The Formation and Disruption of Black Hole Jets

Astrophysics and Space Science Library

More information about this series at
http://www.springer.com/series/5664

The Formation and Disruption of Black Hole Jets

Astrophysics and Space Science Library

More information about this series at
http://www.springer.com/series/5664

Ioannis Contopoulos • Denise Gabuzda •
Nikolaos Kylafis
Editors

The Formation and Disruption of Black Hole Jets

Editors
Ioannis Contopoulos
Academy of Athens
Athens, Greece

Denise Gabuzda
Physics Department
University College Cork
Cork, Ireland

Nikolaos Kylafis
Department of Physics
University of Crete
Heraklion, Greece

ISSN 0067-0057 ISSN 2214-7985 (electronic)
ISBN 978-3-319-10355-6 ISBN 978-3-319-10356-3 (eBook)
DOI 10.1007/978-3-319-10356-3
Springer Cham Heidelberg New York Dordrecht London

Library of Congress Control Number: 2014954253

Cover illustration: A 3D rendering of a numerical simulation (by Tchekhovskoy - chapter 2 of this book) of a magnetically-arrested disk around a black hole with Kerr parameter a = 0.99. Dynamically-important magnetic fields are twisted by the rotation of a Black Hole at the center of an accretion disk. The azimuthal magnetic field component clearly dominates the jet structure.

Printed on acid-free paper

Springer is part of Springer Science+Business Media (www.springer.com)

Preface

Since the discovery of the first optical jet in the galaxy Messier 87 (M87) in 1918, and the discovery of the first 'micro-quasar' in the Galactic source GRS 1915+105 in 1994, it has become clear that relativistic outflows are a ubiquitous feature in both stellar black hole X-ray binaries and active galactic nuclei (AGNs).

The properties of the radio jets in Galactic black holes depend on the X-ray spectral state and history of the source. Steady, compact radio jets are emitted in the hard X-ray state; in contrast, the jets become eruptive as the sources move toward the soft state, and then disappear completely when they reach the soft state. Models for this pattern have been proposed, and there is general agreement about the nature of the accretion disk around the black hole, but a complete and consistent physical picture of the appearance and disappearance of the radio jets is not yet available. On the other hand, accreting supermassive black holes show a clear division into a minority that exhibit radio jets (radio-loud) and a majority that do not (radio-quiet). There have also been some hints of possible connections between the birth of superluminal knots in radio-loud AGNs and dips of their X-ray flux, suggesting a similar phenomenology to that observed in black hole X-ray binaries. However, overall it remains unclear what determines the presence or absence of radio jets in AGNs.

One thing does seem clear, though: much of the physics governing these two types of relativistic outflows must be common, but acting on very different spatial and temporal scales. It is believed that, in both cases, magnetic fields play a fundamental role in the formation and powering of the jets, but the study of how they interact with the strong gravitational field of the central black hole and generate highly relativistic collimated outflows is a formidable problem of modern astrophysics.

We tried to address all of the above issues during a one-day Special Session on 'The Formation and Disruption of Jets in Black-Hole Binaries and AGNs' that we organized as part of the 2012 European Week of Astronomy and Space Sciences (Rome, July 6, 2012). Our experience with this Session provided the motivation for the present volume. Our aim has been to present reviews of the varied phenomenology regarding the radio to X-ray spectra of stellar binaries and the properties of AGN jets on the wide range of scales through which they propagate, as well as recent theoretical efforts to understand the physical mechanisms that contribute to the origin of black hole jets on all scales. We have given particular emphasis to the role and the origin of the black hole magnetosphere and the magnetic fields that drive, collimate, and accelerate the jets.

This project has been an exciting opportunity for us to try to put together a consistent, unified physical picture of the formation and disruption of jets in accreting black hole systems. New observational and theoretical results are piling up every day, so the contents of this volume only represent our current best ideas. Time will tell how close our present understanding is to reality.

We thank all the contributing authors for their efforts, and the Springer Senior Editor of Physics and Astronomy Ramon Khanna for his trust and support.

Athens, Greece — Ioannis Contopoulos
Cork, Ireland — Denise Gabuzda
Heraklion, Greece — Nick Kylafis

Contents

Chapter 1
Jets at Birth and Death

D. Lynden-Bell

Abstract The collimation of jets by magnetic fields is shown to need external pressure which, due to the stabilised pinch effect, may be considerably less than the magnetic pressure at the core of the jet, but must then operate over a much larger area. This article concentrates on the extreme situation where magnetism dominates within the jet.

1.1 Introduction

It is a surprising fact that swirling disks of conducting fluid around young stars, dying stars, quasars and micro-quasars (Mirabel 1999) generate highly collimated jets perpendicular to the plane of the motion. These jets sometimes remain well collimated for distances of hundreds or even tens of thousands of times the size of the disks that generate them. Sometimes these jets elongate at speeds comparable to c, the velocity of light, but where the jets originate from stars in formation they have far smaller speeds of 100 or 200 km/s. Thus there is a clear correlation between the maximum circular velocity in the accretion disk (or the escape speed from the central object) and the velocities observed in their jets. Whereas Curtis discovered the optical jet in the galaxy Messier 87 in 1918 (Curtis 1918), the jet phenomenon was discovered by radio astronomers studying the large lobes of radio galaxies. Early examples were found by Hargrave and Ryle (1974) studying Cygnus A from Cambridge while much higher resolution maps were given later by Carilli and Barthel (1996) (Fig. 1.1) using the very large array (VLA) in New Mexico.

D. Lynden-Bell (✉)
Institute of Astronomy, Madingley Road, CB3 0HA Cambridge, UK

Clare College, Cambridge, UK
e-mail: dlb@ast.cam.ac.uk

I. Contopoulos et al. (eds.), *The Formation and Disruption of Black Hole Jets*,
Astrophysics and Space Science Library 414,
DOI 10.1007/978-3-319-10356-3_1

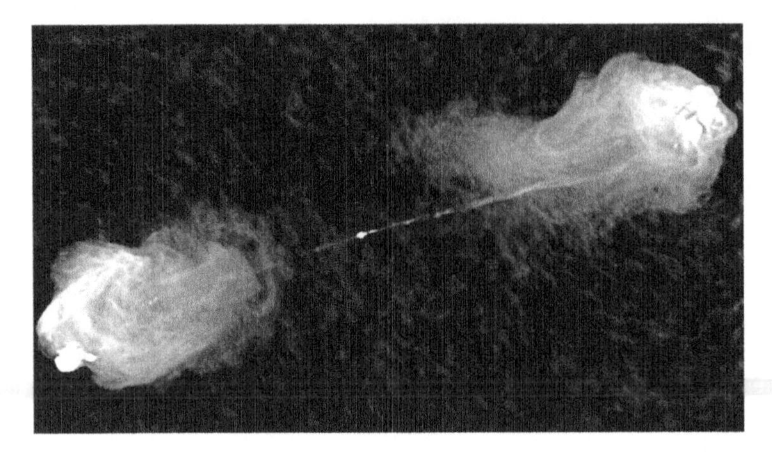

Fig. 1.1 Cynus A showing both radio lobes and the jets feeding them, VLA radiogram Carilli and Barthel (Produced by the Very Large Array)

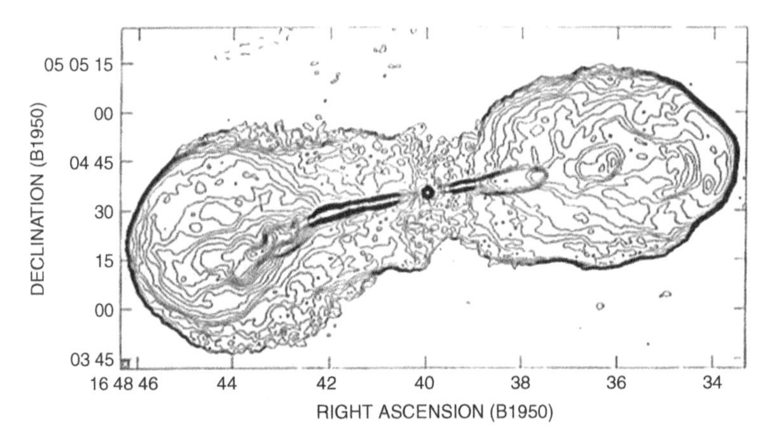

Fig. 1.2 Hercules A shows both the strong central jet and a surrounding magnetised cocoon with a sharp edge (From Gizani and Leahy (2003) and Produced by the Very Large Array)

Contours of the emission from Hercules A were given by Gizani and Leahy see Fig. 1.2. A particularly beautiful example emanating from the galaxy NGC 6258 was discovered by Baldwin (Waggett et al. 1977) in his 6C survey of low surface brightness objects and Cohen and Readhead (1979) resolved the jet down to parsec scales using very long baseline interferometry (VLBI), see Fig. 1.3. Even before jets were found both Scheuer (1974) and Blandford and Rees (1974) argued that jets were needed as a means of maintaining the radio emission of the lobes.

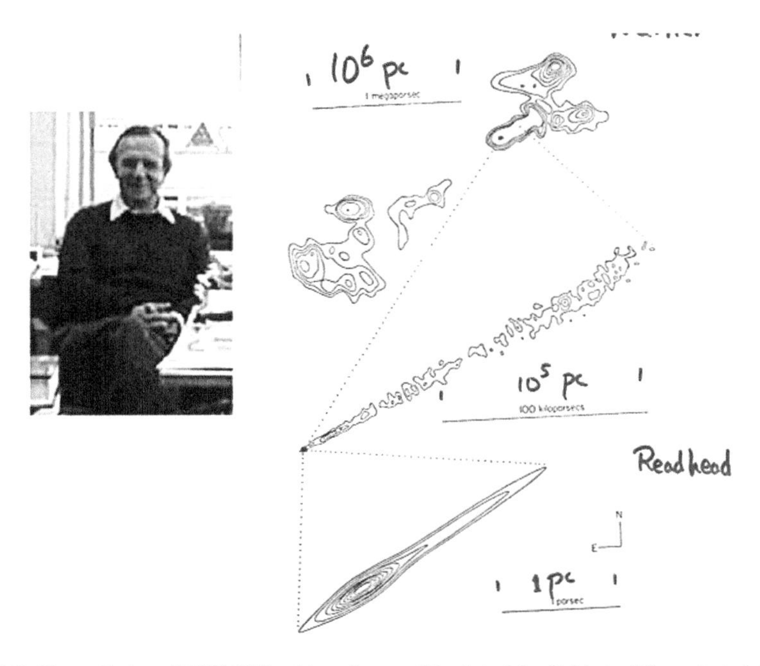

Fig. 1.3 The radio jet of NGC 6251 with a picture of the late John Baldwin (Waggett et al. 1977) and *inset* the VLBI picture of the jet close to the source (Cohen and Readhead 1979). Note the similarity of the jet structure over scales ranging by a factor of 1,00,000!

In these radio galaxies the main emission mechanism is synchrotron radiation from highly relativistic electrons spiralling around magnetic field lines. The strong linear polarisation predicted for this mechanism is observed. It is not too hard to find mechanisms to accelerate electrons given shocks and re-connecting magnetic fields, so it is likely that swirling disks produce magnetic fields and these in turn produce some very fast electrons that radiate as they gyrate about the field. However it is far from clear that the fields generated should be collimated into narrow jets. There has been much work on the theory of the jets themselves see for example work by Blandford and Payne (1982), Ouyed and Pudritz (1999), Appl and Camenzind (1993), and Gammie et al. (2003). Lovelace (1976) was perhaps the first in producing a magnetically collimated model but he put in an ambient initially uniform magnetic field ab initio so the final collimation along that field was no surprise. Contopoulos (1995) found interesting MHD solutions with flows along the magnetic fields, see also Bogovalov and Tsinganos (2003). Here I shall give a personal account of how my ideas have evolved concentrating on how twisted magnetic fields behave and what collimates the jets. I started with the belief that there might be a gun barrel inside the central object (Lynden-Bell 1978) but

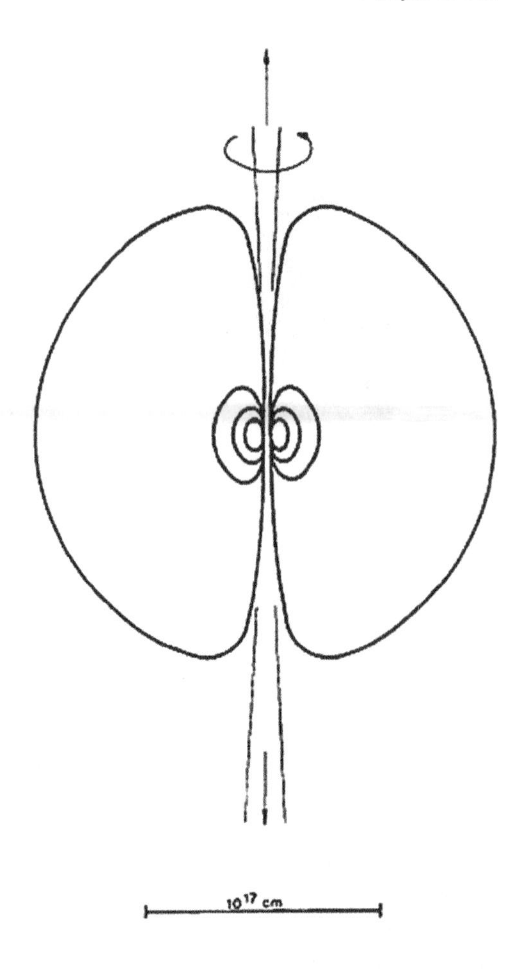

Fig. 1.4 After core collapse to a black hole material of greater angular momentum can not reach the axis, so a barrel forms

while something of that type may be present I now favour magnetic collimation, with the barrel itself being magnetic (Fig. 1.4). The job of a theorist is to create understanding. I shall not hesitate to simplify the problem to its bare essentials. We are interested in what I deem to be a strongly magnetic phenomenon so we shall assume that in the jet itself the magnetic forces dominate over any residual gas pressure or any inertial pressures. Nevertheless outside the jet's magnetic cavity we shall be forced later to assume that there is a coronal medium whose pressure and inertia can not be neglected. By contrast to the magnetic dominance in the jet, we shall assume that the central body and the accretion disk around it are so massive and so conducting that their inertial motions dominate, dragging the magnetic field with their fluid motion. It is the build up of magnetic energy due to differential motions that must provide the driving force for the jets. An early paper that considered this

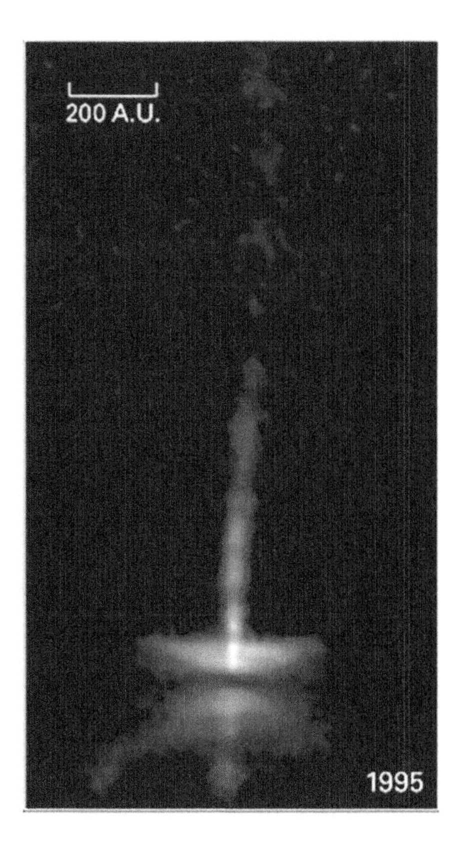

Fig. 1.5 Hubble picture of a jet from a forming star Herbig-Haro 30. The star is obscured by the dark band of the accretion disc (From the Hubble Space Telescope)

build up was Hoyle and Ireland (1960) and their instability was later exploited by Parker (1966). We shall first gain some insight into the behaviour of the anisotropic Maxwell stresses by working out the energies of magnetic configurations and their equilibria. We shall then progress to problems with non-relativistic motions and only finally consider relativistic effects. Non-relativistic jets from young stellar objects show strong collimation and are remarkably similar in form to the relativistic jets of radio sources. See the jet from HH30 shown in Fig. 1.5. It is not clear that the basic mechanism of jet formation requires relativistic motion. My belief is that the basic mechanism is the same and the differences are caused by the extreme velocities that make the highly relativistic cases more spectacular and more visible. Perhaps the first paper to produce a convincing jet from first principles was that by LeBlanc and Wilson (1970) who computed a core collapse in a star with a magnetic field The winding of the field produced a double ended blow-out along the axis. Just such a mechanism is now enviewed as the origin of the gamma-ray bursts from rapidly spinning supernovae (Fig. 1.6).

Fig. 1.6 Hubble picture of jets from young stars in Carina (From the Hubble Space Telescope)

1.2 Magnetic Energy and the Pinch Effect

The magnetic energy may be written as a volume integral $E_m = \int \mathbf{B}^2/(8\pi)dV$. If we expand the whole configuration by a factor λ while preserving the magnetic flux through each fluid element then $\mathbf{B} \propto \lambda^{-2}$, $dV \propto \lambda^3$ so $E_m \propto \lambda^{-1}$. Thus the magnetic energy decreases under a uniform expansion. Speaking anthropomorphically the field wants to expand and must be held in check by other mechanical forces such as gravitation and external gas pressure.

Suppose now that the magnetic field dies away at large height, z, and all the boundary conditions defining the magnetic configuration via the foot-points of the magnetic field in the central body and the accretion disk lie below some plane $z =$ const. Now imagine a virtual displacement in which the slice between heights z and $z + \delta z$ is thickened to occupy the region z to $z + \lambda\delta z$ (Fig. 1.7). We can do this by imagining the part of the system above the slice to be lifted a little but otherwise unchanged. Flux freezing within the slice ensures that B_x and B_y will decrease by a factor $1/\lambda$ while B_z will be unchanged. Since the volume of the slice increases by

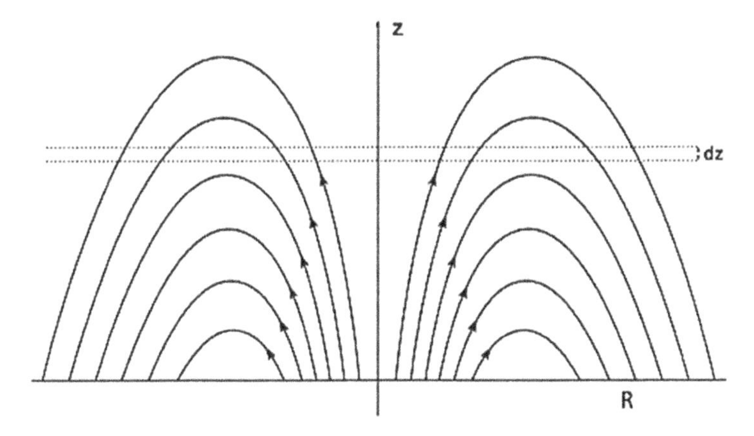

Fig. 1.7 Slice through a magnetic configuration. Only the meridional components of the field are drawn (Figure 2 from page 1361 of my paper Lynden-Bell (2003))

a factor λ we find the magnetic energy of the field in the slice behaves as

$$\Delta E_m=(8\pi)^{-1}[\lambda^{-1}\int\int(B_x^2+B_y^2)dxdy\delta z+\lambda\int\int B_z^2dxdy\delta z]=\lambda^{-1}\Delta W_{xy}+\lambda\Delta W_z \tag{1.1}$$

Thus as λ increases from unity the magnetic energy will decrease if $\Delta W_{xy} > \Delta W_z$ but will increase if the inequality is reversed. Since such a displacement could occur spontaneously any magnetically dominated equilibrium must have

$$\int\int(B_x^2+B_y^2)dxdy=\int\int B_z^2dxdy \tag{1.2}$$

over any plane above all the field's foot-points. This condition becomes more interesting in cylindrical polar coordinates (R,ϕ,z) when it reads $\int\int(B_R^2+B_\phi^2)Rd\phi dR = \int\int B_z^2Rd\phi dR$. Now imagine a field attached to a central body that is contracting and rotating faster than its surroundings but suppose those are massive enough to hold back the field attached to them. The differential rotation will inevitably wind up the poloidal magnetic flux so that an equal toroidal flux is generated by each turn of the central object relative to its surroundings. After N turns the toroidal flux above the disk will be N times the poloidal flux there. With the toroidal flux continually increasing and a fixed poloidal flux I was led to think that $W_\phi = (8\pi)^{-1}\int\int\int B_\phi^2RdRd\phi dz$ would soon outstrip the magnetic energy in the z component of the field W_z, so that the equilibrium condition (1.2) integrated over height would be violated. Then λ will increase dynamically so driving a vertical extension of the field. With great expectations that these simple ideas would lead to the long jets observed, I sought in 1979 a more precise model that I could calculate exactly. The resulting calculation so shattered my dream that I did not write up the model, nor did I return to the jet problem for 15 years. However before continuing that story I should clear up the apparent discrepancy between the general theorem that magnetic fields have a tendency to expand, with Ampere's finding that parallel currents attract which is vividly illustrated in the pinch effect.

Consider first a purely toroidal field confined within height Z between two co-axial cylinders of radii $b_0 < b$ and b. Minimising the magnetic energy at fixed flux $F_\phi = \int_0^Z \int_{b_0}^b B_\phi dRdz$ we find $B_\phi = B_b b/R$, $F_\phi = B_b bZ \ln(b/b_0)$ and $W_\phi = \frac{1}{4} B_b^2 b^2 Z \ln(b/b_0) = \frac{1}{4} F_\phi^2 / [Z \ln(b/b_0)]$. The magnetic pressure on the inner cylinder is $\frac{B_b^2 b^2}{8\pi b_0^2}$, so for b_0 small it is much greater than the pressure $\frac{B_b^2}{8\pi}$ that must be applied by the outer cylinder to maintain equilibrium. Thus the toroidal magnetic flux is here acting as a pressure amplifier giving a much larger pressure on smaller area. This is the pinch effect. However now consider a uniform lateral contraction at constant flux, keeping b/b_0 constant. The magnetic energy, W_ϕ, does not change, there is no release of magnetic energy, rather the external work done by the pressure of the outer cylinder is balanced by the work done in compressing the inner cylinder. Under a uniform expansion in all three directions the magnetic energy does decrease due to the increase of Z so the main push is directed upwards. However note that an increase in b with both Z and b_0 held fixed does decrease W_ϕ in keeping with the work done by the magnetic pressure on the outer cylinder. Likewise a decrease in b_0 with the others held fixed releases magnetic energy in keeping with the work done on the inner cylinder via the magnetic pressure there. In summary the primary push of this toroidal field configuration is along z but there is also a push to increase $\ln(b/b_0)$ which can give very large pressures over small areas when b_0 is small. These effects are significantly modified by the presence of a longitudinal field. This is well illustrated by the constant pitch, force-free field of Gold and Hoyle (1958) in which,

$$B_z = \frac{B_0}{1+\kappa^2 R^2}; \quad B_\phi = \kappa R B_z; \quad B^2 = \frac{B_0^2}{1+\kappa^2 R^2}; \tag{1.3}$$

In a cylinder of height Z and radius b the magnetic fluxes are

$$F_\phi = \int\int B_\phi dRdz = \tfrac{1}{2} B_0 Z \kappa^{-1} \ln(1+\kappa^2 b^2) : \tag{1.4}$$

$$F_z = \int B_z 2\pi R dR = \pi B_0 \kappa^{-2} \ln(1+\kappa^2 b^2); \quad f = F_\phi / F_z = \kappa Z/(2\pi) \tag{1.5}$$

The lines of force are given by $dz/B_z = Rd\phi/B_\phi$ so $d\phi/dz = B_\phi/(RB_z) = \kappa$, which justifies our claim that the field has constant pitch. The magnetic pressure on the boundary is $p = \frac{B_b^2}{8\pi} = \frac{B_0^2}{8\pi(1+1/\kappa^2 b^2)}$, however the central magnetic pressure is now limited to $B_0^2/(8\pi)$. Writing the energy of the whole configuration, including the $\int pdV$ work in making the magnetic cavity against a constant external pressure, in terms of the fluxes, and the sizes of b and Z,

$$W = \int \mathbf{B}^2/(8\pi)dV + pV = \tfrac{1}{8}Z\kappa^{-2}B_0^2\ln(1+\kappa^2b^2) + pV = \frac{F_\phi^2}{2Z\ln[1+(2\pi fb/Z)^2]} + \pi b^2 Zp. \tag{1.6}$$

Minimising W with respect to b^2 merely gives restatement of external pressure balance but that gives us the radius b in terms of the fluxes, the pressure and Z

$$b = \frac{Zu}{2\pi f}; \qquad (1+u^2)[\ln(1+u^2)]^2 = \frac{2\pi f^2}{Z^4 p} \tag{1.7}$$

with b so determined, we replace it by writing $u = 2\pi fb/Z$ to obtain $W = F_\phi^2/[2Z\ln(1+u^2)] + \frac{1}{4}\pi^{-1}pZ^3u^2/f^2$. Since we have already minimised over b we see that $\partial W/\partial u = 0$. We notice that the magnetic push to larger Z is now opposed by the pressure term. To see how this works in reality, we now take our cylinder and bend it around into a torus of major radius $a \gg b$, which remains the minor radius. Following early work by Weiss (1964) we shall consider this toroidal magnetic configuration as a flux rope in the equatorial region of the Sun and put $p = p(a)$ as would be appropriate in the Sun's interior. Now $Z = 2\pi a$ and the equilibrium condition $\partial W/\partial a = 0$ gives,

$$\frac{F_\phi^2}{4\pi a^2\ln(1+u^2)} = \frac{2\pi^2u^2}{f^2}\frac{d(pa^3)}{da} \tag{1.8}$$

Evidently pa^3 must increase with a for any equilibrium to be possible. However a further restriction follows since we also have $\partial W/\partial u = 0$ which gives

$$\frac{F_\phi^2}{4\pi a[\ln(1+u^2)]^2(1+u^2)} = \frac{2\pi^2pa^3}{f^2} \tag{1.9}$$

On division we see that,

$$(1+u^{-2})\ln(1+u^2) = d\ln(pa^3)/d\ln a. \tag{1.10}$$

The expression on the left increases from one at $u = 0$, so there can only be an equilibrium solution if $d\ln(pa^3)/d\ln a \geq 1$ that is pa^2 must increase outwards. In the Sun this only occurs at radii less than a fifth of the Solar radius. Evidently in regions in which the Sun's pressure decreases more rapidly than $1/a^2$ there is no equilibrium solution and the toroidal loop will expand with both b and a increasing until it bursts out of the Sun and expands into the solar wind. Now suppose that our twisted flux tube is in a region where pa^2 increases and that $p \propto a^s$ with $s < 2$ Then there will be a solution of Eq. (1.10) for u. Then from Eq. (1.9) a is determined by,

$$a^{4-s} = \frac{F_z^2 f^4}{8\pi^5(pa^s)[\ln(1+u^2)]^2(1+u^2)} \tag{1.11}$$

Since $4 > 4 - s > 2, a$ will increase with f. So increasing the twisted flux while keeping the longitudinal flux F_z constant causes an increase in a. Twisting the flux helps the tube to float.

1.3 Time-Dependent Force-Free Magnetic Fields

1.3.1 Grad-Safranov Equation Solved: Collimation Lost

We shall assume axial symmetry. Then since $\mathbf{divB} = 0$ the field takes the form

$$\mathbf{B} = (2\pi)^{-1}\nabla P \times \nabla\phi + RB_\phi\nabla\phi. \tag{1.12}$$

Here ϕ is the azimuthal angle about the axis in cylindrical coordinates (R, ϕ, z) and $P(R, z)$ gives the magnetic flux through a circle about the axis of radius R at height z so $P = 0$ on the axis. Evidently P is constant along lines of force and defines axially symmetric tubes on which they lie. It is called the poloidal flux function. Its maximum value gives the total poloidal flux F. The poloidal components of the magnetic field are given by $B_z = (2\pi R)^{-1}\partial P/\partial R$; $B_R = -(2\pi R)^{-1}\partial P/\partial z$; The form of Eq. (1.12) guarentees that the field has no divergence and its curl is

$$\mathbf{curl\,B} = -(2\pi)^{-1}[\nabla^2 P - (\nabla \ln R^2.\nabla P)]\nabla\phi + \nabla(RB_\phi) \times \nabla\phi. \tag{1.13}$$

The Maxwell equation reads $\mathbf{curl\,B} = 4\pi\mathbf{j} + c^{-1}\partial\mathbf{E}/\partial t$. The final term is Maxwell's displacement current which is important for rapidly time varying fields as occur in radiation and light waves but can be neglected here. As we are assuming that magnetism dominates within the jet there is nothing to resist the $\mathbf{j} \times \mathbf{B}$ force density. Thus for non-relativistic force-free motion $\mathbf{curl\,B}$ must be parallel to $\mathbf{B}$. Comparing Eqs. (1.12) and (1.13) and using the axial symmetry $\nabla(RB_\phi) = \beta'/(2\pi)\nabla P$ Which implies that $\beta' = \beta'(P) = d\beta(P)/dP$. Then without loss of generality we may take $B_\phi = (2\pi R)^{-1}\beta(P)$. Comparing the $\nabla\phi$ terms in (1.12) and (1.13) we now find

$$\nabla^2 P - \nabla \ln R^2.\nabla P = -\tfrac{1}{2}d/dP[\beta(P)]^2; \quad 4\pi\mathbf{j} = \mathbf{curl\,B} = d\beta(P)/dP\,\mathbf{B}. \tag{1.14}$$

The first is known as the Grad-Safranov equation. Equations (1.14) and (1.12) together with their boundary conditions are sufficient to determine the magnetic field. Those equations do not involve any time derivatives. Thus for non-relativistic force-free time-dependent problems the time dependence is dictated by the changes of the boundary conditions e.g. the winding caused by differential rotation and the system speeds kinematically through a sequence of configurations that could be equilibria if the boundary conditions ceased to change. To get this important result we had to neglect the displacement current but this is justified when the motions are non-relativistic. We also had to assume that the magnetic field is so dominant in the jet that it is force-free. This has the effect that the Alfve'n speed is c, the velocity of light, and since all our velocities are small compared with c the system can be taken to respond instantaneously and so pass from one equilibrium to the next. Equilibria can be determined from the principle of minimal potential energy embodied in the so-called energy principles. These principles are now seen to extend to time-dependent problems obeying the conditions discussed above. We would like to see how a magnetic configuration in a highly conducting plasma responds to continual twisting of the field's foot-points. The most natural problem of this type would be a field like Saturn's but with the outer feet of the field lines attached to a conducting massive accretion disk taking the place of Saturn's rings. However, to make the problem more amenable to exact calculation, it is simpler if both feet are on a sphere of radius a where the field has its outgoing magnetic flux rotating relative to the ingoing flux. This problem can be exactly solved for a dipole, (Lynden-Bell and Boily 1994) but to make it closer to the accretion disk problem we take the case of an axially symmetrical initially quadrupolar field whose flux goes out at high latitudes and returns at lower latitudes but is constrained not to cross the equatorial plane by the presence of the conducting accretion disk. For radii $r \gg a$ we may expect the field to take up a form that is independent of the value of a. Such considerations suggest that there should exist solutions that are proportional to some negative power of r. In spherical polar coordinates with $\mu = \cos\theta$ the GS equation (1.14) reads,

$$\frac{\partial^2 P}{\partial r^2} + \frac{(1-\mu^2)\partial^2 P}{r^2 \quad \partial\mu^2} = -\beta'\beta \tag{1.15}$$

We look for solutions of the form $P = \Pi(r)f(\mu)$. By division by P and differentiation we can prove that either (i) $\beta'\beta$ is linear in P or else (ii) Π is a power of r and $\beta'\beta$ is a power of P. The linear case has been explored but it leads to solutions that oscillate as a function of r and have field-lines in the different oscillations that do not connect back to small r. Such solutions are not related to our problem of continual twisting so we consider case (ii). Setting $\Pi \propto (r/a)^{-l}$ where l is a real number, not generally an integer, we find there are only solutions if $\beta'\beta \propto P^{1+2/l}$. Then all the terms have an $r^{-(2+l)}$ dependence and the function $f(\mu)$ has to obey,

$$(1-\mu^2)d^2 f/d\mu^2 = -l(l+1)f - C(1+1/l)f^{1+2/l} \tag{1.16}$$

where C is a constant. This equation was studied and solved by Lynden-Bell and Boily (1994), hereafter LBB (Lynden-Bell and Boily 1994). Looking for solutions in which the flux extends beyond a few a we need l to be small as $P \propto r^{-l}$. By choice of the proportionality constants in Π and β we can normalise f so that its maximum is one. With l small we see that both $l(l+1)f$ and $f^{1+2/l}$ are small everywhere except close to that maximum (just because f is raised to a high power). Thus $d^2 f/d\mu^2$ is small everywhere except near that maximum at say $\mu = \mu_m$. To exploit this we write $1-\mu^2 = 1-\mu_m^2 - (\mu^2 - \mu_m^2)$ in Eq. (1.16) and then multiply

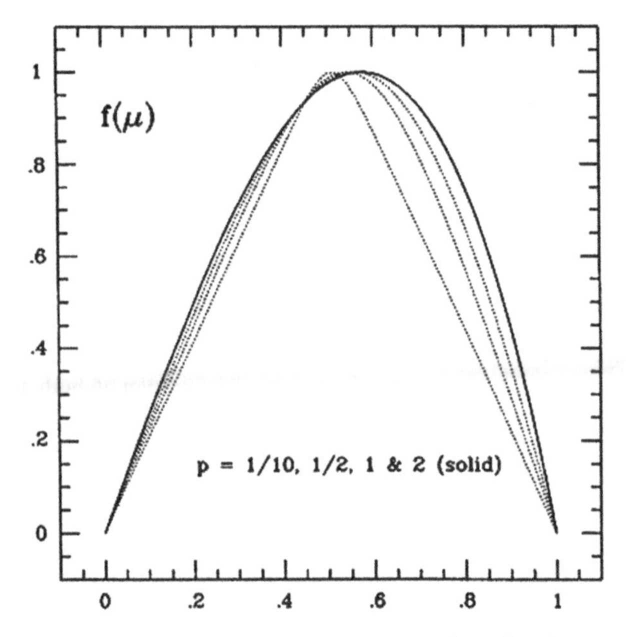

Fig. 1.8 The angular profile function for twist angles of 180°, 120°, 80°, and 0°. At 208° it becomes an isoseles triangle (Figure 2 from page 150 of Lynden-Bell and Boily (1994))

by $2df/d\mu$ and integrate from μ_m to μ obtaining,

$$(1-\mu_m^2)(df/d\mu)^2 = \gamma_m^2 - \gamma(f)^2 + I(\mu);\ \gamma(f)^2 = l(l+1)f^2 + Cf^{2+2/l};\ \gamma_m = \gamma(1); \tag{1.17}$$

$$I(\mu) = \int_{\mu_m}^{\mu} (\mu^2 - \mu_m^2)\frac{d}{d\mu}(\frac{df}{d\mu})^2 d\mu. \tag{1.18}$$

Near μ_m $I(\mu)$ is of order $(\mu - \mu_m)^3$ which is smaller than the other terms while away from there it contains $d^2 f/d\mu^2$ which we have shown to be small. Thus I is small everywhere and may be treated by perturbation theory. Neglecting it LBB showed that the solution for small l is $f = 1 - \alpha \ln(\cosh u)$; $\alpha = 1/(d \ln \gamma/df)_{f=1}$; $u = k(\mu - \mu_m)$: $k = \gamma_m/(\alpha\sqrt{1-\mu_m}) \gg 1$. $u_m = k\mu_m$. When I is included via perturbation theory this solution is amended to $f = 1 - \alpha \ln \cosh(u + \Delta u)$ where $\Delta u = \frac{1}{2k^2(1-\mu_m^2)}[2u_m(\ln \cosh u - \frac{1}{6}\tanh^2 u) + (u - \tanh u)\ln 4 - 3(\ln 4 - \frac{1}{2})\tanh^3 u]$. Imposition of the boundary conditions that $f = 0$ at $\mu = 1$ and at $\mu = 0$ gives $\mu_m = \frac{1}{2} + \Delta\mu_m$. Where $\Delta\mu_m$ gives the shift in the maximum of f seen in Fig. 1.8.

As the toroidal flux increases there is more magnetic energy in the field that has to be held in equilibrium. This leads to an expansion of the field, see Fig. 1.9. What I found surprising was that after a twist of only $2\pi/\sqrt{3}^{\,c} = 208°$, just over

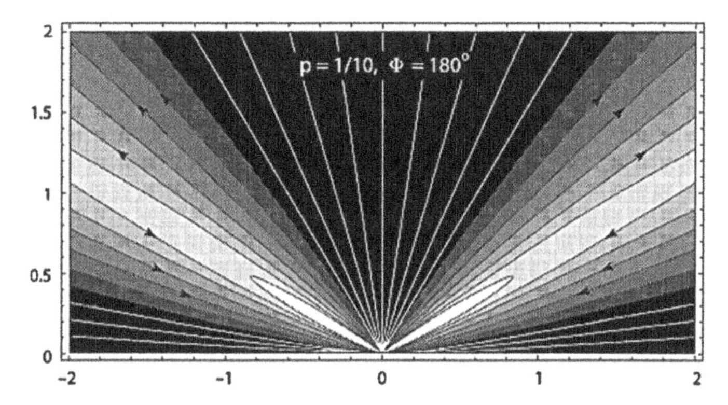

Fig. 1.9 At a twist of 180° the field lines reach out toward ∞. The flux function $P \propto r^{-1/10}$ (Figure 1 from page 1362 of my paper Lynden-Bell (2003))

half a turn the field becomes radial and extends to infinity. The couple (or torque) necessary to produce the twist increases up to a twist of 80° but then it diminishes until it vanishes at 208°. Thereafter the outward flux can be rotated without effort. The outgoing flux having reached to infinity is effectively disconnected from the incoming flux and our W_ϕ no longer increases with each turn, so our argument for collimation falls! Instead of my hoped-for narrow jet I obtained a whole hemisphere! The behaviour above was essentially predicted by Aly (1984) who pointed out that a radial field with the same fluxes at the foot-points has a finite total energy since $(8\pi)^{-1} \int \mathbf{B}^2 dV \propto \frac{1}{2} \int_a^\infty r^{-4} r^2 dr = \frac{1}{2} a^{-1}$ which is finite and depends on the lower radius a even when the flux extends to infinity. In the limiting case of 208°, $\mu_m = 1/2$, so exactly half the solid angle of the sky is filled with outgoing field and the other half above the disk has incoming field. This behaviour is true also when the lower boundary is shifted from $\mu = 0$ the equator to $\mu = -1$ the south pole. Then the initially dipolar field evolves after only half a turn of the northern hemisphere relative to the southern into a split monopole with the outward radial field of the north separated from the radial inward field of the south by a sheet in which the current circulates azimuthally. In all such cases twisting the field does not of itself lead to collimation. Similar results hold when the solution is confined by a cone. Again the solid angle is equally split between outgoing and incoming field in the limiting case but the twist needed to reach that limit increases as the confining cone narrows. Such solutions are collimated by the imposition of the cone i.e. not by physics but by fiat!

1.3.2 Collimation Regained

In the solution given above we took no account of the pressure that any coronal medium would exert. The flux expanded to infinity without having to do work in making a magnetic cavity. However even in the limiting case the magnetic fields fall as r^{-2} so the magnetic pressures and stresses fall as r^{-4} at $100a$ the pressures will be down by 10^{-8}, at $1{,}000a$ by 10^{-12}. Thus we are wrong in claiming that the magnetic pressures will dominate all the way to infinity; the ambient pressure must eventually intervene. To demonstrate its effect we first make a simple minded model to give us an orientation as to which terms are important. Consider a field within a cylinder of radius b and height Z both of which are free to vary. Since P must vanish on the axis and at the boundaries we take it to be

$$P = 4F(R/b)^2[1 - (R/b)^2](1 - z/Z) \tag{1.19}$$

The total poloidal flux is F. After N turns of the outgoing flux relative to the incoming flux the toroidal flux will be NF and this must pass through the meridional section of area bZ so the mean toroidal field over that area will be $\overline{B}_\phi = NF/(bZ)$ (Lynden-Bell 2003). A rough estimate of the energy in this component of the field is $W_\phi = (8\pi)^{-1} \int \overline{B}_\phi^2 dV = \frac{1}{8} N^2 F^2/Z$. Adding that to the energies in the poloidal components $W_R = F^2/(24\pi^2 Z)$ and $W_z = F^2 Z/(3\pi^2)b^2$ and the energy stored in making the magnetic cavity $W_p = \pi b^2 Z$ we get the configuration energy

$$W = W_R + W_\phi + W_z + W_p = \frac{F^2}{24\pi^2}[(1+3\pi^2 N^2)/Z + 8Z/b^2] + \pi b^2 Zp \tag{1.20}$$

Minimizing over b we find that at equilibrium $W_p = W_z$, so $b = (\frac{F^2}{3\pi^3 p})^{1/4}$ and the energy principle becomes the minimization of $W = W_R + W_\phi + 2W_z$. Now minimizing over Z we find $W_R + W_\phi = 2W_z$ which is, but for the factor 2, what we got previously when we neglected the pressure term. This equality now gives us $Z = \frac{1}{4} b\sqrt{1 + 3\pi^2 N^2} \to \frac{1}{8}\sqrt{3}\, b\, \Omega t$. So the collimation Z/b grows linearly with the number of turns and when we replace N by $\Omega t/(2\pi)$ the top of the jet advances at constant speed $\frac{1}{8}\sqrt{3} V_c$. Here $\Omega b = V_c$ may be related to the circular velocity in the accretion disk. Thus even at this crude level we have a correlation of jet speed with something related to the circular velocity of the disk. Looking at the orders of magnitude we see the W_R is small compared with W_ϕ. This is because the cylinder becomes tall and no radial flux is added. Thus the radial flux is spread over an ever increasing area. By contrast the ϕ flux increases and this forces the field up, so W_z increases when the toroidal flux increases. The above field is not force-free. Had we put in many free parameters and taken care that each line of force had the right prescribed twist then minimisation would have lead us to a force-free field but our aim was to get orders of magnitude not to get a precise model. This model has taught us that W_R can be neglected in finding the overall shape of the magnetic cavity, that the radius b is primarily determined by lateral pressure balance and that the structure is tall and expands upwards once lateral pressure balance has been achieved. Finally the upward expansion is not due to any gradient in the pressure of the surrounding medium since it occurs in a medium of uniform pressure. We now give an idealised but exact time-dependent solution of the force-free equations that expands into a

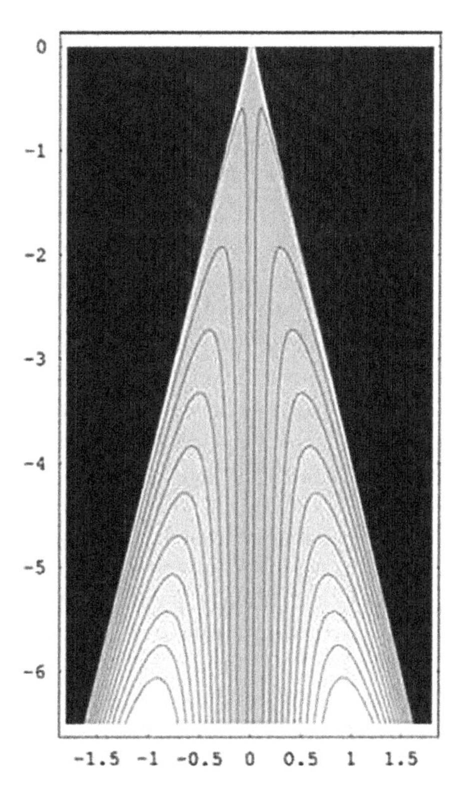

Fig. 1.10 The field-lines of the Dunce's cap model. Only the meridional field is shown. As the toroidal flux is increased the model expands vertically uniformly, with the apex moving at constant speed (Part of figure 6 from page 1182 of my paper Lynden-Bell (2006))

medium of constant pressure. The Dunce's Cap model (Lynden-Bell 2006; Mestel 2012), is particularly interesting as it is the natural limit of a pinch stabilised by longitudinal field. Surprisingly it is described in spherical polar coordinates (r, θ, ϕ) centred at the moving apex of the jet with theta measured from the downward vertical axis that points toward the central object. p is the external pressure which is assumed to be constant even when the magnetic cavity is expanding. $\mu = \cos\theta$ and the boundary of the cavity is at some not large angle $\theta_m < 15°$; $\mu_m = \cos\theta_m$. Thus the magnetic cavity is in the form of a cone pointing down from the jet's apex. See Fig. 1.10. As the jet advances this cone becomes taller and thinner its base at the accretion disc being of fixed cross section. The poloidal flux function and $\beta(P)$ are,

$$P = 4\pi\sqrt{2\pi p}\, r^2 f(\mu, \mu_m); \qquad \beta(P) = 4\sqrt{\frac{(1+\mu_m)}{(1+\mu_m^2)}}\sqrt{2\pi^3 p}\ P^{1/2}; \quad (1.21)$$

$$f = \frac{(1-\mu^2)}{(1+\mu_m^2)}[1 - \frac{1+\mu_m}{1+\mu} + \tfrac{1}{2}(1+\mu_m)\ln|\frac{(1+\mu)(1-\mu_m)}{(1-\mu)(1+\mu_m)}|]. \quad (1.22)$$

It is readily seen that $P = 0$ both at the axis $\mu = 1$ and at μ_m. For opening angles at the jet apex of less than 30° $1 - \mu_m = \Delta < 0.03$ so we may expand in powers of Δ and write $1 - \mu = (\theta/\theta_m)^2\Delta$ with a maximum error less than 0.6 %. The total poloidal flux arising from the accretion disk at a distance Z below our origin is $F = 2\pi\sqrt{2\pi p}\ \theta_m^2 Z^2/e$ where $e = 2.718281828$ is the base of Napierian logarithms. Neglecting a term of order θ_m^3 the total toroidal flux is $F_\phi = 2\pi\sqrt{2p}Z^2\theta_m$ so the flux ratio $F_\phi/F = \frac{e}{\sqrt{\pi}\,\theta_m}$. Now $F_\phi/F = N = \Omega t/(2\pi)$ so θ_m decreases as t^{-1}. However F is constant and the footprint of the flux on the disk is the radius $b = Z\theta_m$ which is also fixed, so $Z \propto \theta_m^{-1}$. Thus Z again increases like t. In fact $\dot{Z} = 1/(2e\sqrt{(\pi)})V_c$ and as before we write $V_c = \Omega b$. In the frame fixed at the accretion disc we may now calculate the electric field $\mathbf{E} = -\mathbf{v} \times \mathbf{B}/c$ where $\mathbf{v} = \hat{\mathbf{z}}\dot{Z}/Z$. The magnetic fields in this solution are of special interest, all components are independent of r. This is obvious for the field along the generators of the θ_m cone since there the magnetic pressure has to balance the constant external pressure p but it s true everywhere within the conical cavity. The fields close to the axis demonstrate the extreme pinch. B_r varies as $-\ln\theta$ and B_ϕ as $\sqrt{\|ln\theta|}$. Thus the field lies along the axis although both become infinite there. Despite this the infinities are so weak that the total energy remains finite. At the apex of the cone the field along the axis bends right over to produce the constant fields along the cone's generators. See Fig. 1.5. It could be the very strong field along the axis, seen here, that is responsible for the extreme narrowness of the jets observed in synchrotron radiation, as that emission is often taken to be proportional to the fourth power of the field (two from the field's energy density and two more from the density of relativistic electrons) To generate the exact field structure described above the field lines have to be turned with a particular velocity at the accretion disk. Furthermore the poloidal flux has to emerge from that disc with the profile given by the P given above and in reality the ambient coronal pressure is hardly likely to be independent of height. For all those reasons much more general solutions are needed which are not dictated by mathematical simplicity but by the physics of the real problem. Attack via the Grad-Safranov equation is then much less attractive. The function $\beta(P)$ though related to the twisting of the field lines is NOT the same as the function $\Phi(P)$ which gives the total angle twisted between the foot-points of the field lines labelled by P. The relationship between them is not known until the solution is given. Furthermore the shape of the domain in which the equation is to be solved i.e. the magnetic cavity, is also determined by the solution. Solution of an ill-defined equation over an ill-defined domain is not a good starting point!

1.3.3 Minimum Energy Solution via an Adaptable Trial Function

We used minimum energy to get initial order of magnitude estimates at the start of Sect. 1.3.2 but there we crudely demanded that the magnetic cavity be a cylinder, albeit one whose dimensions could be varied, and we prescribed both the form of the poloidal and toroidal fields within it. Our aim now is to remove all such restrictions by allowing the magnetic cavity freedom to take any axially symmetrical shape, allowing the poloidal field freedom of form and allowing the field lines to have any prescribed twist about the axis $\Phi(P)$ between their foot points at $z = 0$. We shall also allow the external pressure to be a prescribed function of height $p(z)$. However, in line with our order of magnitude calculation, we shall assume that the jet cavity is tall and thin and that the components of magnetic field that are radial from the axis are small everywhere except possibly near the top of the cavity where they have to oppose the ambient pressure. It is perhaps surprising that we can find a trial function with all these freedoms which nevertheless gives analytically soluble equations for the minimisation (Lynden-Bell 2006). When pressure varies with height the energy required to make the magnetic cavity is no longer pV but is rather $W_p = \int p(z)A(z)dz$ where $A = \pi[R_m(z)]^2$ is the cross-sectional area of the cavity at height z and R_m is its radius there. We again consider expanding vertically the slice of the system between z and $z + \delta z$ but now the work done against pressure in raising the upper part bodily depends on height. As a result our former theorem, cf. Eqs. (1.2) and below (1.20), is modified to read

$$< B_R^2 > + < B_\phi^2 > \quad = \quad < B_z^2 > +8\pi \left[p(z) + A(z)^{-1} \int_z^{top} A(z')(dp/dz')\, dz' \right] \tag{1.23}$$

The final integral from z up is negative when p gets smaller as height increases. The averages $< - >$ are over the areas A at height z. We now consider the same slice but expand it uniformly in its own plane via a displacement field $\boldsymbol{\xi} \propto \mathbf{R}$. In doing this we consider the virtual work done by the magnetic stresses $B_R B_z/(4\pi)$ both at z and $z + \delta z$. As in our discussion of the pinch effect the varying of the B_ϕ field does not change the energy but work is done against the pressure at R_m. Setting $W_s = (4\pi)^{-1} \int B_R B_z R^2 d\phi dR$, we obtain the exact result

$$< B_z^2 >= 8\pi p(z) - 4\pi[A(z)]^{-1} dW_s/dz. \tag{1.24}$$

Now $\Phi(P, t)$ will increase linearly with t due to the differential rotation of the source. Hence $\overline{\Phi} = \overline{\Omega} t$ so from Eq. (1.31) we see that $Z^2\sqrt{8\pi p(Z)} \propto (\overline{\Omega} t)^2$. This gives us the generalisation of the earlier constant speed law found for the constant pressure case (Fig. 1.11). However our former example showed us that the radial fields were small so the W_s term may be neglected and we have approximately $< B_z^2 >= 8\pi p(z)$. We now need an estimate of $< B_\phi^2 >$ at height z in terms of the known twist function $\Phi(P)$ that expresses the angle turned by the inner foot-point relative to the outer one. Let $Z(P)$ be the maximum height of the surface of constant P. Where the poloidal field lines are primarily along the jet we think of each having a twist per unit height but to get the total twist this will be supplemented by a contribution from near the top of that line where it lies more radially. If the latter contributes $\lambda\Phi(P)$ to the twist then the twist per unit height along the sides of the jet will be $(1 - \lambda)\Phi(P)/Z(P)$ Now each element dP of toroidal flux would produce a toroidal flux dP if twisted once around the axis so

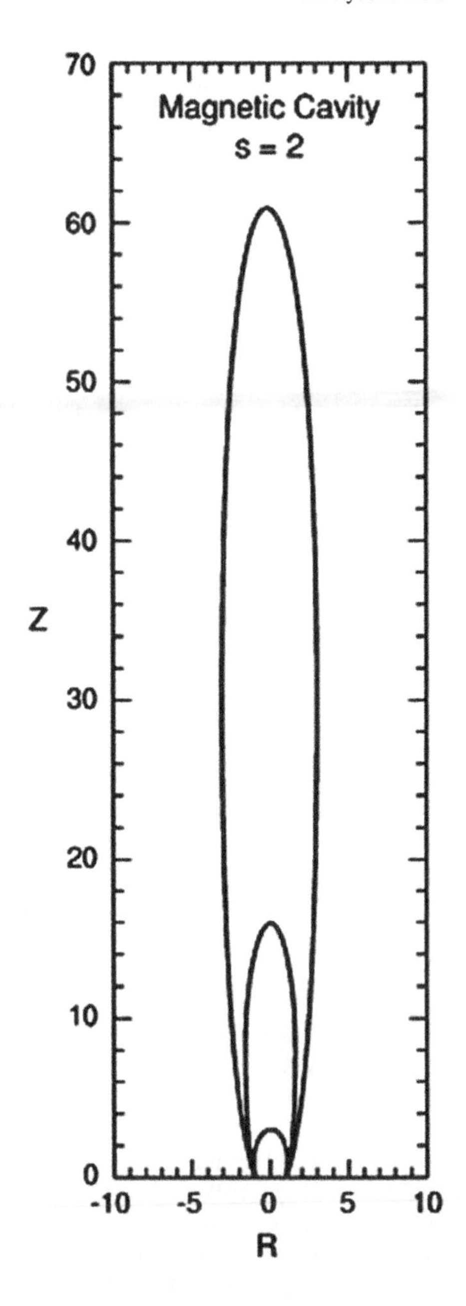

Fig. 1.11 Shape of the magnetic cavity when the external pressure $p \propto (1+z)^{-2}$ at three times in the ratio 1 : 4 : 8. The initial expansion at constant speed is replaced by constant acceleration once the z^{-2} region is reached

it produces a toroidal flux $(2\pi)^{-1}\Phi(P)dP$ when twisted through the angle $\Phi(P)$. Now consider the toroidal flux through the toroidal slot between z and $z + \delta z$ of area $R_m \delta z$. Defining $\overline{B}_\phi(z) = R_m^{-1} \int_0^{R_m} B_\phi dR$ at constant z, this flux $R_m B_\phi$ will consist of a contribution from the flux per unit height of all field lines that reach higher than z and another contribution from all field lines that have maxima within the slot. If $P_m(z)$ is the maximum that P achieves at height z then only those field lines with $P < P_m$ will get higher Their contribution to the flux in the slot is then $(1 - \lambda) \int_0^{P_m(z)} [\Phi(P)/Z(P)] dP \delta z$. The field lines that have their maxima within the slot have P values between P_m and $P_m + \delta P_m$ where $\delta P_m = (dP_m/dz)\delta z$. These will contribute a toroidal flux $\lambda(2\pi)^{-1}\Phi(P_m)(-dP_m/dz)\delta z$ to the slot. The negative sign arrises because P_m decreases at greater heights. Hence

$$R_m \overline{B}_\phi = (2\pi)^{-1}[(1 - \lambda) \int_0^{P_m} \Phi/Z dP + \lambda \Phi(P_m)(-dP_m/dz)] \tag{1.25}$$

The two contributors $\Phi(P_m)(-dP_m/dz)$ and $\int_0^{P_m} \Phi/Z dP$ have the same dimension and depend on the same quantities; we shall call their typical ratio g and use an average value in what follows. The magnetic energy of the ϕ component of the magnetic field in the volume between z and $z + \delta z$ is $(8\pi)^{-1} \int_0^{R_m} B_\phi^2 \, 2\pi R dR \delta z = \frac{1}{8} < B_\phi^2 > R_m^2 \delta z$. Because B_ϕ will have some profile as a function of R we shall write $< B_\phi^2 >= J^2 \overline{B}_\phi^2$ where J is a dimensionless number that depends on the profile. Nevertheless we shall take J to be independent of height as it would be in a self-similar jet. Then

$$W_\phi = (8\pi)^{-1} \int B_\phi^2 dV = \frac{1}{8} J^2 \int \left[(1 - \lambda) \int_0^{P_m} \Phi/Z \, dP + \lambda \Phi(P_m)(-dP_m/dz) \right]^2 dz. \tag{1.26}$$

We treat λ as independent of z and use the ratio g to obtain, writing $\tilde{J} = J(1 - \lambda + g\lambda)/(4\pi)$,

$$W_\phi = \frac{1}{2} \tilde{J}^2 \int [\int_0^{P_m} (\Phi/Z) dP]^2 dz = \frac{1}{2} (\tilde{J}^2/g) \int [\Phi(P_m)(-dP_m/dz) \int_0^{P_m} (\Phi/Z) dP] dz. \tag{1.27}$$

On integration by parts $\int [\int_0^{P_m} (\Phi/Z) dP]^2 dz = 2 \int [\Phi(P_m)(-dP_m/dz) \int_0^{P_m} (\Phi/Z) dP]$ dz so we deduce that $g = \frac{1}{2}$ and

$$W_\phi = (32\pi^2)^{-1} J^2 (1 - \tfrac{1}{2} \lambda)^2 \int [\int_0^{P_m} (\Phi/Z) dP]^2 dz \tag{1.28}$$

To complete our variational principle we need to express W_z in terms of the functions $P_m(z)$, $R_m(z)$ that we vary. At any height, z,

$$\pi R_m^2 < |B_z| >= \int_0^{R_m} |B_z| 2\pi R dR = \int_0^{R_m} |\partial P/\partial R| dR = 2 P_m(z), \tag{1.29}$$

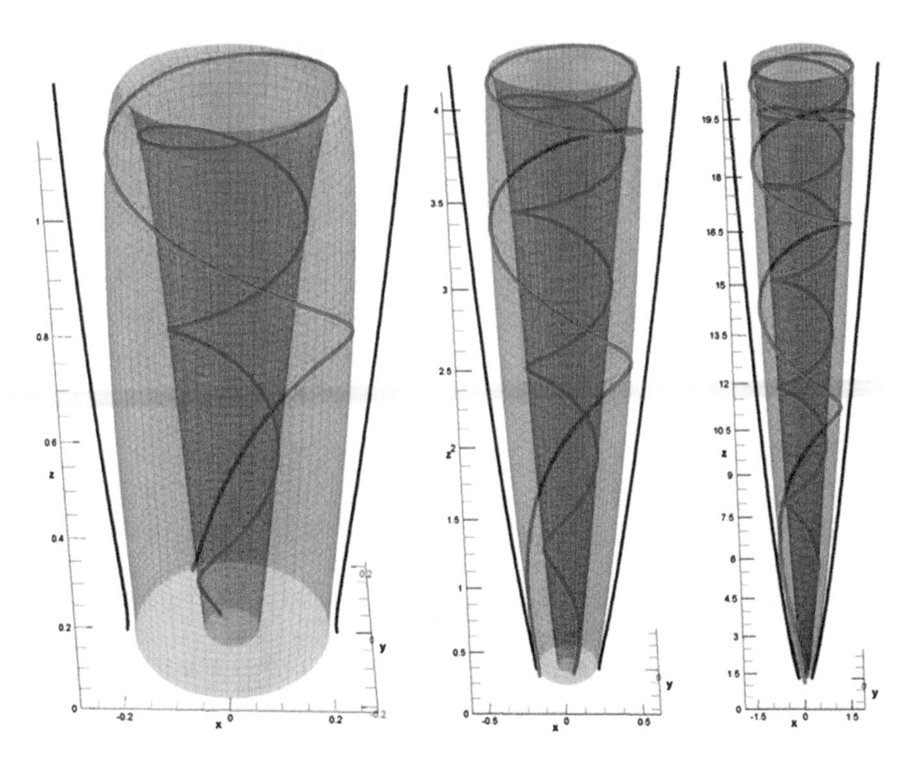

Fig. 1.12 One field line in *blue* comes from an inner foot-point where the twisting is more rapid. It lies on the constant P surface, turns *red* at the top and descends to the outer foot-point. The whole configuration gets much taller at later times so the figures to the *right* have been shrunk. The pressure falls as $(1+z)^{-3}$ and the configurations are shown at dimensionless times 30, 60 and 90. The *heavy line* shows the edge of the magnetic cavity (Figure 4 from page 413 of Sherwin and Lynden-Bell (2007))

since P has one maximum and is zero at both ends. Hence $< |B_z| >= 2P_m(z)R_m^{-2}$ (Fig. 1.12). However it al We now put $I^2 =< B_z^2 > /(< |B_z| >)^2$ and obtain $W_z = \frac{1}{2}I^2 \int P_m^2/R_m^2 dz$. Gathering all the energy contributions and neglecting W_R in line with our earlier discussion

$$W = W_\phi + W_z + W_p = \frac{J^2(1-\frac{1}{2}\lambda)^2}{32\pi^2} \int \left[\int_0^{P_m} (\Phi/Z)dP \right]^2 dz + \\ + \frac{1}{2}I^2 \int (P_m^2/R_m^2)dz + \int p(z)\pi R_m^2 dz. \quad (1.30)$$

Now $Z(P)$ is directly related to $P_m(z)$ through the property that $P_m(Z(P)) = P$; so when we vary $P_m(z)$, $Z(P)$ automatically varies too. In the above expression for W the functions to be freely varied are $P_m(z)$ and $R_m(z)$. The functions $\Phi(P,t)$, $p(z)$ are given as they define the problem to be solved while J, I, λ are fixed constants which we now determine by comparison with known exact solutions the Dunce's cap and the linear solution in a cylinder $P = A\exp(-\kappa z)k_1 R J_1(k_1 R)$. Here J_1 is the Bessel function $k_1 b$ is its first zero where b is the radius of the cylinder, $2\pi R B_\phi = \beta(P) = \alpha P$ and $\kappa^2 = \alpha^2 - k_1^2$. From these self similar solutions we find in the order Dunce's cap first, $I = e/2 = 1.359, 1.179$; $J = \sqrt{2/\pi} = 0.800, 1.098$. We evaluated the value of λ by comparison with exact results from the linear model which gave values between 0.47 and 0.62 so the mean $\lambda = 0.55$ is an appropriate value to use along with $I = 1.27, J = 0.95$.

We are now in a position to perform the variation and determine the dynamic shape of the magnetic cavity and the fields within it. Varying $R_m(z)$ we find that the last two terms in W must be equal and at each value of z, $R_m(z) = \sqrt{I\ P_m(z)}/[2\pi p(z)]^{1/4}$. We can now remove the penultimate term and replace it by doubling the final one which we now write in the form $\int_0^F \Pi(P_m)dP_m$ where $\Pi(P_m)$ is the function $\int_0^z 4I\ \sqrt{8\pi p(z')}dz$ re-expressed as a function of P_m rather than z, so $\Pi(P_m) = \int_0^{Z(P_m)} 4I\sqrt{8\pi p(z)}dz$. Finally minimising W over all choices of $P_m(z)$ or equivalently $Z(P_m)$ we find

$$Z^2\sqrt{8\pi p(Z)} = \frac{[J(1-\frac{1}{2}\lambda)]^2}{16\pi I} P_m \overline{\Phi}^2, \tag{1.31}$$

where $\overline{\Phi}^2 = \Phi(P_m)P_m^{-1}\int_{P_m}^F \Phi(P)dP$.

Now $\Phi(P,t)$ will increase linearly with t due to the differential rotation of the source. Hence $\overline{\Phi} = \overline{\Omega}t$ so from Eq. (1.31) we see that $Z^2\sqrt{8\pi p(Z)} \propto (\overline{\Omega}t)^2$. This gives us the generalisation of the earlier constant speed law found for the constant pressure case.

However it also tells us something of great interest that relates back to the floating flux tube in the Sun discussed in Sect. 1.2. We notice that with $\overline{\Phi}$ increasing there will continue to be solutions of (1.31) for Z provided that $Z^2\sqrt{8\pi p(Z)}$ increases with Z. But what happens if at some point $p(Z)Z^4$ has a maximum? There all equilibria cease. Beyond there with no equilibrium to go to the system must evolve at a speed determined by its own inertia i.e. at relativistic speeds that we can no longer treat by energy principles. Inside the Sun such a maximum occurs at about 0.3 of a solar radius. Could it be that the relativistic jets expand so fast because the effective pressures have fallen faster than Z^{-4}? Lorentz transformation of any solution will give another solution so a relativistic version of the Dunce's hat may well be attainable, but I shall not attempt that here.

1.4 Remark on the C-K Cosmic Battery

Some 15 years ago Contopoulos and Kasanas (1998) proposed a cosmic battery that acts on disks circulating around central sources of radiation. This battery may build significant magnetic fields even without further dynamo amplification, but it may also generate seed fields. C-K showed that their battery produces currents that circulate with the disc, so the magnetic fields within the current loop have magnetic moments aligned with spins $\mathbf{M}.\boldsymbol{\Omega} > 0$. They argued that this prediction can distinguish the fields so generated from those made by normal dynamo action. It was the symmetry breaking predicted by this model that led me to look for symmetry breaking in the Astronomical objects themselves. I found further tentative evidence (Lynden-Bell 2013) favouring symmetry breaking in the sense predicted. This is interesting but in no way proves the efficacy of their particular mechanism. I give below order of magnitude estimates of the sort of fields it might produce in a steady state. These estimates ignore winds, accretion flows, time dependent induction etc. all of which modify the results in any detailed modelling of a particular system. In the C-K battery radiation from the central source falls on the ionised inner wall of the circulating disc. The radiation is Thomson scattered by electrons. Relative to the fixed frame the scattered radiation on average moves forward like the electrons which experience a back reaction as a result. An electron exposed to a radiation flux $L/(4\pi r^2)$ scatters an energy $L\sigma_T/(4\pi r^2)$ per second and loses the forward momentum $[L\sigma_T/(4\pi r^2 c)](V/c)$ per second. Here V is the circulation velocity of the inner edge of the disk and the Thomson scattering cross-section is $\sigma_T = (8\pi/3)(e^2/mc^2)^2$, but the protons scatter $(1{,}836)^2$ times less, so the effect on them can be neglected. The force difference means that the electrons circulate slightly slower than the protons thus causing an electric current in the direction of the rotational velocity. The EMF of this cosmic battery is $1/e$ times the integral of the force around the circuit which gives the EMF = $[L\sigma_T/(4\pi r^2 ec)](V/c)(2\pi r)$. The current only runs on the surface layer one optical depth, $1/(n\sigma_T)$, thick (where n is the number density of electrons in the layer) and over the width to $2h$ of the circulating disc. If the electrical conductivity is σ the resistance in the circuit is $\mathcal{R} = 2\pi r/(A\sigma)$ where the cross-sectional area is $A = 2h/(n\sigma_T)$. In a steady state the current will be, EMF/$\mathcal{R}$ = $[L/(4\pi r^2 ec)](V/c)(\sigma/n)2h$. The magnetic field at the centre will be $B_0 = 2\pi I/r$ while that close to the disk will be $2I/h = [L/(\pi r^2 ec)](\sigma/n)(V/c)$. To evaluate such fields we need the conductivity of plasma to a current moving across the field that it generates. This is given by $\sigma = [e^2 n\tau/(mc)]/[1 + (eB\tau/mc)^2]$ where τ is the time between collisions of electrons and protons. Now the expressions for B_d and B_0 involve σ/n so it is only through the dependence of τ on n that the electron density influences the fields generated. For $eB\tau/(mc) < 1$ this has the paradoxical effect that lowering the electron density increases the current and the field. Lowering n does not change the total number of charge carriers as the radiation sinks in deeper, increasing the thickness of the current-carrying layer. Lowering n increases the time between collisions with protons which have the same number density, so the resistance in the circuit is lowered. However once τ gets larger than $mc/(eB)$ then σ decreases as τ increases, so further lowering of the electron density is counter productive. We therefore put τ equal to its optimum value and deduce that $\sigma/n = e/(2B)$. Now within the current carrying layer b falls from B_d through zero. So putting $B = \frac{1}{2}B_d$ we have $\sigma/n = e/B_d$. Inserting this formula into our expression for B_d we find the simple formula,

$$B_d = \left[\frac{L}{\pi r^2 c}\right]^{\frac{1}{2}} \left[\frac{V_c}{c}\right]^{1/2} \tag{1.32}$$

from which most details of the disk have disappeared. We now consider some examples:

1. Sun-grazing disk $L = 4 \times 10^{33}$ ergs/s.$r = 7 \times 10^{10}$ cm. $M = M_{sun}$ $B_d = 0.1$ Gauss.
2. Quasar disk $L = 10^{46}$ ergs/s $r = 4 \times 10^{13}$ cm $M = 10^8 M_{sun}$ $B_d = 6{,}000$ Gauss.
3. Proto-Jupiter $L = 4\pi r^2 \sigma_{Stefan} T^4$ $T = 10^4$ K $V = 40$ km/s $B_d = 0.1$ Gauss.

References

Aly, J.J.: On some properties of force-free magnetic fields in infinite regions of space. ApJ **283**, 349 (1984)

Appl, S., Camenzind, M.: The structure of relativistic MHD jets: a solution to the nonlinear Grad-Shafranov equation. A&A **270**, 71 (1993)

Blandford, R.D., Rees, M.J.: A 'Twin-exhaust' model for double radio sources. MNRAS **169**, 395 (1974)

Blandford, R.D., Payne, D.G.: Hydromagnetic flows from accretion discs and the production of radio jets. MNRAS **199**, 883 (1982)

Bogovalov, S., Tsinganos, K.: Shock formation at the magnetic collimation of relativistic jets. MNRAS **337**, 553 (2003)

Carilli, C.L., Barthel, P.D.: Cygnus A. AAR **7**, 1 (1996)

Cohen, M.H., Readhead, A.C.S.: Misalignment in the radio jets of NGC 6251. ApJ **233**, 101 (1979)

Contopoulos, J.: Force-free self-similar magnetically driven relativistic jets. ApJ **450**, 616 (1995)

Contopoulos, I., Kasanas, D.: A cosmic battery. ApJ **508**, 859 (1998)

Curtis, H.D.: Descriptions of 762 nebulae and clusters (NGC 4486, p. 31). Pub. Lick Obs. **13**, 11–42 (1918)

Gammie, C.F., McKinney, J.C., Gabor, T.: HARM: a numerical scheme for general relativistic magnetohydrodynamics. ApJ **589**, 444 (2003)

Gizani, N.A.B., Leahy, J.P.: A multiband study of Hercules A – II. Multifrequency very large array imaging. MNRAS **342**, 399 (2003)

Gold, T., Hoyle, F.: On the origin of solar flares. MNRAS **120**, 89 (1958)

Hargrave P.J., Ryle M.: Observations of Cygnus A with the 5-km radio telescope. MNRAS **166**, 305.(1974)

Hoyle F., Ireland, J.G.: Note on the transference of angular momentum within the galaxy through the agency of a magnetic field. MNRAS **120**, 173 (1960)

LeBlanc, J.M., Wilson, J.R.: A numerical example of the collapse of a rotating magnetized star. ApJ **161**, 541 (1970)

Lovelace, R.V.E.: Dynamo model of double radio sources. Nature **262**, 649 (1976)

Lynden-Bell, D.: Gravity power. Phys. Scr. **17**, 185 (1978)

Lynden-Bell, D.: On why discs generate magnetic towers and collimate jets. MNRAS **341**, 1360 (2003)

Lynden-Bell, D.: Magnetic jets from swirling discs. MNRAS **369**, 1167 (2006)

Lynden-Bell, D.: Magnetism along spin. Observatory **133**, 266 (2013)
Lynden-Bell, D., Boily, C.: Self-similar solutions up to flashpoint in highly wound magnetostatics. MNRAS **341**, 1360 (1994)
Mestel, L.: Stellar Magnetism, p. 549, 2nd edn. Oxford University Press, Oxford/New York (2012)
Mirabel, T.F., Rodriguez, L.F.: Sources of relativistic jets in the galaxy. ARAA **37**, 409 (1999)
Ouyed, R., Pudritz, R.E.: Numerical simulations of astrophysical jets from Keplerian discs – III. The effects of mass loading. MNRAS **309**, 233 (1999)
Parker, E.N.: The dynamical state of the interstellar gas and field. ApJ **145**, 811 (1966)
Scheuer, P.A.G.: Models of extragalactic radio sources with a continuous energy supply from a central object. MNRAS **166**, 513 (1974)
Sherwin, B.D., Lynden-Bell, D.: Electromagnetic fields in jets. MNRAS **378**, 409 (2007)
Waggett, P.C., Warner, P.J., Baldwin, J.E.: NGC 6251, a very large radio galaxy with an exceptional jet. MNRAS **181**, 465 (1977)
Weiss, N.O.: Magnetic flux tubes and convection in the Sun. MNRAS **128**, 225 (1964)

Chapter 2
Relativistic Jets in Stellar Systems

Elena Gallo

Abstract Albeit their nature remains elusive, relativistic, collimated outflows of energy and particles appear to be a nearly ubiquitous feature of accreting black holes. As evidence accumulates for a dominant role of the jet in dissipating the liberated accretion power, questions around their powering mechanism and even composition remain unanswered. In this chapter, I will describe the main observational properties of relativistic jets from black hole X-ray binaries, with a particular emphasis on recent developments around three main topics: (i) the role and relative importance of the accretion flow, relativistic jet and equatorial wind; (ii) the existence of global luminosity-luminosity correlation(s) in quiescent and hard state black hole X-ray binaries, and their interpretation(s); (iii) (ways of estimating) the total jet power, and its relation to black hole spin.

2.1 Review of Observations

2.1.1 Black Hole X-Ray Binary Outbursts

Though tens of thousands of such systems are thought to exist throughout the Galaxy (Fender et al. 2013), black hole X-ray binaries become detectable to most ground- and space-based telescopes when they enter an outburst phase, that is, when their luminosity increases by several orders of magnitude at all wavelengths. There is general agreement that black hole X-ray binary outbursts are triggered by a relatively sudden increase of the accretion disk viscosity, caused in turn by a rise in the ionization degree of hydrogen in the disk (see Lasota 2001, and reference therein for a review, and Coriat et al. 2011 for a recent test of the disk instability model). In what follows, I will give a very brief description of our phenomenological understanding of the properties of outbursting black hole

E. Gallo (✉)
Department of Astronomy, University of Michigan, 500 Church St., Ann Arbor, MI 48104, USA
e-mail: egallo@umich.edu

I. Contopoulos et al. (eds.), *The Formation and Disruption of Black Hole Jets*,
Astrophysics and Space Science Library 414,
DOI 10.1007/978-3-319-10356-3_2

X-ray binaries. For comprehensive reviews, the reader is directed to Remillard and McClintock (2006) and Fender (2006), on X-ray states and relativistic jets, respectively.

2.1.2 *Accretion Modes and Relativistic Jets I: Hard State Steady Jets*

A useful and practical tool to visualize the evolution of the global properties of black hole X-ray binaries across a typical outburst is to rely on the so-called hardness-intensity diagram, shown schematically in Fig. 2.1. Plotted here is the X-ray spectral evolution of a prototypical system throughout an outburst (GX339–4), in the form of X-ray hardness (typically 2–10 keV over 0.1–2 keV count rate) as a function of integrated X-ray luminosity (or more simply count rate).

Black hole X-ray binary systems spend the overwhelming majority of their lifetime in a 'quiescent' state, with X-ray luminosities $\lesssim 10^{30-32}$ erg s^{-1}. This stage corresponds to the bottom right corner of the diagram in Fig. 2.1. Once an outburst

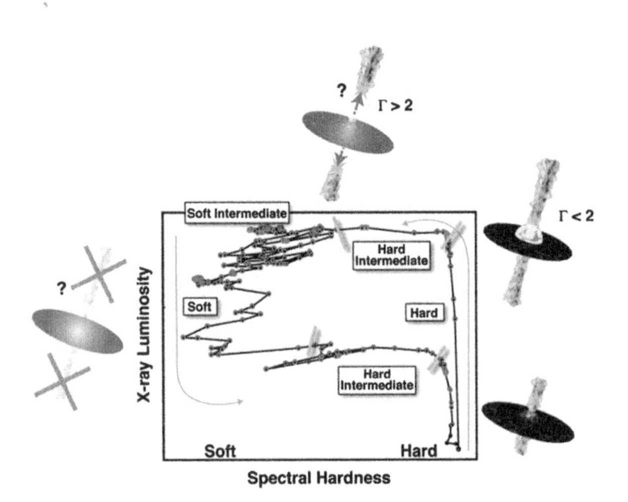

Fig. 2.1 This sketch illustrates our basic understanding of how different accretion modes of black hole X-ray binaries, as traced by the X-ray continuum, are known to map into different 'varieties' of relativistic outflows, traced by the radio-IR emission. Specifically, persistent, synchrotron emitting jets with flat-to-inverted radio-IR spectra are routinely observed in hard X-ray states; such 'steady jets' are quenched once the transition to the soft state is made. During the hard-to-soft X-ray state transition, bright radio flares are often observed, arguably in relation to the ejection of highly relativistic material. Fast moving 'blobs', i.e. the flaring jets, are then observed moving away from the binary system in opposite directions after the transition to the soft state has taken place (sometimes with apparent superluminal velocities); they expand adiabatically, display optically thin radio spectra, and only last a few days-weeks; they are thus refereed to as 'flaring jets' (From: http://www.issibern.ch/teams/proaccretion/)

begins, the source X-ray luminosity can rise up to $\sim 10^{38}$ in timescales of days. All throughout this phase the broadband X-ray spectrum is well represented by a power-law with photon index $\Gamma \sim 2.1 - 1.6$, with the spectrum gradually hardening as the system increases in luminosity (Plotkin et al. 2013; Reynolds et al. 2014). The X-ray spectrum also shows a cutoff around a few tens of keV, and is generally ascribed to thermal Comptonization of thermal disk photons off of a population of hot electrons/positrons within a compact 'corona' that enshrouds the inner accretion disk. Strong X-ray variability (up to 40 % r.m.s. in the Fourier frequency range 0.01–100 Hz) is typically associated with the hard state (although the X-ray spectral evolution is not dramatic over this phase, it is likely accompanied by more substantial variations in the ultraviolet band, which is unfortunately unaccessible for most Galactic X-ray binaries).

Hard/quiescent state black hole X-ray binaries are associated to weak but persistent radio emission, with a flat-to-inverted ($\alpha \sim 0 - 0.5$, where flux density is proportional to ν^{α}) spectrum. In analogy with compact extragalactic radio sources, the relative flux steadiness, flat spectrum, low degree of polarization and high brightness temperature indicate that this emission originates in a (nearly-) continuously replenished, partially self-absorbed, synchrotron-emitting outflow, a.k.a. jet. This interpretation has been directly confirmed for two (bright) systems; for both GRS 1915+105 and Cyg X-1, the compact, persistent, hard state radio source has been resolved into a $\sim$10 A.U., highly collimated jet with very long based interferometry (Stirling et al. 2001; Dhawan et al. 2000, respectively).

2.1.3 Accretion Modes and Relativistic Jets II: Hard-to-Soft Transition – Flaring Jets

As the source reaches an X-ray luminosity of $\sim 10^{37}$ erg s^{-1}, it is very likely to make a transition to the soft state (albeit many 'failed' state transition have been reported for multiple systems). The hard-to-soft state transition can take place in the face of relatively little change in the broadband X-ray luminosity, while significant evolution is observed in the power density spectrum (see, e.g., Casella et al. 2005 for a thorough study of the broad band noise and quasi-periodic oscillations during state transitions). For the purpose of this review, the most important feature of the state transition is its association with bright, radio flare(s) (occurring sometimes singly and sometimes in sequences). These events are thought to be associated with ejections of synchrotron emitting, adiabatically expanding plasmons, which are (occasionally) later resolved as highly relativistic 'flaring', or ballistic, jets, i.e., moving away from the binary system in opposite directions out to thousands of A.U., and fading away over time scales of days or weeks (Mirabel and Rodríguez 1999; Fender et al. 1999; Miller-Jones et al. 2012).

As discussed in Fender and Gallo (2014), "there have been a number of attempts to associated the 'moment of jet launch' with an associated change or event in the accretion flow [...], in particular with the occurrence of the type-B quasi-periodic oscillations, but no convincing one-to-one relation could be firmly established and whether or not there exists a key signature of the moment of launch remains unclear". The possibility remains that the flaring jet(s) may be a more complex phenomenon resulting from rapid – but not instantaneous – changes in the inner disk/outflowing gas properties, and the suppression of the hard X-ray power law. Nevertheless, there has been considerable interest around these events, in that the peak luminosity of the hard-to-soft state transition radio flares has been adopted by Narayan and McClintock (2012) as a proxy for the total jet power, and shown to correlate with the inferred spin parameter value for a handful of black hole X-ray binaries. This is further discussed in Sect. 2.3.

2.1.4 Accretion Modes and Relativistic Jets III: Soft State Winds

The X-ray spectrum of black hole X-ray binaries in the soft state (left portion of the diagram in Fig. 2.1) is well represented by a blackbody-like component which peaks around 1 keV, combined with a steep power law that accounts for less than 10 % of the integrated luminosity. In this state, the overall X-ray variability drops to below 5 %. All these features are well understood in terms of an optically thick, geometrically thin accretion disk extending very close to innermost stable circular orbit of the black hole, with a residual corona of thermal and quasi-thermal particles responsible for a minimal level of Comptonization. The core radio, mm and near-infrared emission all drop below detectable levels in the soft state (Fender et al. 1999; Russell et al. 2011), strongly indicative that the core jet emission has switched off.

A major, recent breakthrough in understanding X-ray states and their connection with outflows is represented by the work of Ponti et al. (2012), who demonstrated that accretion disk winds (revealed by absorption lines in high resolution X-ray spectra) appear to be uniquely observed in *edge-on soft-states*, thus indicating a broad equatorial geometry (see Fig. 2.2; see also Miller et al. 2006, 2008; Neilsen and Lee 2009; King et al. 2013).

In summary, observations indicate a *bimodal regime, whereby, as they transition from the hard to the soft state, black hole X-ray binaries move from a jet-dominated regime with no evidence (so far) of strong winds, to the opposite regime, where the onset of a strong wind coincides with the quenching the steady, relativistic radio jet.* As noted by Ponti et al., however, these two regimes are not likely to tap into the same energy reservoir, with the soft state wind likely being more mass-loaded

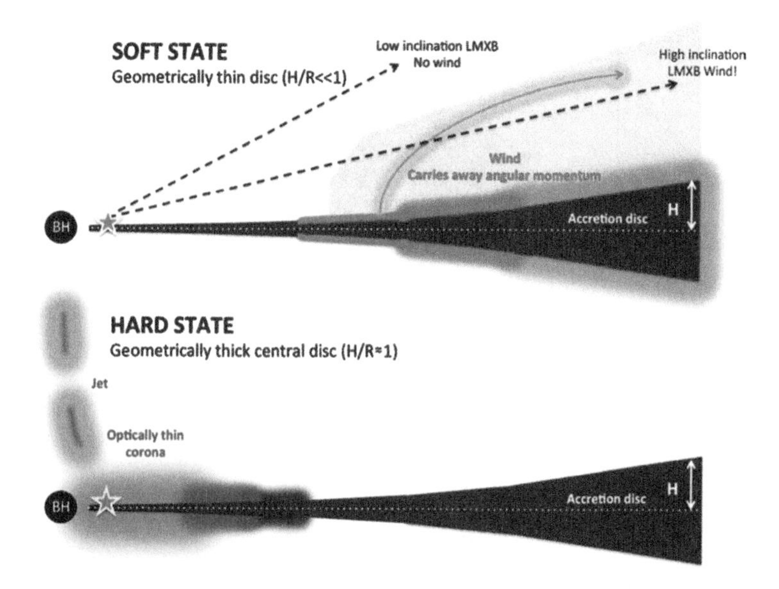

Fig. 2.2 Schematic illustrating the geometry of X-ray winds from soft state black hole X-ray binaries. This follows from the realization that high inclination (dipping) sources show absorption lines every time they are in the soft state and upper limits in the hard states, whereas lines are never detected in low inclination (non dipping) sources (From Ponti et al. (2012; see also their figure 2))

and yet carrying less kinetic energy than the steady hard state jet (Fender et al. in preparation).

2.1.5 Accretion Modes and Relativistic Jets IV: Caveats and Other Recent Developments

As noted by Fender and Gallo (2014), whereas there is general agreement around the physical mechanism that is responsible for the onset of the outburst, the associated 'spectral hysteresis' effect is far from being understood. Figure 2.3, from Dunn et al. (2010), serves to illustrate the broad range of observed luminosities for the hard-to-soft vs. soft-to-hard state transitions for a sample of 24 black hole X-ray binaries observed by the Rossi X-ray Timing Explorer over a period of 13 year. In a recent model put forward by Begelman and Armitage (2014), this hysteresis pattern is driven by an increase in turbulent stress in a disk that is threaded by a net magnetic field, combined with the ability of geometrically thick (but not thin) disks to advect such a field in the radial direction. In this framework, the transition to the soft stare

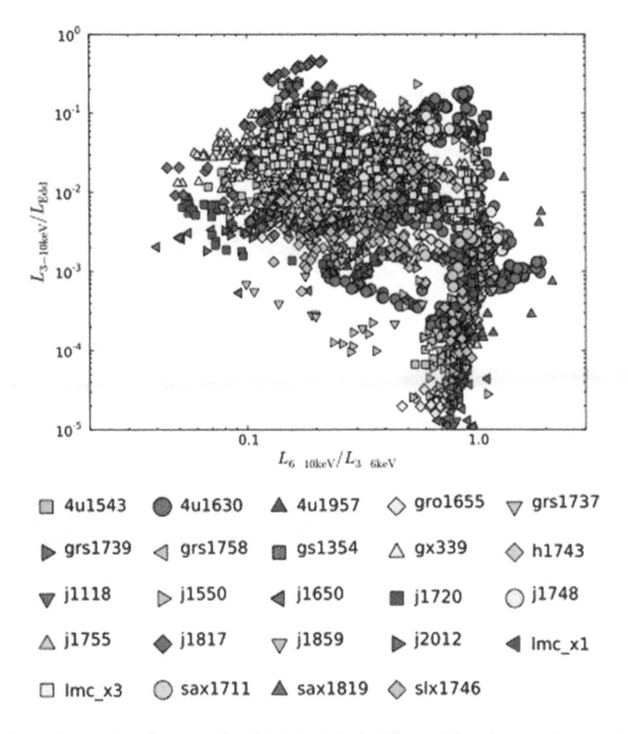

Fig. 2.3 Hardness-intensity diagram for 24 black hole X-ray binaries as observed by the Rossi X-ray Timing Explorer over a period of 13 years (From Dunn et al. (2010))

occurs when the total accretion power luminosity, which in turn is proportional to the second power of the alpha parameter, approaches the Eddington luminosity.

The same hysteresis behavior can also be visualized via a (model-dependent, unlike the hardness-intensity-diagram)'reflection-intensity' diagram (see Fig. 2.4), where the flux in the hard X-ray power law component is shown as a function of the flux in the reflection component during the outburst evolution. Overall, the hard state (in blue) is characterized by a weak reflection component; at the same time, though, the ratio of reflection to power law component (i.e., the reflection fraction) increases as the source luminosity rises. In contrast, the soft state (in red) is found to be reflection-dominated. In the hard state, the reflection fraction can be increased by progressively decreasing the size of the inner radius of the accretion disk, i.e., by moving the disk closer and closer to the innermost stable circular orbit. In contrast, the scale-height of the corona above the disk decreases in the soft state, leading to an increase in the fraction of back-scattered, hard X-ray photons, and hence of the reflection fraction. The role of the jet in this picture remains to be explored.

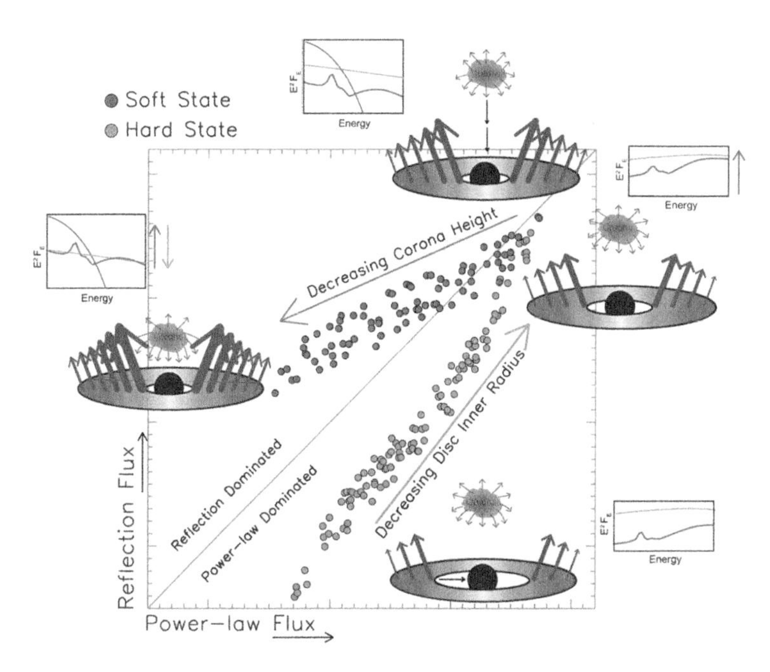

Fig. 2.4 Reflection-intensity diagram of the black hole X-ray binary GX339–4 as observed by the Rossi X-ray Timing Explorer during three of its outbursts. The evolution of the power law and reflection fraction are interpreted in terms of two main varying parameters: the extent of the inner accretion disk radius and the height of the Comptonizing corona above/below the disk (From Plant et al. (2014))

2.2 The Radio/X-Ray Domain of Hard and Quiescent State Black Hole X-Ray Binaries

Coordinated radio and X-ray monitoring of hard and quiescent state black hole X-ray binaries have proven to be a powerful observational tool for studying the connection between accretion and the production of steady jets in these systems. A tight and repeating correlation between the radio and X-ray luminosity (with L_X being proportional to $L_r^{0.7\pm0.1}$) was first established for the black hole X-ray binary GX339-4 in the hard state by Corbel et al. (2003), and later confirmed with data from 7 outbursts over a period of 15 year by Corbel et al. (2013) (with a revised slope of 0.62 ± 0.01; see Fig. 2.5). A similar relation, with slope in the range 0.5–0.7 and holding over several orders of magnitude in L_X, has also been established for the black hole X-ray binary V404 Cyg (Gallo et al. 2003, 2005; Corbel et al. 2008), and, more recently, XTE J1118+480 (Gallo et al. 2014; see Fig. 2.6). A great deal of work has been carried out to identify possible causes for the different normalisations among different systems; however, to date no convincing evidence has been reported for a dependence of the normalisation on orbital parameters, relativistic beaming,

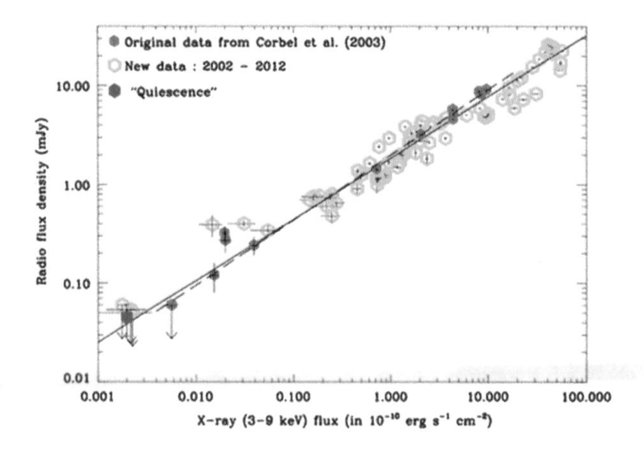

Fig. 2.5 Radio-X-ray correlation for the black hole X-ray binary GX339-4 in hard state: $L_X \propto L_r^{0.62\pm0.01}$. (Data are taken from 7 different outbursts spanning over 15 year. From Corbel et al. (2013))

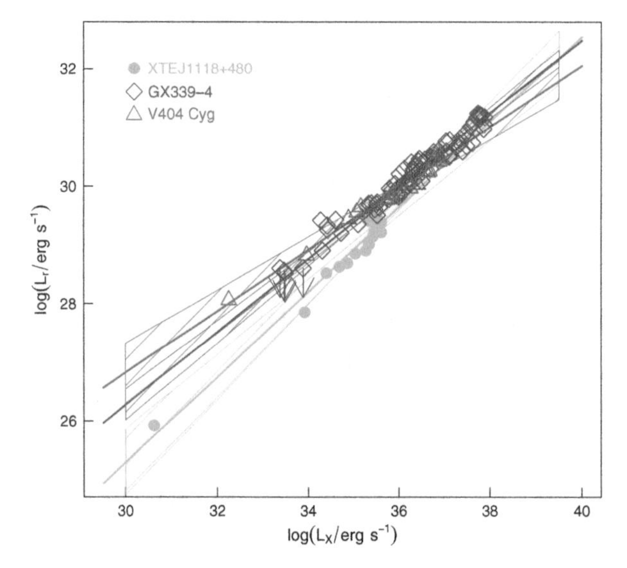

Fig. 2.6 Radio-X-ray correlation for the black hole X-ray binaries GX339-4 (in *blue*, same data set as in Fig. 2.5), V404 Cyg (*red*) and XTE J1118+480 (*green*), with slopes: 0.62±0.04, 0.52±0.07, and 0.72 ± 0.09, respectively (From Gallo et al. (2014))

black hole spin and/or black hole mass (see Soleri and Fender 2011; Gallo et al. 2012, 2014).

Additionally, despite the tightness of the relation for the three individual sources, its 'universality' (cf. Gallo et al. 2003) has recently come under severe scrutiny, and rightly so. Figure 2.7 summarizes the current state of the problem by assembling what is likely the most complete data collection as of today (data

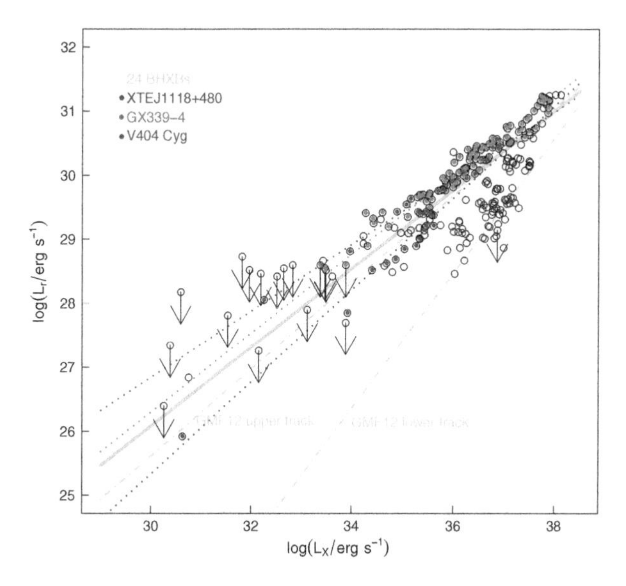

Fig. 2.7 The radio/X-ray luminosity plane of 24 hard and quiescent state black hole X-ray binaries. The *dashed grey lines* indicate the best-fit relations to the upper and lower tracks as identified in the clustering analysis of Gallo et al. (2012). Also highlighted are the three sources for which a tight non-linear correlation has been reported over a wide range in X-ray luminosity, i.e. GX339–4, V404 Cyg and XTE J1118+480 (shown individually in Fig. 2.6). The best-fitting slope for the whole data set is $\beta = 0.61 \pm 0.03$, with an intrinsic scatter of 0.31 ± 0.03 dex (From Gallo et al. (2014))

from Gallo et al. 2003, 2012, 2014; Corbel et al. 2003, 2008, 2013, and references therein). The existence of a cluster of 'outliers' at X-ray luminosities between $\sim 10^{35-37}\,\mathrm{erg\,s^{-1}}$ is apparent, so much so that Gallo et al. (2012) suggested the existence of statistically significant, separate tracks (labeled as upper and lower track; shown in gray in Fig. 2.7).

Regardless of whether a second track or a genuinely large scatter to a single relation provides a more meaningful description of the data, particularly at high X-ray luminosities, it is worth stressing that at least two sources have been observed declining along the (allegedly) steeper, lower track to then re-join the upper track. For example, the black hole X-ray binary H1743–22, as observed during the decline of its 2008 outburst (Jonker et al. 2010; Miller-Jones et al. 2012; Miller et al. 2012), started off as under-luminous in the radio band during the initial decay phase ($10^{36} \lesssim L_X \lesssim 10^{38}\,\mathrm{erg\,s^{-1}}$), proceeded to make a nearly-horizontal excursion toward lower X-ray luminosities (between $10^{36} \lesssim L_X < \lesssim 10^{36}\,\mathrm{erg\,s^{-1}}$), and finally reached a comparable radio luminosity level (for the same L_X) as GX339–4 and V404 Cyg (somewhere below $\simeq 10^{34}\,\mathrm{erg\,s^{-1}}$).

Coriat et al. (2011) interpreted the behavior of H1743−22 and other radio-quiet systems as due to the temporary onset radiatively efficient accretion. In a more theoretical work, Meyer-Hofmeister and Meyer (2014) argue that thermal photons from a weak inner accretion disk could be responsible for enhancing the photon bath available for Comptonization, and hence the hard X-ray flux, in bright hard states. Within this model, the condensation of optically thin accreting gas into an inner, keplerian disk is expected above a critical mass accretion rate once thermal conduction and Compton cooling are properly taken into account in the energy balance equations (for reasonable values of the viscosity parameter the threshold can be set around 10^{-3} times the Eddington limit; Meyer et al. 2007; Liu et al. 2006, 2007). Such disk would cease to exist at low accretion rate, as also indicated by observations (Miller et al. 2006; Reis et al. 2010; Reynolds 2013). In the context of this model, the 'radio quiet' track is thus better described as 'X-ray bright'.

Shortly after Corbel et al. (2003) reported on the non-linear radio/X-ray luminosity correlation for GX339-4 and Gallo et al. (2003) extended the analysis to a larger sample of systems, claiming the existence of a universal correlation for hard state black hole X-ray binaries, Merloni et al. (2003) and, independently, Falcke et al. (2004) presented the first evidence that, with the addition of a mass term, a similar scaling holds for super-massive black holes in radiatively inefficient AGN. Since, the so-called 'Fundamental plane of black hole activity' has become a standard tool for estimating the nature of compact radio and X-ray sources of unknown mass (see, e.g., Mïller and Gültekin 2011; Gültekin et al. 2014), and is generally taken as strong evidence for a common physics driving the jet-accretion coupling across the mass scale.

2.2.1 X-Ray/Optical-IR Correlation

Mounting evidence supports the claim that, while in the hard state, the steady jet extends all the way up to IR and often optical frequencies (Fender 2001; Chaty et al. 2003; Gallo et al. 2007; Russell et al. 2006, 2010, 2013b; Brocksopp et al. 2010; Malzac et al. 2004; Hynes et al. 2004, 2006, 2009; Casella et al. 2010). In a comprehensive work, Russell et al. (2006) assembled nearly-simultaneous IR-optical and X-ray observations of 33 X-ray binaries (black holes and neutron stars) and estimated the relative contributions of various IR/optical emission processes (companion star, direct disk emission, irradiated disk emission, jet) as a function of the X-ray luminosity. Again, evidence for a positive, non-linear correlation was found, of the form $L_{\rm opt-IR} \propto L_{\rm X}^{0.6}$, extending all the way from the peak of the hard state down to the quiescent regime. The radio/IR correlation is shown in Fig. 2.8.

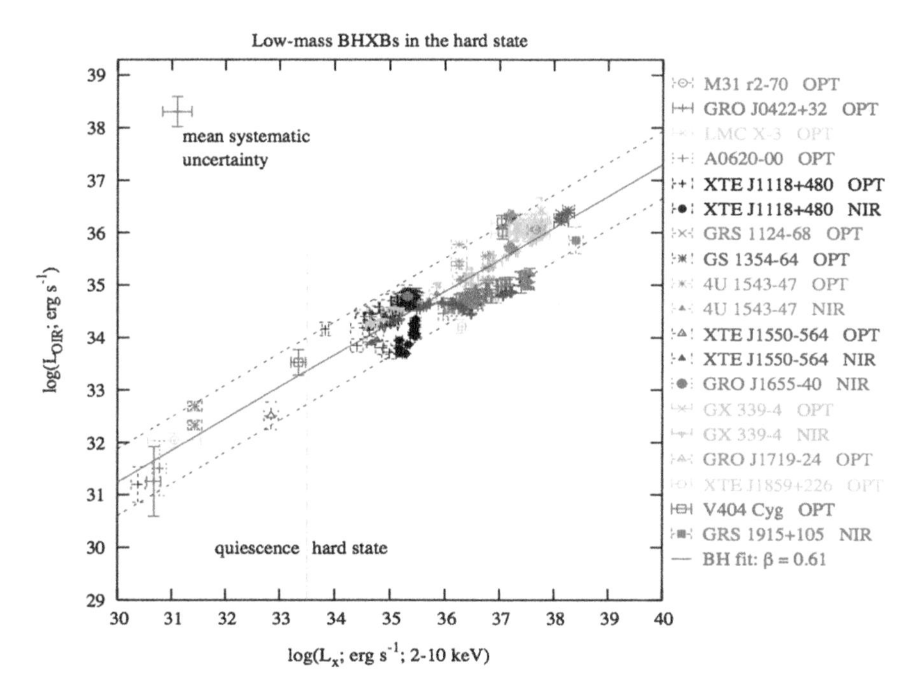

Fig. 2.8 The X-ray/optical-IR luminosity correlation for hard state black hole X-ray binaries, with $L_{\mathrm{opt-IR}} \propto L_{\mathrm{X}}^{0.6}$ (From Russell et al. (2006))

2.3 Relation Between Jet Power and Black Hole Spin

The powering mechanism of relativistic jets remains one of the most important and elusive questions in high energy astrophysics. For the purpose of modeling the evolution of galaxies and their nuclear super-massive black holes, the more powerful jets, i.e., those of radio galaxies, are often *assumed* (e.g., Sikora et al. 2007) to be powered by black hole spin, i.e. via the 'Blandford-Znajek' mechanism (Blandford and Znajek 1977). In contrast, Seyfert galaxies and lower luminosity jets in general may be powered by differential rotation coupled with magnetic fields ('Blandford-Payne' mechanism; Blandford and Payne 1982). However (for super-massive black hole at least), there is no compelling observational evidence for this to be the case.

In this respect, albeit the much lower statistics, black hole X-ray binaries offer the advantage of a neater environment (and higher count rate) to comfortably apply X-ray spectral fitting techniques for measuring the temperature, and thus the extent, of the inner accretion disk, by fitting either the reflection and iron line complex in bright hard states (hereafter reflection-fitting), or the thermal continuum in soft states (hereafter continuum-fitting; see Reynolds et al. 2014, and McClintock et al. 2013, respectively, for recent, detailed reviews; notice that the latter method can only be applied to stellar mass black holes, as the thermal disk peaks in the UV band for super-massive black holes in AGN). This translates into a measurement of the black hole spin parameter, a, which – in principle – can be compared against jet power, P_j to test the presence of a relation of the form, $P_j \propto a^2$, predicted for spin-powered jets.

2.4 Measures of Jet Power

Though several modeling uncertainties are at play in estimating the black hole spin parameter via fitting techniques, the uncertainties that affect jet power estimates for steady jets are undeniably much more severe. Simply put, the steady jet radio emission is known to be at best a poor indicator of total (kinetic plus electromagnetic) jet power.

This is readily apparent from a number of considerations; as clearly shown in Fig. 2.7, the radio luminosity of steady hard state jets varies by orders of magnitude during an outburst cycle, while the black hole spin, obviously, remains constant. To add to this, the radiative efficiency of the synchrotron process is known to be lower than a few per cent at most (Fender 2001), making any determination of jet power based on radio luminosity alone (either integrated or single frequency) highly unreliable (this might not be the case for flaring jets, though, as discussed below).

In order to circumvent this well know problem, and test for the presence of a relation between jet power and spin parameter in a sample of black hole X-ray binaries with measured spin parameter, Fender et al. (2010) adopted the relative normalization of the radio/X-ray correlation for individual sources as a proxy for their total jet power. No evidence for a positive correlation emerged. This result can be easily interpreted as due the large uncertainties in jet power, and/or, to a second extent, spin determinations (see however King et al. 2013, who report on a marginally significant correlation between the mass-scaled radio luminosity and spin parameter across a sample of 11 stellar mass black holes and 37 super-massive black holes in Seyfert galaxies). Alternatively, taken at face value, it argues against the black hole rotational energy as the main supply of power for steady jets.

The jet power proxy adopted by Fender et al. (2010) was admittedly "susceptible to errors resulting from poor sampling of events, uncertainties in Doppler boosting, assumptions about equipartition, etc." Ideally, radio lobes and/or cavities, of the kind of those that are observed around powerful radio galaxies or brightest center galaxies in galaxy clusters, provide the best (and possibly only) reliable diagnostics for the jet power, allowing for a direct measurement of the amount of work exerted by the jets on the surrounding medium. However, primarily due to the highly under-dense environment of black hole X-ray binaries (compared to AGN), such jet-ISM interaction regions are very rare among X-ray binaries (Heinz 2002). In fact, only three such structures are known in the Galaxy, and only one of those surrounds a dynamically confirmed black hole accretor, i.e. Cyg X-1 (Gallo et al. 2005).

The above considerations are largely centered around steady jets. In a recent work however, Narayan and McClintock (2012) claimed evidence for a positive correlation between spin parameter – as measured from continuum fitting – and jet power *for flaring (or ballistic) jets*. The correlation, shown in Fig. 2.9, appears to be consistent with a scaling of the form $P_{\rm j} \propto \Omega_H^2$, as predicted for 'Blandford-Znajek' jet-powering (dashed line, where $\Omega_H = a(c^3/2GM)/(1+\sqrt{1-a^2})$, is the angular frequency of the black hole horizon[1]). Here, the (mass- and distance-scaled) peak 5 GHz luminosity of the bright radio flare associated with the hard-to-soft state transitions (Sect. 2.1.3), S_ν, is taken as proxy for jet power: $P_{\rm j} = D^2(\nu S_\nu)/M_{\rm BH}$, where $M_{\rm BH}$ is the black hole mass and D its distance. Unlike the radio/X-ray correlation normalization adopted by Fender et al. (2010) for steady jets, this is indeed a model-independent quantity.

In a subsequent work, however, Russell et al. (2013a) argued against there being a significant correlation. The core of the controversy is illustrated in Fig. 2.10, most notably the bottom left panel (see Fig. 2.9 for a direct comparison with Narayan and McClintock 2012). Compared to Narayan and McClintock (2012), Russell et al. (2013a) include several additional data points that, taken at face value, do not fit the spin-powering relation. The interested reader who wishes to form an unbiased opinion on whether the above-mentioned points ought to be included in the analysis, or not, is directed to the discussion in Russell et al. (2013a), Narayan

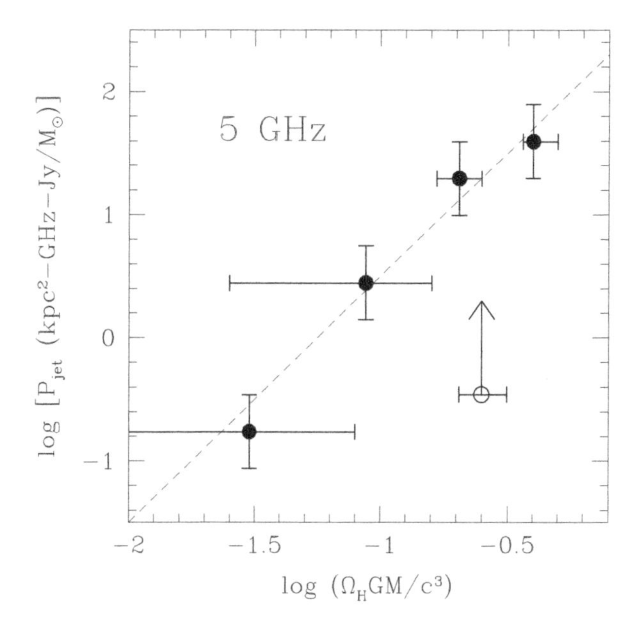

Fig. 2.9 Jet power vs. angular frequency of the black hole horizon, Ω_H, for a sample of five black hole X-ray binaries. The *dashed line* represents a scaling of the form $P_{\rm j} \propto \Omega_H^2$, as expected from Blandford and Znajek (1977) (From Narayan and McClintock (2012))

---

[1]The 'standard' $P_{\rm j} \propto a^2$ scaling relation is formally correct only for slowly spinning black holes; see Tchekhovskoy et al. (2011) for a full relativistic treatment.

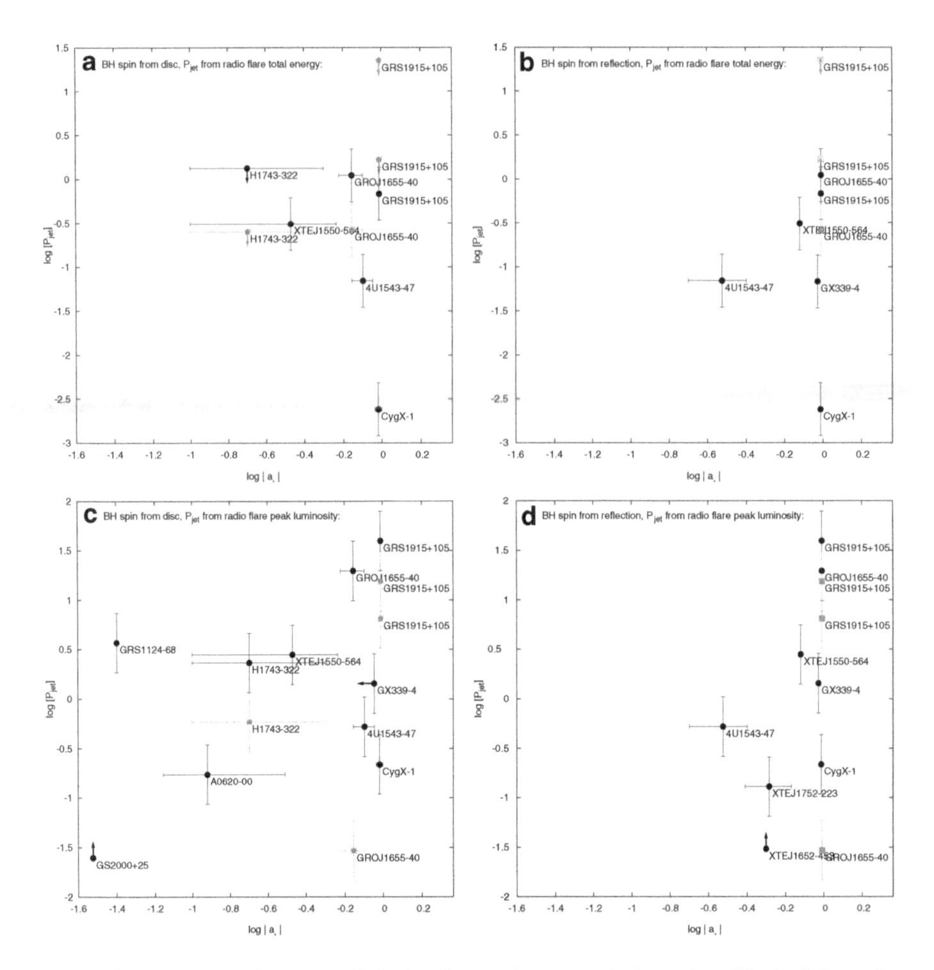

Fig. 2.10 Black hole spin measured via the disk continuum method are plotted in the *left panels*; black hole spin measured via the reflection method are plotted in the *right panels*. The jet powers estimated from total energy arguments (following Fender et al. 2010 for flaring, ballistic jets) and peak radio luminosity (following Narayan and McClintock 2012) are shown in the *upper* and *lower panels*, respectively. The *bottom left panel* allows a direct comparison with Narayan and McClintock (2012) and Fig. 2.9 (From Russell et al. (2013a))

and McClintock (2012) and Steiner et al. (2013), respectively (with regard to the right panels of Fig. 2.10, it is important to keep in mind the fact that the reflection-fitting method does not return a correlation with jet power that is consistent with the spin-powering mechanism should not be taken as evidence in favor, or against, either method).

As recognized by both groups of authors, much of the uncertainty revolves around the small number of observations of Eddington-limited systems (see discussion in Steiner et al. 2013). A further source of disagreement – this one more conceptual than practical – has to due with the implicit assumption, by Narayan and McClintock (2012), that the average duration of the bright radio flare does not vary significantly between different outbursts of the same source (this assumption is at the base of equation B15 in Steiner et al. 2013), which in turn is presented as the physical motivation for using the peak flare luminosity as proxy for jet power. While new data will likely settle the debate around this potentially groundbreaking discovery over the next few years, a word of caution is in order for the moment, as any extraordinary (important) discoveries ought to be supported by extraordinary (robust) evidence.

2.4.1 Multi-wavelength Spectral Modeling and Jet Power

Extrapolating the integrated radio luminosity to total (i.e. radiative plus kinetic) jet power relies on a number of assumptions, most importantly on the location of the optically-thin-to thick jet break, and the location of the cooling break. Standard synchrotron theory predicts the jet break frequency, ν_b, here defined as the location where the jet partially self-absorbed spectrum becomes optimally thin, to scale with the mass accretion rate and black hole mass (Heinz and Sunyaev 2003; Markoff et al. 2001, 2003, 2005). It follows that, under the reasonable assumption that the magnetic field value does not vary by orders of magnitude, ν_b is expected to scale with X-ray luminosity as $L_{\rm X}^{1/3}$. Based on a systematic literature search for coordinated multi-wavelength data of hard and quiescent state black hole X-ray binaries, however, Russell et al. (2013b) found no evidence for such a relation.

More recently, high cadence, multi-wavelength monitoring of the 2011 outburst of MAXI J1836–194 (Russell et al. 2014), showed even more severe discrepancies between the data and the expected scaling of break frequency with $L_{\rm X}$, suggesting a much higher degree of complexity in the jet spectral energy distribution evolution.

As discussed in greater detail in Fender and Gallo (2014), this multi-wavelength campaign also led to the first (albeit indirect) observational inference of the high energy synchrotron cooling break for this system ($3.2\times10^{14} \lesssim \nu_c \lesssim 4.5\times10^{14}$ Hz). For X-ray binaries, the cooling break is expected to move from the UV to the hard X-ray band as the outburst luminosity rises towards bright X-ray states (Pe'er and Markoff 2012), while the reported value for MAXI J1836–194 falls into the optical band, implying (if correct) a substantial reduction of the radiative jet power.

The importance of obtaining high quality, multi-wavelength coverage for understanding the interplay between accretion and relativistic outflows has been highlighted in several instances throughout this review. One last example to illustrate the power of this 'multi-messenger' approach is offered by the recent, multi-wavelength campaign targeting the high Galactic latitude black hole X-ray binary XTE J1118+480 in a quiescent system, close to 10^{-9} times its Eddington luminosity. Along with the nearby system in A0620-00 (Gallo et al. 2006), XTE J1118+480 is the only other system for which we can probe whether the jet extends all the way into quiescence, and, if so, if and how the hard state and quiescent

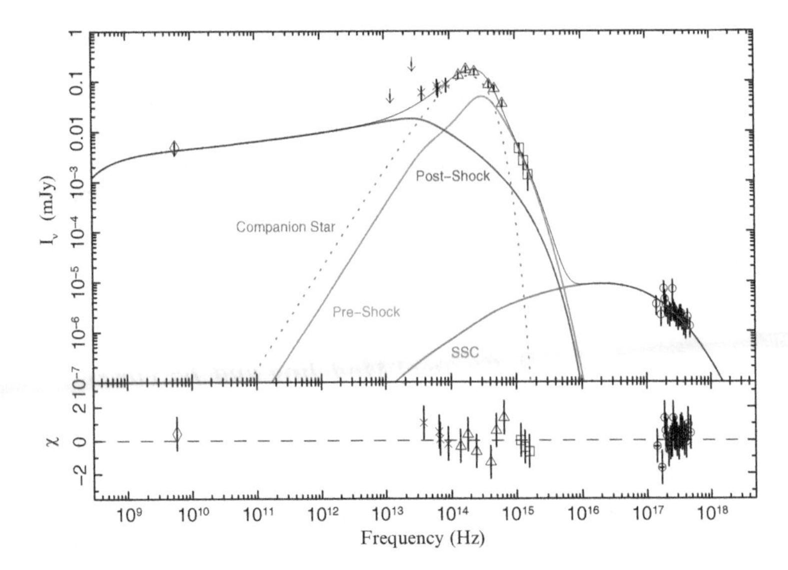

Fig. 2.11 Broadband spectral energy distribution of the black hole X-ray binary XTE J1118+480 in quiescence ($L_X/L_{Edd} \simeq 10^{-8.5}$). Fitting the data with a multi-zone jet model (Markoff et al. 2005) indicates that the jet is only weakly accelerated in this regime. As a consequence, non-thermal particle do not contribute substantially to the high energy radiation emissions and the X-ray spectrum arises from synchrotron self-Compton (From Plotkin et al. (2014))

jets differ in any physically meaningful way. The simultaneous radio (Very Large Array), NIR (William Herschel Telescope), UV (Swift) and X-ray (Chandra) data of XTE J1118+480 are shown in Fig. 2.11, along with non-simultaneous FIR data (from Spitzer and WISE). Our best-fitting jet model (cf. Markoff et al. 2005 and references therein) suggests that jet particle acceleration is weaker in quiescence compared to the hard state (see Markoff et al. 2001 for the same model fit to the hard state spectrum of XTE J1118+480). The jet base is also less magnetically dominated and more compact in quiescence, in agreement with broadband modeling for the stellar mass black hole in A0620-00 (Gallo et al. 2007), as well as the Galactic Center super-massive black hole, Sgr A*, and the super-massive black hole in M81 (Markoff et al. 2008).

A consistent picture is thus starting to emerge whereby, in quiescence, black hole jets that are too weakly accelerated for non-thermal particles to contribute significant amounts of X-rays. Conversely, the jet acceleration and magnetization appear to increase as the overall luminosity rises, leading to a significant contribution from optically thin synchrotron to the high energy radiation in the canonical hard state.

Acknowledgements The author wishes to thank all her long-term collaborators in this field, and in particular: Dave Russell, James Miller-Jones, Rob Fender, Sera Markoff, Peter Jonker, Jeroen Homan, Stephane Corbel and Rich Plotkin.

References

Begelman, M.C., Armitage, P.J.: A mechanism for hysteresis in black hole binary state transitions. ApJ **782**, 18 (2014). doi:10.1088/2041-8205/782/2/L18

Blandford, R.D., Payne, D.G.: Hydromagnetic flows from accretion discs and the production of radio jets. MNRAS **199**, 883–903 (1982)

Blandford, R.D., Znajek, R.L.: Electromagnetic extraction of energy from Kerr black holes. MNRAS **179**, 433–456 (1977)

Brocksopp, C., Jonker, P.G., Maitra, D., Krimm, H.A., Pooley, G.G., Ramsay, G., Zurita, C.: Disentangling jet and disc emission from the 2005 outburst of XTE J1118+480. MNRAS **404**, 908–916 (2010). doi:10.1111/j.1365-2966.2010.16323.x

Casella, P., Belloni, T., Stella, L.: The ABC of low-frequency quasi-periodic oscillations in black hole candidates: analogies with Z sources. ApJ **629**, 403–407 (2005). doi:10.1086/431174

Casella, P., Maccarone, T.J., O'Brien, K., Fender, R.P., Russell, D.M., van der Klis, M., Pe'Er, A., Maitra, D., Altamirano, D., Belloni, T., Kanbach, G., Klein-Wolt, M., Mason, E., Soleri, P., Stefanescu, A., Wiersema, K., Wijnands, R.: Fast infrared variability from a relativistic jet in GX 339-4. MNRAS **404**, 21–25 (2010). doi:10.1111/j.1745-3933.2010.00826.x

Chaty, S., Haswell, C.A., Malzac, J., Hynes, R.I., Shrader, C.R., Cui, W.: Multiwavelength observations revealing the evolution of the outburst of the black hole XTE J1118+480. MNRAS **346**, 689–703 (2003). doi:10.1111/j.1365-2966.2003.07115.x

Corbel, S., Nowak, M.A., Fender, R.P., Tzioumis, A.K., Markoff, S.: Radio/X-ray correlation in the low/hard state of GX 339-4. A&A **400**, 1007–1012 (2003). doi:10.1051/0004-6361:20030090

Corbel, S., Körding, E., Kaaret, P.: Revisiting the radio/X-ray flux correlation in the black hole V404 Cyg: from outburst to quiescence. MNRAS **389**, 1697–1702 (2008). doi:10.1111/j.1365-2966.2008.13542.x

Corbel, S., Coriat, M., Brocksopp, C., Tzioumis, A.K., Fender, R.P., Tomsick, J.A., Buxton, M.M., Bailyn, C.D.: The 'universal' radio/X-ray flux correlation: the case study of the black hole GX 339-4. MNRAS **428**, 2500–2515 (2013). doi:10.1093/mnras/sts215

Coriat, M., Corbel, S., Prat, L., Miller-Jones, J.C.A., Cseh, D., Tzioumis, A.K., Brocksopp, C., Rodriguez, J., Fender, R.P., Sivakoff, G.R.: Radiatively efficient accreting black holes in the hard state: the case study of H1743-322. MNRAS **414**, 677–690 (2011). doi:10.1111/j.1365-2966.2011.18433.x

Dhawan, V., Mirabel, I.F., Rodríguez, L.F.: AU-scale synchrotron jets and superluminal ejecta in GRS 1915+105. ApJ **543**, 373–385 (2000). doi:10.1086/317088

Dunn, R.J.H., Fender, R.P., Körding, E.G., Belloni, T., Cabanac, C.: A global spectral study of black hole X-ray binaries. MNRAS **403**, 61–82 (2010). doi:10.1111/j.1365-2966.2010.16114.x

Falcke, H., Körding, E., Markoff, S.: A scheme to unify low-power accreting black holes. Jet-dominated accretion flows and the radio/X-ray correlation. A&A **414**, 895–903 (2004). doi:10.1051/0004-6361:20031683

Fender, R.P.: Powerful jets from black hole X-ray binaries in low/hard X-ray states. MNRAS **322**, 31–42 (2001). doi:10.1046/j.1365-8711.2001.04080.x

Fender, R.: Jets from X-Ray Binaries, 1st edn., pp. 381–419. Cambridge Astrophysics Series, No. 39. Cambridge, UK: Cambridge University Press (2006)

Fender, R., Corbel, S., Tzioumis, T., McIntyre, V., Campbell-Wilson, D., Nowak, M., Sood, R., Hunstead, R., Harmon, A., Durouchoux, P., Heindl, W.: Quenching of the radio jet during the X-ray high state of GX 339-4. ApJ **519**, 165–168 (1999). doi:10.1086/312128

Fender, R.P., Gallo, E., Russell, D.: No evidence for black hole spin powering of jets in X-ray binaries. MNRAS **406**, 1425–1434 (2010). doi:10.1111/j.1365-2966.2010.16754.x

Fender, R.P., Maccarone, T.J., Heywood, I.: The closest black holes. MNRAS **781** (2013). doi:10.1093/mnras/sts688

Fender, R., Gallo, E.: Space science reviews (2014). doi:10.1007/s11214-014-0069-z, (arXiv:1407.3674)

Gallo, E., Fender, R.P., Pooley, G.G.: A universal radio-X-ray correlation in low/hard state black hole binaries. MNRAS **344**, 60–72 (2003). doi:10.1046/j.1365-8711.2003.06791.x

Gallo, E., Fender, R.P., Hynes, R.I.: The radio spectrum of a quiescent stellar mass black hole. MNRAS **356**, 1017–1021 (2005). doi:10.1111/j.1365-2966.2004.08503.x

Gallo, E., Fender, R.P., Miller-Jones, J.C.A., Merloni, A., Jonker, P.G., Heinz, S., Maccarone, T.J., van der Klis, M.: A radio-emitting outflow in the quiescent state of A0620-00: implications for modelling low-luminosity black hole binaries. MNRAS **370**, 1351–1360 (2006). doi:10.1111/j.1365-2966.2006.10560.x

Gallo, E., Migliari, S., Markoff, S., Tomsick, J.A., Bailyn, C.D., Berta, S., Fender, R., Miller-Jones, J.C.A.: The spectral energy distribution of quiescent black hole X-ray binaries: new constraints from spitzer. ApJ **670**, 600–609 (2007). doi:10.1086/521524

Gallo, E., Miller, B.P., Fender, R.: Assessing luminosity correlations via cluster analysis: evidence for dual tracks in the radio/X-ray domain of black hole X-ray binaries. MNRAS **423**, 590–599 (2012). doi:10.1111/j.1365-2966.2012.20899.x

Gallo, E. et al.: MNRAS **445**, 290 (2014)

Gültekin, K., Cackett, E.M., King, A.L., Miller, J.M., Pinkney, J.: Low-mass AGNs and their relation to the fundamental plane of black hole accretion. ApJ **788**, 22 (2014). doi:10.1088/2041-8205/788/2/L22

Heinz, S.: Radio lobe dynamics and the environment of microquasars. A&A **388**, 40–43 (2002). doi:10.1051/0004-6361:20020402

Heinz, S., Sunyaev, R.A.: The non-linear dependence of flux on black hole mass and accretion rate in core-dominated jets. MNRAS **343**, 59–64 (2003). doi:10.1046/j.1365-8711.2003.06918.x

Hynes, R.I., Charles, P.A., Garcia, M.R., Robinson, E.L., Casares, J., Haswell, C.A., Kong, A.K.H., Rupen, M., Fender, R.P., Wagner, R.M., Gallo, E., Eves, B.A.C., Shahbaz, T., Zurita, C.: Correlated X-ray and optical variability in V404 Cygni in quiescence. ApJ **611**, 125–128 (2004). doi:10.1086/424005

Hynes, R.I., Robinson, E.L., Pearson, K.J., Gelino, D.M., Cui, W., Xue, Y.Q., Wood, M.A., Watson, T.K., Winget, D.E., Silver, I.M.: Further evidence for variable synchrotron emission in XTE J1118+480 in outburst. ApJ **651**, 401–407 (2006). doi:10.1086/507669

Hynes, R.I., Bradley, C.K., Rupen, M., Gallo, E., Fender, R.P., Casares, J., Zurita, C.: The quiescent spectral energy distribution of V404 Cyg. MNRAS **399**, 2239–2248 (2009). doi:10.1111/j.1365-2966.2009.15419.x

Jonker, P.G., Miller-Jones, J., Homan, J., Gallo, E., Rupen, M., Tomsick, J., Fender, R.P., Kaaret, P., Steeghs, D.T.H., Torres, M.A.P., Wijnands, R., Markoff, S., Lewin, W.H.G.: Following the 2008 outburst decay of the black hole candidate H 1743-322in X-ray and radio. MNRAS **401**, 1255–1263 (2010). doi:10.1111/j.1365-2966.2009.15717.x

King, A.L., Miller, J.M., Reynolds, M.T., Gültekin, K., Gallo, E., Maitra, D.: A distinctive disk-jet coupling in the lowest-mass Seyfert, NGC 4395. ApJ **774**, 25 (2013). doi:10.1088/2041-8205/774/2/L25

Lasota, J.-P.: The disc instability model of dwarf novae and low-mass X-ray binary transients. New Astron. Rev. **45**, 449–508 (2001). doi:10.1016/S1387-6473(01)00112-9

Liu, B.F., Meyer, F., Meyer-Hofmeister, E.: An inner disk below the ADAF: the intermediate spectral state of black hole accretion. A&A **454**, 9–12 (2006). doi:10.1051/0004-6361:20065430

Liu, B.F., Taam, R.E., Meyer-Hofmeister, E., Meyer, F.: The existence of inner cool disks in the low/hard state of accreting black holes. ApJ **671**, 695–705 (2007). doi:10.1086/522619

Malzac, J., Merloni, A., Fabian, A.C.: Jet-disc coupling through a common energy reservoir in the black hole XTE J1118+480. MNRAS **351**, 253–264 (2004). doi:10.1111/j.1365-2966.2004.07772.x

Markoff, S., Falcke, H., Fender, R.: A jet model for the broadband spectrum of XTE J1118+480. Synchrotron emission from radio to X-rays in the low/hard spectral state. A&A **372**, 25–28 (2001). doi:10.1051/0004-6361:20010420

Markoff, S., Nowak, M., Corbel, S., Fender, R., Falcke, H.: Exploring the role of jets in the radio/X-ray correlations of GX 339-4. A&A **397**, 645–658 (2003). doi:10.1051/0004-6361:20021497

Markoff, S., Nowak, M.A., Wilms, J.: Going with the flow: can the base of jets subsume the role of compact accretion disk coronae? ApJ **635**, 1203–1216 (2005). doi:10.1086/497628
Markoff, S., Nowak, M., Young, A., Marshall, H.L., Canizares, C.R., Peck, A., Krips, M., Petitpas, G., Schödel, R., Bower, G.C., Chandra, P., Ray, A., Muno, M., Gallagher, S., Hornstein, S., Cheung, C.C.: Results from an extensive simultaneous broadband campaign on the underluminous active nucleus M81*: further evidence for mass-scaling accretion in black holes. ApJ **681**, 905–924 (2008). doi:10.1086/588718
McClintock, J.E., Narayan, R., Steiner, J.F.: Black hole spin via continuum fitting and the role of spin in powering transient jets. Space Sci. Rev. (2013). doi:10.1007/s11214-013-0003-9
Merloni, A., Heinz, S., di Matteo, T.: A fundamental plane of black hole activity. MNRAS **345**, 1057–1076 (2003). doi:10.1046/j.1365-2966.2003.07017.x
Meyer, F., Liu, B.F., Meyer-Hofmeister, E.: Re-condensation from an ADAF into an inner disk: the intermediate state of black hole accretion? A&A **463**, 1–9 (2007). doi:10.1051/0004-6361:20066203
Meyer-Hofmeister, E., Meyer, F.: The relation between radio and X-ray luminosity of black hole binaries: affected by inner cool disks? A&A **562**, 142 (2014). doi:10.1051/0004-6361/201322423
Mïller, J.M., Gültekin, K.: X-ray and radio constraints on the mass of the black hole in swift J164449.3+573451. ApJ **738**, 13 (2011). doi:10.1088/2041-8205/738/1/L13
Miller, J.M., Raymond, J., Homan, J., Fabian, A.C., Steeghs, D., Wijnands, R., Rupen, M., Charles, P., van der Klis, M., Lewin, W.H.G.: Simultaneous Chandra and RXTE spectroscopy of the microquasar H1743-322: clues to disk wind and jet formation from a variable ionized outflow. ApJ **646**, 394–406 (2006). doi:10.1086/504673
Miller, J.M., Reynolds, C.S., Fabian, A.C., Cackett, E.M., Miniutti, G., Raymond, J., Steeghs, D., Reis, R., Homan, J.: Initial measurements of black hole spin in GX 339-4 from Suzaku spectroscopy. ApJ **679**, 113–116 (2008). doi:10.1086/589446
Miller, J.M., Raymond, J., Fabian, A.C., Reynolds, C.S., King, A.L., Kallman, T.R., Cackett, E.M., van der Klis, M., Steeghs, D.T.H.: The disk-wind-jet connection in the black hole H 1743-322. ApJ **759**, 6 (2012). doi:10.1088/2041-8205/759/1/L6
Miller-Jones, J.C.A., Sivakoff, G.R., Altamirano, D., Coriat, M., Corbel, S., Dhawan, V., Krimm, H.A., Remillard, R.A., Rupen, M.P., Russell, D.M., Fender, R.P., Heinz, S., Körding, E.G., Maitra, D., Markoff, S., Migliari, S., Sarazin, C.L., Tudose, V.: Disc-jet coupling in the 2009 outburst of the black hole candidate H1743-322. MNRAS **421**, 468–485 (2012). doi:10.1111/j.1365-2966.2011.20326.x
Mirabel, I.F., Rodríguez, L.F.: Sources of relativistic jets in the galaxy. Ann. Rev. A&A **37**, 409–443 (1999). doi:10.1146/annurev.astro.37.1.409
Narayan, R., McClintock, J.E.: Observational evidence for a correlation between jet power and black hole spin. MNRAS **419**, 69–73 (2012). doi:10.1111/j.1745-3933.2011.01181.x
Neilsen, J., Lee, J.C.: Accretion disk winds as the jet suppression mechanism in the microquasar GRS 1915+105. Nature **458**, 481–484 (2009). doi:10.1038/nature07680
Pe'er, A., Markoff, S.: X-ray emission from transient jet model in black hole binaries. ApJ **753**, 177 (2012). doi:10.1088/0004-637X/753/2/177
Plant, D.S., Fender, R.P., Ponti, G., Muñoz-Darias, T., Coriat, M.: Revealing accretion on to black holes: X-ray reflection throughout three outbursts of GX 339-4. MNRAS **442**, 1767–1785 (2014). doi:10.1093/mnras/stu867
Plotkin, R.M., Gallo, E., Jonker, P.G.: The X-ray spectral evolution of galactic black hole X-ray binaries toward quiescence. ApJ **773**, 59 (2013). doi:10.1088/0004-637X/773/1/59
Plotkin, R.M. et al.: MNRAS (2014, submitted)
Ponti, G., Fender, R.P., Begelman, M.C., Dunn, R.J.H., Neilsen, J., Coriat, M.: Ubiquitous equatorial accretion disc winds in black hole soft states. MNRAS **422**, 11 (2012). doi:10.1111/j.1745-3933.2012.01224.x
Reis, R.C., Fabian, A.C., Miller, J.M.: Black hole accretion discs in the canonical low-hard state. MNRAS **402**, 836–854 (2010). doi:10.1111/j.1365-2966.2009.15976.x

Remillard, R.A., McClintock, J.E.: X-ray properties of black-hole binaries. Ann. Rev. A&A **44**, 49–92 (2006). doi:10.1146/annurev.astro.44.051905.092532

Reynolds, C.S.: Measuring black hole spin using X-ray reflection spectroscopy. Space Sci. Rev. (2013). doi:10.1007/s11214-013-0006-6

Reynolds, M.T., Reis, R.C., Miller, J.M., Cackett, E.M., Degenaar, N.: The quiescent X-ray spectrum of accreting black holes. MNRAS **441**, 3656–3665 (2014). doi:10.1093/mnras/stu832

Russell, D.M., Fender, R.P., Hynes, R.I., Brocksopp, C., Homan, J., Jonker, P.G., Buxton, M.M.: Global optical/infrared-X-ray correlations in X-ray binaries: quantifying disc and jet contributions. MNRAS **371**, 1334–1350 (2006). doi:10.1111/j.1365-2966.2006.10756.x

Russell, D.M., Maitra, D., Dunn, R.J.H., Markoff, S.: Evidence for a compact jet dominating the broad-band spectrum of the black hole accretor XTE J1550-564. MNRAS **405**, 1759–1769 (2010). doi:10.1111/j.1365-2966.2010.16547.x

Russell, D.M., Miller-Jones, J.C.A., Maccarone, T.J., Yang, Y.J., Fender, R.P., Lewis, F.: Testing the jet quenching paradigm with an ultradeep observation of a steadily soft state black hole. ApJ **739**, 19 (2011). doi:10.1088/2041-8205/739/1/L19

Russell, D.M., Gallo, E., Fender, R.P.: Observational constraints on the powering mechanism of transient relativistic jets. MNRAS **431**, 405–414 (2013a). doi:10.1093/mnras/stt176

Russell, D.M., Markoff, S., Casella, P., Cantrell, A.G., Chatterjee, R., Fender, R.P., Gallo, E., Gandhi, P., Homan, J., Maitra, D., Miller-Jones, J.C.A., O'Brien, K., Shahbaz, T.: Jet spectral breaks in black hole X-ray binaries. MNRAS **429**, 815–832 (2013b). doi:10.1093/mnras/sts377

Russell, T.D., Soria, R., Miller-Jones, J.C.A., Curran, P.A., Markoff, S., Russell, D.M., Sivakoff, G.R.: The accretion-ejection coupling in the black hole candidate X-ray binary MAXI J1836-194. MNRAS **439**, 1390–1402 (2014). doi:10.1093/mnras/stt2498

Sikora, M., Stawarz, Ł., Lasota, J.-P.: Radio loudness of active galactic nuclei: observational facts and theoretical implications. ApJ **658**, 815–828 (2007). doi:10.1086/511972

Soleri, P., Fender, R.: On the nature of the 'radio-quiet' black hole binaries. MNRAS **413**, 2269–2280 (2011). doi:10.1111/j.1365-2966.2011.18303.x

Steiner, J.F., McClintock, J.E., Narayan, R.: Jet power and black hole spin: testing an empirical relationship and using it to predict the spins of six black holes. ApJ **762**, 104 (2013). doi:10.1088/0004-637X/762/2/104

Stirling, A.M., Spencer, R.E., de la Force, C.J., Garrett, M.A., Fender, R.P., Ogley, R.N.: A relativistic jet from Cygnus X-1 in the low/hard X-ray state. MNRAS **327**, 1273–1278 (2001). doi:10.1046/j.1365-8711.2001.04821.x

Tchekhovskoy, A., Narayan, R., McKinney, J.C.: Efficient generation of jets from magnetically arrested accretion on a rapidly spinning black hole. MNRAS **418**, 79–83 (2011). doi:10.1111/j.1745-3933.2011.01147.x

Chapter 3
Launching of Active Galactic Nuclei Jets

Alexander Tchekhovskoy

Abstract As black holes accrete gas, they often produce relativistic, collimated outflows, or jets. Jets are expected to form in the vicinity of a black hole, making them powerful probes of strong-field gravity. However, how jet properties (e.g., jet power) connect to those of the accretion flow (e.g., mass accretion rate) and the black hole (e.g., black hole spin) remains an area of active research. This is because what determines a crucial parameter that controls jet properties—the strength of large-scale magnetic flux threading the black hole—remains largely unknown. First-principles computer simulations show that due to this, even if black hole spin and mass accretion rate are held constant, the simulated jet powers span a wide range, with no clear winner. This limits our ability to use jets as a quantitative diagnostic tool of accreting black holes. Recent advances in computer simulations demonstrated that accretion disks can accumulate large-scale magnetic flux on the black hole, until the magnetic flux becomes so strong that it obstructs gas infall and leads to a magnetically-arrested disk (MAD). Recent evidence suggests that central black holes in jetted active galactic nuclei and tidal disruptions are surrounded by MADs. Since in MADs both the black hole magnetic flux and the jet power are at their maximum, well-defined values, this opens up a new vista in the measurements of black hole masses and spins and quantitative tests of accretion and jet theory.

3.1 Introduction

Black holes (BHs) of all sizes produce relativistic jets, one of the most spectacular manifestations of BH accretion. Figure 3.1 illustrates that jet-producing accretion systems span nearly 10 orders of magnitude in BH mass: from stellar-mass BHs in BH binaries (BHBs, Fig. 3.1c) and gamma-ray bursts (GRBs, Fig. 3.1d, e), with masses $M_{\rm BH} \sim$ few $\times\, M_\odot$, to supermassive BHs in active galactic nuclei (AGN, Fig. 3.1a), with masses $M_{\rm BH} \sim 10^{6-10} M_\odot$, where $M_\odot \approx 2 \times 10^{33}$ g is a solar mass. If a single mechanism is at work across the entire BH mass scale, it should be scale

---

A. Tchekhovskoy (✉)
Lawrence Berkeley National Laboratory, 1 Cyclotron Rd, Berkeley, CA 94720, USA
e-mail: atchekho@berkeley.edu

I. Contopoulos et al. (eds.), *The Formation and Disruption of Black Hole Jets*,
Astrophysics and Space Science Library 414,
DOI 10.1007/978-3-319-10356-3_3

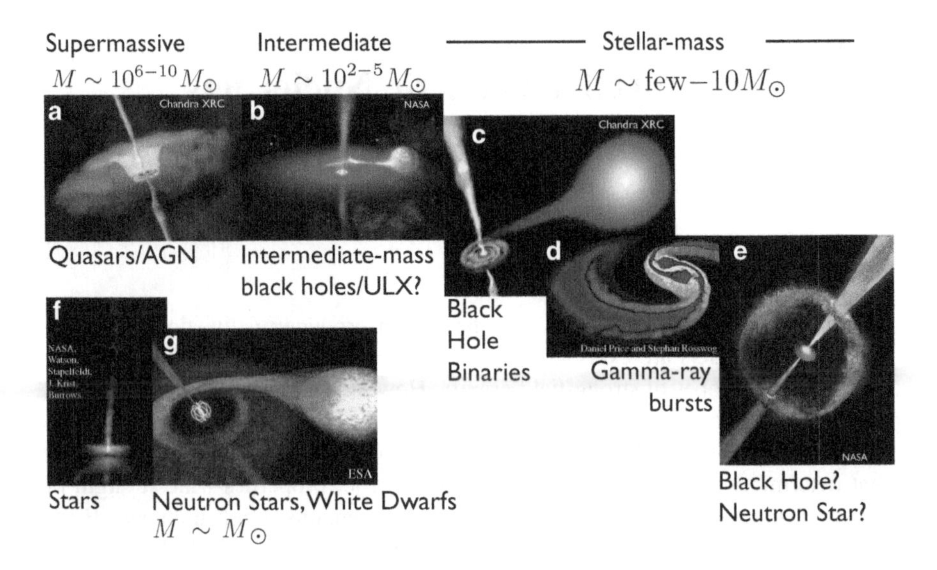

Fig. 3.1 Black holes of all sizes produce jets. BHs come in two broad categories: supermassive, with masses ranging between millions and billions of solar masses, and stellar-mass BHs, with masses ranging from a few to tens of solar masses. Supermassive BHs are found at the centers of AGN (panel (**a**)), and stellar-mass BHs are found in binary systems (panel (**c**)), or formed as a result of binary neutron star mergers (panel (**d**)) and core collapse of massive stars that is thought to give rise to GRBs (panel (**e**)). Recent evidence suggests there is a third class of *intermediate-mass* BHs with masses bridging the mass gap (Hui and Krolik 2008; Farrell et al. 2009; Davis et al. 2011; Webb et al. 2012; Straub et al. 2014). To be fair to non-BH systems, the presence of an event horizon is not a necessity for producing jets: neutron stars (panels (**e**) and (**g**)) and white dwarfs (panel (**g**)), as well as normal stars (panel (**f**)), also produce jets

invariant. Magnetic fields are a promising agent for jet production because they are abundant in astrophysical plasmas and because the properties of magnetically-powered jets scale trivially with BH mass (Blandford and Znajek 1977; Chiueh et al. 1991; Heinz and Sunyaev 2003; Tchekhovskoy et al. 2008, 2009, 2010).

How are jets magnetically launched? Figure 3.2 shows a cartoon depiction of this. Consider a vertical magnetic field line attached on one end to a perfectly conducting sphere, which represents the central compact object, and on the other end to a perfectly conducting "ceiling", which represents the ambient medium (panel a). As the sphere is spinning at an angular frequency Ω, after N turns the initially vertical field line develops N toroidal field loops (panel b; we assume the ceiling is a distance $z_{\rm ceil} \ll c/\Omega$ away from the central object, an assumption we later relax). This magnetic spring pushes against the ceiling due to the pressure of the toroidal field. As more toroidal loops form and the toroidal field becomes stronger, the spring pushes away the ceiling and accelerates any plasma attached to it along the rotation axis, forming a jet (panels (c) and (d) in Fig. 3.2, see the caption for details; after the ceiling is pushed away, the final state is independent of $z_{\rm ceil}$). In the case when the central body is a BH, which does not have a surface, the rotation of space-time

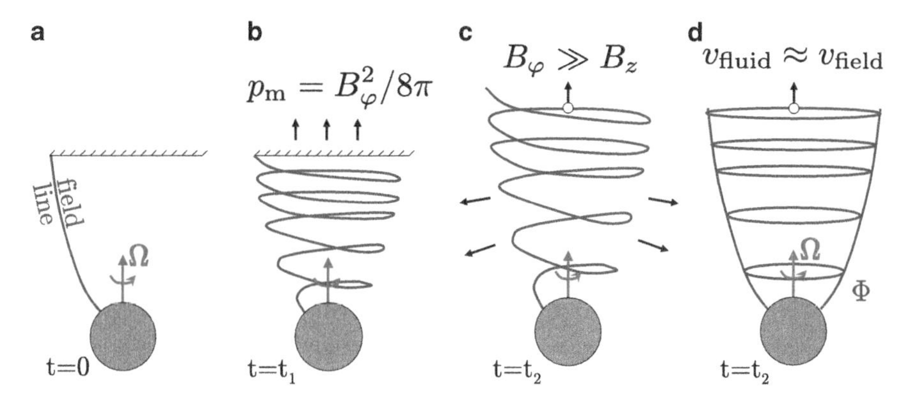

Fig. 3.2 Illustration of jet formation by magnetic fields. Panel (**a**): Consider a purely poloidal (i.e., toroidal field vanishes, $B_\varphi = 0$) field line attached on one end to a stationary "ceiling" (which represents the ambient medium and is shown with *hashed horizontal line*) and on the other end to a perfectly conducting sphere (which represents the central BH or neutron star and is shown with *gray filled circle*) rotating at an angular frequency Ω. Panel (**b**): After N rotations, at time $t = t_1$, the initially purely poloidal field line develops N toroidal loops. This magnetic spring pushes against the "ceiling" with an effective pressure $p_m \sim B_\varphi^2/8\pi$ due to the toroidal field, B_φ. As time goes on, more toroidal loops form, and the toroidal field becomes stronger. Panel (**c**): At some later time, $t = t_2$, the pressure becomes so large that the magnetic spring, which was twisted by the rotation of the sphere, pushes away the "ceiling" and accelerates the plasma attached to it along the rotation axis, forming a jet. Asymptotically far from the center, the toroidal field is the dominant field component and determines the dynamics of the jet. Panel (**d**): It is convenient to think of the jet as a collection of toroidal field loops that slide down the poloidal field lines and accelerate along the jet under the action of their own pressure gradient and tension (hoop stress). The rotation of the sphere continuously twists the poloidal field into new toroidal loops at a rate that, in steady state, balances the rate at which the loops move downstream. As we will see below (Sects. 3.3–3.4), the power of the jet is determined by two parameters: the rotational frequency of the central object, Ω, and the radial magnetic flux threading the object, Φ

causes the rotation of the field lines, and jets form in a similar fashion via a process referred to as Blandford-Znajek mechanism (BZ, hereafter) (Blandford and Znajek 1977).

3.2 Physical Description of Highly Magnetized Plasmas

To describe the motion of magnetized plasma on a curved space-time of a BH from first-principles is a formidable task. The relativistic analog of second Newton's law "$\mathbf{F} = m\mathbf{a}$" in the absence of gravity takes the form:

$$\rho_c \mathbf{E} + \mathbf{j} \times \mathbf{B} = \rho \frac{\mathrm{d}(\gamma \mathbf{v})}{\mathrm{d}t}, \tag{3.1}$$

where all quantities are measured in the lab frame: ρ is mass density and ρ_c is electric charge density, $\mathbf{E}$ and $\mathbf{B}$ are electric and magnetic field strengths, $\mathbf{j}$ is the electric current density, and γ and $\mathbf{v}$ are the Lorentz factor and velocity of the plasma. To close the system, we complement Eq. (3.1) with Maxwell's equations, $\nabla \cdot \mathbf{E} = 4\pi\rho_c$, $\partial\mathbf{E}/\partial t = c\nabla \times \mathbf{B} - 4\pi\mathbf{j}$, and $\partial\mathbf{B}/\partial t = -c\nabla \times \mathbf{E}$. For simplicity, in Eq. (3.1) we dropped non-magnetic forces (e.g., the thermal pressure term) on the left hand side.

In order to make progress, simplifications are necessary. The first simplification that is usually made is the assumption that the fluid is *ideal*, or infinitely conducting. That is, in the fluid frame the Ohm's law takes the form: $\mathbf{j}' = \sigma\mathbf{E}'$ with $\sigma = \infty$, where the prime symbols indicate quantities measured in the fluid frame. Since $\mathbf{j}'$ is finite, the electric field in the frame of the fluid vanishes: $\mathbf{E}' \propto \mathbf{E} + \mathbf{v} \times \mathbf{B}/c = 0$. This gives us the *ideal magnetohydrodynamics* (ideal MHD) approximation.

For highly magnetized plasmas even further simplification is possible. A particularly useful approach is to utilize a *force-free* approximation. It works well for the cases when magnetic field is so strong that the effects of inertia of plasma particles attached to the field lines as well as of pressure forces are negligible. This amounts to neglecting the right-hand side in Eq. (3.1). The resulting equation states that the left-hand side of Eq. (3.1), the force, vanishes. Hence, the name: *force-free* approximation.

Due to space constraints we will not be able to describe the details of ideal MHD and force-free approaches. We will only mention that both approaches can be generalized to a curved space-time of a spinning BH, and the resulting sets of equations can be solved either analytically or numerically, with examples of such solutions given below.

3.3 Extraction of BH Rotational Energy via Magnetic Field

Magnetic Field Configuration Consider a rotating BH threaded with a force-free magnetic field. The simplest magnetic field configuration of this type is a BH with a nonzero magnetic monopole charge, as illustrated in Fig. 3.3a. While such a configuration may seem unrealistic—after all, magnetic monopoles do not exist!—this not quite so. Energetically, this is equivalent to a *split-monopole* configuration, in which magnetic field direction reverses in the equatorial plane, as shown in Fig. 3.3b. The split-monopole has no monopole charge on the BH: the magnetic field is sourced by an equatorial current sheet. The modern thinking is that the current sheet represents the current carried by the plasma in a razor-thin accretion disk orbiting the BH (Blandford and Znajek 1977). As we will see below, this is indeed a good description of reality (see Fig. 3.9b). An even more realistic field configuration is a parabolic one, illustrated in Fig. 3.3c: it also has an equatorial current sheet, but now the field lines thread not only the BH but also the sheet. This configuration is qualitatively similar to what is found in numerical BH accretion-jet simulations, as we will see below (see, e.g., Fig. 3.9b).

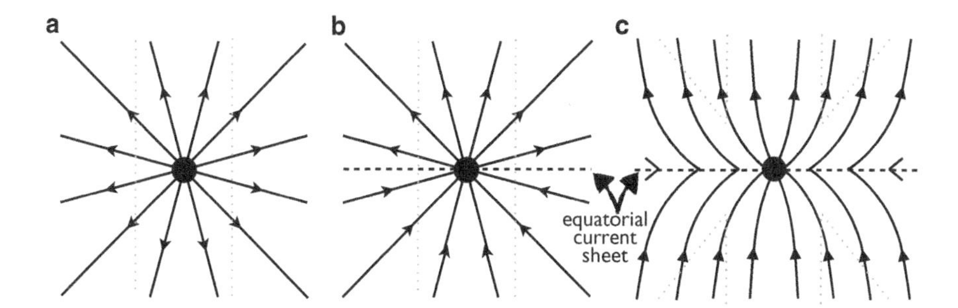

Fig. 3.3 Magnetic field configurations around a BH. First, consider a non-rotating BH. Panel (**a**) shows a BH, indicated with the *black filled circle*, threaded with *monopolar* (radial) magnetic field B_r. This is the simplest configuration for computing BH power output. Since it implies the presence of a nonzero magnetic (monopole) charge on the hole, by analogy with the electric charge, this configuration is a solution to force-free equations. Whereas it might seem that this solution is artificial (because magnetic monopoles do not exist), with a small modification we can convert it to a physical solution. Panel (**b**) shows a BH with a *split-monopolar* magnetic field. In this configuration, the magnetic field is also radial but changes direction across the equatorial plane. Unlike panel (**a**), here the radial magnetic field is monopole-free and is sourced by an equatorial current sheet, which is shown with the *black dashed line*. The modern thinking is that the current sheet represents the current carried by the plasma in a razor-thin accretion disk orbiting the BH (Blandford and Znajek 1977). (As we will see below, in jet-producing BHs, accretion disks are thought to be geometrically-thick (Sect. 3.5), therefore their current is a distributed current rather than a singular current (see Fig. 3.9b).) Due to the equatorial symmetry, the magnetic fields are in force-balance across the current sheet; therefore, just like the configuration in panel (**a**), what we have here is also a solution to force-free equations. Now suppose the BH is spinning. Due to rotation, each field line winds up, develops an azimuthal field component B_φ (not shown here) and rotates at an angular frequency $\Omega_{\rm F}$ (which can vary from one field line to another but is a constant along each of the field lines). The rotation brings about a characteristic length scale, $R_{\rm LC} = c/\Omega_{\rm F}$, that is indicated by *gray vertical dotted lines* and is called the *light cylinder* (LC, Eq. 3.4). Panel (**c**) shows a BH with a parabolic magnetic field. This configuration is more realistic than the ones in panels (**a**) and (**b**): it is closer to what is found in global numerical simulations of accreting BHs (as we will later; see, e.g., Fig. 3.9b). In this configuration the field lines thread not only the BH but also the current sheet. The field lines threading the sheet can be thought of as being powered by the rotation of the razor-thin disk, which is represented by the current sheet. Since angular frequency $\Omega_{\rm F}$ differs from one field line to another in this configuration (e.g., disk field lines can rotate slower than BH field lines), the LC does not have a cylindrical shape

Black Hole "Hairs" A *nonrotating* BH is charaterized by two "hairs": mass $M_{\rm BH}$ and charge $Q_{\rm BH}$. The charge of astrophysical BHs is thought to be negligibly small to affect the gravity of the BH,[1] thus we will set $Q_{\rm BH} = 0$ for the rest of our discussion. In addition to $M_{\rm BH}$, an astrophysical *rotating* BH is characterized by the value of its angular momentum $J_{\rm BH}$, or, equivalently, dimensionless BH spin, $a = J_{\rm BH}/J_{\rm max}$. Here $r_g = GM_{\rm BH}/c^2$ is BH gravitational radius, $J_{\rm max} = M_{\rm BH} r_g c$ is the maximum angular momentum a BH of mass $M_{\rm BH}$ can have, and c is the speed

[1]If a BH were strongly charged, it would attract oppositely-charged particles, which would neutralize the BH charge.

of light. Thus defined, BH spin varies from 0 (nonspinning BHs) to 1 (maximally spinning BHs).

Rotation of Black Hole and Magnetospheric Structure BH rotation causes the inertial frames to be dragged about the BH at an angular frequency, $\Omega \approx \Omega_{\rm H} \times (r/r_{\rm H})^{-3}$, where

$$\Omega_{\rm H} = \frac{ac}{2r_{\rm H}} \tag{3.2}$$

and $r_{\rm H} = r_g(1 + \sqrt{1-a^2})$ are the angular frequency and radius of BH event horizon, respectively. For convenience we will also use a normalized version of $\Omega_{\rm H}$:

$$\omega_{\rm H} \equiv \frac{\Omega_{\rm H}}{\Omega_{\rm H,max}} \equiv \frac{2r_g\Omega_{\rm H}}{c} \equiv \frac{a}{1+\sqrt{1-a^2}}. \tag{3.3}$$

Frame-dragging attempts to force different parts of the field line to rotate at different frequencies: $\Omega = \Omega_{\rm H}$ near the BH and $\Omega = 0$ at infinity. However, in a steady state every field line must rotate at a single angular frequency.[2] Understandably, this forces a field line to choose an in-between value of Ω, which turns out to be close to the average of the two frequencies, $\Omega_{\rm F} \simeq 0.5\Omega_{\rm H}$ (Blandford and Znajek 1977; Tchekhovskoy et al. 2010).

The rotational frequency introduces a characteristic length scale into the problem,

$$R_{\rm LC} = \frac{c}{\Omega_{\rm F}}, \tag{3.4}$$

which is referred to as the *light cylinder* (LC) radius. It has a clear physical meaning: if a magnetic field line rigidly rotates at an angular frequency $\Omega_{\rm F}$, its rotational velocity reaches the speed of light at a cylindrical radius $R_{\rm LC}$. At the LC, special relativistic effects become important, and all components of electromagnetic field become comparable: $B_\varphi \sim E_\theta \sim B_r \sim \Phi_{\rm BH}/2\pi R_{\rm LC}^2$, where $\Phi_{\rm BH}$ is the magnetic flux through the BH, B_r and B_φ are radial and toroidal magnetic field components, and E_θ is the θ-component of electric field. Note that the LC is of cylindrical shape only if $\Omega_{\rm F} = \text{const}$; if this is not the case, as illustrated in Fig. 3.3c, the shape of the LC can be very different from a cylinder. Yet, it is often referred to by the same name—"light cylinder"—regardless of this.

[2]If this were not the case, differential rotation between different parts of the field line would cause the production of new toroidal magnetic field loops and the violation of steady-state assumption, similar to the process illustrated in Fig. 3.2a, b.

3.4 BH Spindown Power

We can approximately compute the spindown power of a BH as the product of characteristic values of the Poynting flux and the area of the LC:

$$P \sim \frac{c}{4\pi}(\mathbf{E}\times\mathbf{B})_r \times 4\pi R_{\rm LC}^2 = cE_\theta B_\varphi R_{\rm LC}^2 \sim \frac{1}{16\pi^2 c}\Phi_{\rm BH}^2\Omega_{\rm H}^2 = \frac{c}{64\pi^2 r_g^2}\Phi_{\rm BH}^2\omega_{\rm H}^2, \tag{3.5}$$

where we used the fact that $\Omega_{\rm F} \approx 0.5\Omega_{\rm H}$ and that at the LC one has $E_\theta \sim B_\varphi \sim \Phi_{\rm BH}/2\pi R_{\rm LC}^2$. This estimate is within a factor of 2 of a more detailed calculation that gives the 2nd-order accurate expansion of spindown power in powers of $\omega_{\rm H}$ (Tchekhovskoy et al. 2010),

$$P_{\rm BZ2} = \frac{c}{96\pi^2 r_g^2}\Phi_{\rm BH}^2\omega_{\rm H}^2. \qquad \text{(2nd order BZ2 expansion, } a \lesssim 0.95) \tag{3.6}$$

For rapidly rotating BHs, magnetic field lines tend to bunch up toward the rotational axis, which leads to higher order corrections in the expression for jet power relative to the 2nd order expansion (3.6). These corrections are captured by the 6th order accurate expansion (Tchekhovskoy et al. 2010):

$$P_{\rm BZ6} = \frac{\kappa c}{16\pi r_g^2}\Phi_{\rm BH}^2\omega_{\rm H}^2 \times f(\omega_{\rm H}), \qquad \text{(6th order BZ6 expansion, all } a) \tag{3.7}$$

where factor

$$f(\omega_{\rm H}) = 1 + 0.35\omega_{\rm H}^2 - 0.58\omega_{\rm H}^4 \tag{3.8}$$

is a high-spin correction. Here prefactor κ weakly depends on magnetic field geometry, varying from $\kappa \approx 0.045$ for collimating, parabolic magnetic field geometry to $\kappa = 1/6\pi \approx 0.053$ for (split-)monopolar geometry (Tchekhovskoy et al. 2010). Figure 3.4 shows that the second-order BZ2 formula remains accurate for $a \lesssim 0.95$, but over-predicts the power by about 30 % as $a \to 1$.

These results are qualitatively similar to the findings of the pioneering Blandford-Znajek (BZ) work (Blandford and Znajek 1977), but there is difference on a quantitative level. BZ performed an expansion of BH energy loss rate in powers of a, not $\omega_{\rm H}$, and wrote down the following scaling:

$$P_{\rm BZ} = \frac{\kappa c}{48\pi r_g^2}\Phi_{\rm BH}^2 a^2 \qquad \text{(standard BZ expansion, low} - \text{spin limit, } a^2 \ll 1), \tag{3.9}$$

As is clear from Fig. 3.4, this low-spin approximation, which we refer to as the standard BZ formula, remains accurate up to $a \lesssim 0.5$ (Komissarov 2001; Tanabe and Nagataki 2008) and for high spin under-predicts the energy loss rate by a factor

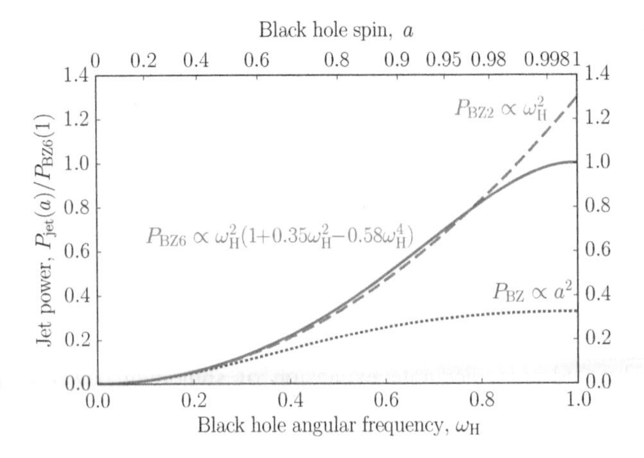

Fig. 3.4 Comparison of various approximations for jet power, $P_{\rm jet}$, versus BH dimensionless angular frequency, $\omega_{\rm H}$ (lower x-axis), and BH spin, a (upper x-axis). All powers are normalized to the maximum achievable power in the BZ6 approximation (Eq. 3.7), $P_{\rm jet} \propto \omega_{\rm H}^2(1 + 0.35\omega_{\rm H}^2 - 0.58\omega_{\rm H}^4)$, which is shown with *solid red line* and which remains accurate for all values of BH spin (Tchekhovskoy et al. 2010). A simpler BZ2 approximation (Eq. 3.6), $P_{\rm jet} \propto \omega_{\rm H}^2$, shown with *green dashed line*, is accurate up to $a \lesssim 0.95$, beyond which it over-predicts the power by about 30 %. The standard BZ approximation (Eq. 3.9; Blandford and Znajek 1977), $P_{\rm jet} \propto a^2$, shown with *blue dotted line*, remains accurate only for moderate values of spin, $a \lesssim 0.5$, beyond which it under-predicts the true jet power by a factor of $\approx$3

of 3. Therefore, when quantitative understanding of BH jet power is required, it is advantageous to use Eq. (3.6) (for $a \lesssim 0.95$) or Eq. (3.7) (for all values of spin).

3.5 When Are Jets Launched by Accreting Black Holes?

The factors that control whether a BH produces jets are not well-understood. Observationally, it is clear that jet production is intimately related to the *spectral state* of the accretion disk (Fender et al. 2004). In Fig. 3.5, we identify 3 such states (see Remillard and McClintock 2006 for a detailed review). We classify them by their normalized luminosity, or the Eddington ratio, $\lambda = L/L_{\rm Edd}$, where L is accretion luminosity and $L_{\rm Edd} \approx 1.2 \times 10^{38} M_{\rm BH}/M_{\odot}$ (erg s^{-1}) is the Eddington luminosity at which the outward radiation force on the electrons balances the inward gravitational force on the ions (see e.g. Frank et al. 2002).

Spectral States of BH Accretion: Thin-disk State We will start our discussion with the simplest and best understood state of accretion: the standard, geometrically-thin disk state, which is also referred to as "high/soft" state or "thermal" state (Shakura and Sunyaev 1973; Novikov and Thorne 1973); for a detailed description of these states and their properties with an emphasis on the accretion flow structure and emission, see Begelman (1985). This state, illustrated in Fig. 3.5b, occurs

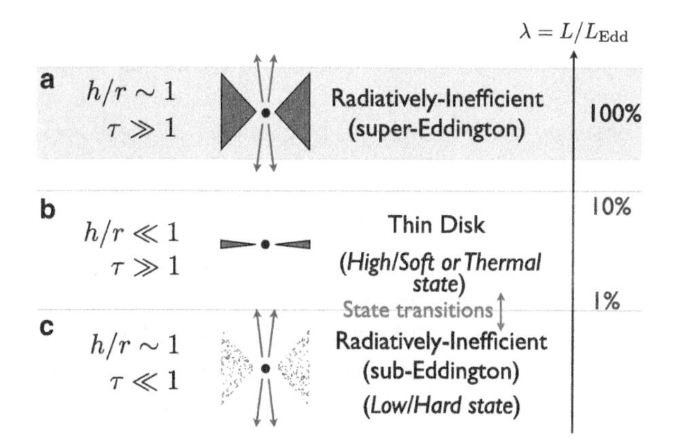

Fig. 3.5 Simplified picture of spectral states of BH accretion. The states are ordered by the value of dimensionless luminosity: the Eddington ratio, $\lambda = L/L_{\rm Edd}$. The BH is shown with a *black circle*, and the accretion disk with a *dark blue wedge*. The presence of jets is indicated with *blue arrows*. For convenience of presentation, we start with panel (**b**), move on to (**c**), and then to (**a**). Panel (**b**): The "thin disk" state—or as it is also referred to, "high/soft" or "thermal" state—occurs between $\sim$1 % and few $\times$ 10 % of $L_{\rm Edd}$ and is the best understood state of all accretion spectral states. Disk gas rotates on Keplerian orbits. Viscosity causes it to gradually lose its angular momentum and very slowly march inward, moving from one Keplerian orbit to another. The gas is optically-thick, $\tau \gg 1$, which means that it radiates as a blackbody. In fact, it takes much longer for the gas to reach the BH than to radiate viscously generated energy, so all viscously generated energy is radiated as the location where it was produced, i.e., the disk is *radiatively-efficient*. This keeps the disk cool and geometrically-thin, $h/r \ll 1$. Panel (**c**): Let us imagine that we start with the "thin disk" state and decrease $\dot{M}$ such that $L \lesssim 0.01 L_{\rm Edd}$. The disk enters the "low/hard" state: disk density drops, and the inner disk becomes optically thin, $\tau \ll 1$, which is illustrated in the figure by the light shading of the disk wedge. This makes it difficult for the disk to cool. Now, the viscously-generated energy, instead of being radiated right away, is locked into the accretion flow, with most of the energy ending up in the BH and only a small fraction escaping as radiation, i.e., the disk is *radiatively-inefficient*. This causes the disk to get hotter and puff up, leading to $h/r \sim 1$. Panel (**a**): Now let us imagine that we start with the "thin disk" state and increase $\dot{M}$ well above Eddington, such that $L \gtrsim L_{\rm Edd}$. In order to accommodate the increased mass supply, disk density, thickness, and radial velocity increase. Disk rotation becomes sub-Keplerian. Due to the higher disk density and thickness, the disk becomes optically-thick, $\tau \gg 1$, and the time it takes for photons to diffuse out of the disk body increases and becomes longer than the time it takes for the gas to reach the BH. This means that most of the photons are locked up inside the accretion flow and end up in the BH, with only a small fraction escaping, i.e., the disk is *radiatively-inefficient*. **Jet launching and spectral state transitions**: Spectral states with geometrically-thick disks, like in panels (**a**) and (**c**), produce jets (indicated with *blue arrows*), whereas geometrically-thin disks do not. In addition to continuous jets discussed above, transient jets can be produced during "hard" to "soft" disk spectral *state transitions* (state c$\rightarrow$b). See text for details

between about one and a few tens of per cent of Eddington, or roughly $L \sim (0.01{-}1)L_{\rm Edd}$ (see e.g. Maccarone 2003). Disk gas rotates on Keplerian orbits. Viscosity causes it to gradually lose its angular momentum and very slowly march inward, moving from one Keplerian orbit to another. The source of viscosity is most likely the magnetorotational instability (MRI, Balbus and Hawley 1991), which amplifies any weak magnetic field in the accretion disk to sub-equipartition levels and transports angular momentum outward and gas inward. The gas is optically-thick, with optical depth $\tau \gg 1$, which means that parcels of gas on different Keplerian orbits radiate as blackbodies at their own temperatures (hence the name "thermal" state). The integrated spectrum is often referred to as "multicolor blackbody" spectrum. Due to the low radial velocity, it takes much longer for the gas to reach the BH than to radiate viscously generated energy, so all viscously generated energy is radiated *locally*, at the location where it is produced, i.e., the disk is *radiatively-efficient*. This keeps the disk cool and geometrically-thin, $h/r \ll 1$. The radiative efficiency of the disk, defined as $\epsilon = L/\dot{M}c^2$, is between 0.05 for non-spinning BHs and $\simeq 0.3$ for nearly maximally-spinning BHs (Novikov and Thorne 1973) (see also Shapiro and Teukolsky 1986; Frank et al. 2002). In many observational applications, authors often set disk radiative efficiency to a characteristic value $\epsilon_{\rm d} = 0.1$ and define Eddington mass accretion rate $\dot{M}_{\rm Edd} = L_{\rm Edd}/\epsilon_{\rm d}c^2 = 10L_{\rm Edd}/c^2$.

Sub-Eddington Thick-disk State Let us imagine that we start with the "thin disk" state and decrease $\dot{M}$ such that $L \lesssim 0.01L_{\rm Edd}$. The disk enters the "low/hard" state illustrated in Fig. 3.5c: disk density drops, and the inner disk becomes optically thin, $\tau \ll 1$ (Esin et al. 1997, 1998). This makes it difficult for the disk to cool. Now, the viscously-generated energy, instead of being radiated right away, is locked into the accretion flow, with most of the energy ending up in the BH and only a small fraction escaping as radiation, i.e., the disk is *radiatively-inefficient*. This causes the disk to get hotter and puff up, causing it to become geometrically-thick, $h/r \sim 1$. See Yuan and Narayan (2014) for a recent review of radiatively-inefficient sub-Eddington accretion.

Super-Eddington Thick-disk State Now let us imagine that we start with the "thin disk" state and increase $\dot{M}$ well above $\dot{M}_{\rm Edd}$, such that $L \gtrsim L_{\rm Edd}$, as illustrated in Fig. 3.5a. In order to accommodate the increased mass supply, disk density, thickness, and radial velocity increase. Disk rotation becomes sub-Keplerian. Due to the higher disk density and thickness, the disk becomes optically-thick, $\tau \gg 1$, and the time it takes for photons to diffuse out of the disk body increases and becomes longer than the time it takes for the gas to reach the BH. This means that most of the photons are locked up inside the accretion flow and end up in the BH, with only a small fraction escaping, i.e., the disk is *radiatively-inefficient*. Note that whereas mass accretion rate can exceed Eddington by essentially any factor (i.e., $\dot{M} \ggg \dot{M}_{\rm Edd}$ is possible), the emerging radiation is always limited by a logarithmic factor times the Eddington luminosity limit, i.e., $L \lesssim {\rm few} \times L_{\rm Edd}$. At the same time, the emission can be beamed into a small solid angle, so an observer exposed to it assuming that the emission is isotropic will incorrectly conclude that the source is a highly super-Eddington emitter (Sądowski et al. 2014; McKinney et al. 2014). There are several observational examples of highly super-Eddington accretion. Gamma-ray bursts, which accrete at $\dot{M} \sim 0.1 M_{\odot}{\rm s}^{-1}$, have $\dot{M}/\dot{M}_{\rm Edd} \sim 10^{13} \ggg 1$. Recent evidence suggests that supermassive BHs can accrete at a respectable $\dot{M}/\dot{M}_{\rm Edd} \sim 100 \gg 1$ (see Sect. 3.10).

Nature of Low Radiative Efficiency Perhaps somewhat surprisingly, both super-Eddington state in Fig. 3.5a and sub-Eddington state in Fig. 3.5c are radiatively-inefficient, but for very different reasons. The super-Eddington accretion flow is radiatively-inefficient because the disk is so optically-thick that it takes longer for a photon to diffuse out of the gas than for the gas fall into the hole. In contrast, the sub-Eddington accretion is radiatively inefficient because viscous dissipation predominantly heats the ions. Due to the low density, the ions do not talk to electrons, which are responsible for radiation. As a result, we end up with a *two-temperature* accretion flow, in which the heat is locked up with the ions, whereas the electrons, responsible for radiation, remain relatively cold (Begelman 1985; Yuan et al. 2003). In our simulations described below, we concentrate on sub- and super-Eddington radiatively-inefficient accretion, and we will neglect radiative cooling.

Jet Launching and Spectral State Transitions Spectral states with geometrically-thick disks, like in Fig. 3.5a, c, produce jets (indicated with blue arrows), as evidenced by observations of AGN and BHBs. Such jets are called *continuous* jets. In contrast, geometrically-thin disks, like in Fig. 3.5b, produce neither jets nor the associated radio emission (Fender et al. 2004; Russell et al. 2011), as seen in BHBs and many AGN. There are competing explanations as to why geometrically thin disks are jet phobic, but no clear winner. As discussed above, geometrically-thin disks have a low radial velocity. Thus, one can argue that magnetic fields diffuse outward faster than the accretion disk drags them inward (Lubow et al. 1994; Guilet and Ogilvie 2012, 2013), so there is no large-scale magnetic flux to power the jets (however, see Rothstein and Lovelace 2008). (This is not a problem for geometrically-thick disks, which have a large radial velocity.) Another possibility is that thin disks do not provide enough collimation to the emerging outflow, as opposed to thick disks. It is possible that not one single factor but a combination of several factors is responsible for the inability of geometrically-thin disks to produce jets.

In addition to continuous jets discussed above, *transient* jets are observed *during* disk spectral "hard to soft" state transitions and are indicated in Fig. 3.5b, c with the two-sided red arrow (see Fender et al. 2004 for more details). During such transitions, disk luminosity L spikes up from $\lesssim 0.01 L_{\rm Edd}$ to $\sim L_{\rm Edd}$. Disk spectrum becomes strongly distorted, presumably by the hot and highly magnetized "corona" that sandwiches the disk, and has little resemblance with the black-body–like spectrum of the standard geometrically-thin disk, until the luminosity drops down to $\sim 0.1 L_{\rm Edd}$. Such outbursts lead to jets that appear as discrete radio-emitting blobs of plasma ejected from the central BH. The power of transient jets is higher than that of continuous jets (Fender et al. 2004). Most of the observational evidence on state transitions comes from BHBs, or microquasars, for which state transitions occur over a period of days, but sometimes cycle over timescales of weeks or months. In AGN, or quasars, observing such state transitions is much harder, since the characteristic time scale, set by the mass of the central BHs, is $\sim 10^7 - 10^8$ times longer (for BH mass of $10^8 - 10^9 M_\odot$) than in BHBs. Thus, if the duration of state transitions scales by the same factor, it is of order of $10^4 - 10^7$ years. Consequently, in any given AGN, we have no chance of observing a state transition from start to finish.

It is presently unclear what causes the production of continuous and transient jets, and there is no agreement on whether they share the same production mechanism. The simplest possibility is that both types of jets are produced by the same BZ-type process (Sect. 3.3) involving the extraction of BH rotational energy, and the differences in their power and timing properties are due to the differences in the supply of BH magnetic flux, $\Phi_{\rm BH}$, that powers the jets. Other suggestions include large-scale magnetic reconnection as the cause of transient jets (Igumenshchev 2009; Dexter et al. 2014). A separate question is what causes the rise in the accretion rate during the state transition, and the answer is presently unclear. It is plausibly related to a global instability of the accretion flow that gives rise to an increased angular momentum transport; such an instability could be driven by temperature-sensitive turbulent transport in the disk (Potter and Balbus 2014) or the accumulation of large-scale magnetic flux (Begelman and Armitage 2014). We will return to the question of state transitions in Sects. 3.7 and 3.8.

3.6 What Sets Jet Power of Accreting Black Holes?

We have shown that BH power is directly proportional to the square of BH magnetic flux, $\Phi_{\rm BH}$, and the square of BH angular frequency, $\Omega_{\rm H}$ (see Eq. 3.7), with small corrections beyond $a \gtrsim 0.95$. In nature, $\Phi_{\rm BH}$ is a free parameter, whose value is poorly observationally constrained. Based on dimensional analysis, we have $\Phi_{\rm BH} \propto \dot{M}^{1/2}$. But what sets the dimensionless ratio,

$$\phi_{\rm BH} = \frac{\Phi_{\rm BH}}{(\langle \dot{M} \rangle r_g^2 c)^{1/2}}, \tag{3.10}$$

which characterizes the degree of inner disk magnetization and controls energy extraction from the BH (Gammie et al. 1999; Komissarov and Barkov 2009; Penna et al. 2010; Tchekhovskoy et al. 2011; Tchekhovskoy and McKinney 2012; McKinney et al. 2012)? Here $\langle \ldots \rangle$ is a time-average. Using $\phi_{\rm BH}$, we define BZ efficiency as BZ6 power (Eq. 3.7) normalized by $\dot{M} c^2$:

$$\eta_{\rm BZ} = \frac{P_{\rm BZ}}{\langle \dot{M} \rangle c^2} \times 100\,\% = \frac{\kappa}{16\pi} \phi_{\rm BH}^2 \omega_{\rm H}^2 f(\omega_{\rm H}) \times 100\,\%, \tag{3.11}$$

where $f(\omega_{\rm H})$ is a high-spin correction given in Eq. (3.8).

The physical meaning of $\eta_{\rm BZ}$ is simple: it is energy investment efficiency into the BH. Let us consider a practical example. Since mass is energy ($E = Mc^2$) and energy is money, suppose you have a hundred dollars (euros, yen, etc.; pick your favorite currency) worth of energy. And suppose you decide to invest it into a BH: you drop the 100 dollars worth of mass-energy into the BH and 20 comes back out. In this case the energy investment efficiency into the BH is $\eta_{\rm BZ} = 20\,\%$. That seems quite low: you just lost 80 dollars! But suppose you get 150 dollars back: that would be a much better outcome. In fact, you would get more out of the BH that you put in. Where would the extra 50 dollars worth of energy come from? It would come from the BH spin energy: BH rotation slows down and the released rotational energy powers the outflow. In a moment we will see how this occurs in a realistic astrophysical setting.

For the rest of the discussion we will make a distinction between (i) outflows powered by the rotation of the central BHs, which we will refer to as the "jets" and whose power we will denote as $P_{\rm jet}$ (which turns out to be very close to $P_{\rm BZ}$), and (ii) outflows powered by the rotation of the accretion disks, which we will refer to as "winds" and whose power we will denote as $P_{\rm wind}$. The sum of the two by definition gives the total outflow power, $P_{\rm outflow} = P_{\rm jet} + P_{\rm wind}$, and the total outflow efficiency,

$$\eta = \frac{P_{\rm outflow}}{\langle \dot{M} c^2 \rangle} \times 100\,\% \tag{3.12}$$

Until recently, general relativistic MHD simulations of jet formation found a rather low jet production efficiency: $\eta_{\rm BZ} \lesssim 20\,\%$, even for nearly maximally spinning BHs (McKinney 2005; De Villiers et al. 2005; Hawley and Krolik 2006; Barkov and Baushev 2011). Moreover, the larger is the large-scale magnetic flux initially present around the BH, the stronger are the jets (McKinney and Gammie 2004; McKinney 2005; Beckwith et al. 2008; McKinney and Blandford 2009; McKinney et al. 2012). Thus, even for a fixed value of BH spin, variations in large-scale poloidal magnetic flux supply are expected to lead to variations in $\eta_{\rm BZ}$: some systems would have no jets at all ($\eta_{\rm BZ} = 0$), some systems would have very strong jets, and the rest would lie in between. A fundamental question emerges: do we expect there to be a limit on how powerful jets can be? If such a limit exists, what does it tell us about the physical processes responsible for jet production?

This is an especially important question since observations suggest high energy efficiency of outflow production (Rawlings and Saunders 1991; Fernandes et al. 2011; Ghisellini et al. 2010; Punsly 2011; McNamara et al. 2011), $\eta \gtrsim 100\,\%$. Can magnetic fields produce jets at such a high efficiency? If yes, then magnetic outflow models are viable. If not, a revision of the models is in order.

Suppose somebody gave us a BH, so its mass and spin are given, and we are interested in extracting its spin energy in the most efficient way possible. According to Eq. (3.6) (or its more accurate version, Eq. (3.7)), to maximize BH power, we need to maximize BH magnetic flux, $\Phi_{\rm BH}$. But what sets the maximum value of $\Phi_{\rm BH}$? To answer this fundamental question, consider a BH immersed into a vertical magnetic field at the center of an accretion disk, as shown in Fig. 3.6. The magnetic pressure force, which can be estimated as magnetic pressure times characteristic surface area, $F_{\rm B} \sim (B^2/8\pi) \times 4\pi r^2 \times (h/r)$, pushes outward on the accretion disk. Clearly, if we removed the disk, the magnetic field would leave the BH due to the "no-hair theorem", which states that an isolated BH can only possess mass, spin, and charge, but not magnetic field or magnetic flux (see Sect. 3.3). Thus, it is the weight of the disk, or the associated force of gravity, $F_{\rm G} \sim G M_{\rm BH} M_{\rm Disk}/r^2$, that keeps the magnetic field from leaving the BH. The disk must be massive enough to keep the magnetic flux on the BH: $F_{\rm B} < F_{\rm G}$. If the opposite is true, i.e. if the magnetic

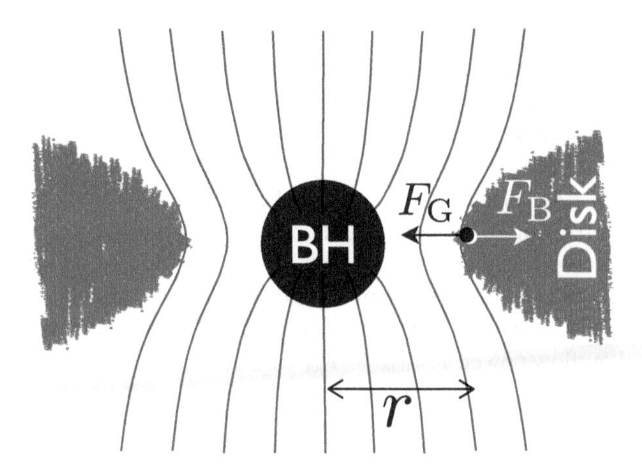

Fig. 3.6 Vertical slice through BH accretion system illustrates how the balance of forces determines the maximum possible field strength on the BH. The BH, shown as a *black circle*, is threaded with vertical magnetic field, whose lines of force are shown with *blue lines*. The magnetic pressure force, F_B, pushes outward on the accretion disk gas, which is shown in *red*. Clearly, if we removed the disk, the magnetic field would leave the BH due to the "no-hair theorem", which states that an isolated BH can only possess mass, spin, and charge, but not magnetic flux (Sect. 3.3). Thus, it is the weight of the disk, or the associated force F_G, that keeps the magnetic field from leaving the BH. The disk must be massive enough to keep the magnetic flux on the BH: $F_B < F_G$. If the opposite is true, i.e. if the magnetic field gets too strong, it pushes parts of the disk away, the excess magnetic field leaves, and the accretion flow enters a *magnetically-arrested disk* (MAD) state (Bisnovatyi-Kogan and Ruzmaikin 1974, 1976; Igumenshchev et al. 2003; Narayan et al. 2003; Igumenshchev 2008; Tchekhovskoy et al. 2011; Tchekhovskoy and McKinney 2012; McKinney et al. 2012). A characteristic size $r \sim$ few $\times\, r_g$ involved into the force balance is indicated

field gets too strong, it pushes parts of the disk away, and the excess magnetic field leaves. The accretion flow then becomes the *magnetically-arrested disk*, or a MAD (Bisnovatyi-Kogan and Ruzmaikin 1974, 1976; Igumenshchev et al. 2003; Narayan et al. 2003; Igumenshchev 2008; Tchekhovskoy et al. 2011; Tchekhovskoy and McKinney 2012; McKinney et al. 2012). In this state the magnetic field on the BH and the jet power are maximum, and we discuss this state in detail below.

Thus, the maximum possible magnetic field strength on the hole is given by the condition $F_B = F_G$. We write disk mass as $M_{\rm Disk} \sim \rho \times (4\pi r^3/3) \times (h/r)$, and get:

$$\frac{B^2}{8\pi} 4\pi r^2 = \frac{GM_{\rm BH}\rho 4\pi r^3/3}{r^2}. \tag{3.13}$$

Now, using mass continuity equation $\dot{M} = 4\pi r^2 \rho v_r \times (h/r)$ to eliminate gas density ρ, where v_r is radial velocity of the infalling gas, we obtain an estimate of field strength at the BH event horizon:

$$B_{\rm MAD} \sim 2 \times 10^4\ [{\rm G}] \left(\frac{L}{0.1 L_{\rm Edd}}\right)^{1/2} \left(\frac{M_{\rm BH}}{10^9 M_\odot}\right)^{-1/2} \left[\frac{(v_r/c) \times (h/r)}{0.05}\right]^{-1/2}, \tag{3.14}$$

where we took $r = r_{\rm H} = 2r_g$ for a non-spinning BH. This magnetic field strength is quite reasonable for AGN (Begelman 1985), so it is possible that at least some AGN can reach the MAD limit. In terms of dimensionless magnetic flux, we obtain:

$$\phi_{\rm MAD} \sim 30 \left[\frac{(v_r/c) \times (h/r)}{0.05} \right]^{-1/2}. \tag{3.15}$$

Here we adopted characteristic values: $v_r \sim c$, because gas falls into the BH at near the speed of light, and $h/r \sim 0.05$, because—as we will see below—strong BH magnetic field squeezes the disk vertically down to $h/r \sim 0.05$ near the BH from $h/r \sim 0.3{-}1$ at large distances. Estimate (3.15) is quite similar to what we will see in the numerical simulations.

MAD vs SANE Initial Conditions We tested the above non-relativistic consideration of force-balance near the BHs with global time-dependent general relativistic MHD accretion disk-jet simulations for different values of BH spin. As is standard, we initialized the simulation with a hydrodynamic gas torus on an orbit around a BH (Chakrabarti 1985; De Villiers and Hawley 2003a), as seen in Fig. 3.7a–d. The gas is in equilibrium under the action of the force of gravity pulling it inward, the centrifugal force pushing it outward, and the thermal pressure gradient balancing the difference between these two forces. That the torus is in equilibrium means that if left alone, it would orbit the BH indefinitely. In order for the gas to accrete, it is standard to insert into the torus a poloidal ($B_\varphi = 0$) magnetic field loop, which is shown in Fig. 3.7a–d with solid black lines. This magnetic field is unstable to the MRI (see Sect. 3.5 and Balbus and Hawley 1991), which drives the accretion of gas and magnetic field on to the BH. We choose a weak magnetic field, with the ratio of gas to magnetic pressures, $\beta = p_{\rm gas}/p_{\rm mag} \geq 100 \gg 1$, so as not to disturb torus' initial equilibrium state and allow the MRI to fully develop.

Clearly, jet efficiency (Eq. 3.11) depends on the amount of large-scale magnetic flux $\Phi_{\rm BH}$, and time-dependent numerical simulations show that the larger the large-scale vertical magnetic flux in the initial torus, the more efficient the jets (McKinney and Gammie 2004; McKinney 2005; Beckwith et al. 2008; McKinney and Blandford 2009; McKinney et al. 2012). To maximize jet efficiency, we would like to populate the torus with a much larger magnetic flux than in previous work. In fact, our goal is for the torus to contain more magnetic flux than the inner disk can push into the BH. For these reasons, we choose a rather large torus capable of holding an extended magnetic flux distribution: the torus extends from $r_{\rm in} = 15r_g$ to $r_{\rm out} = \text{few} \times 10^4 r_g$, with the torus density peaking at $r_{\rm max} = 34r_g$ (Fig. 3.7a–d). We also choose a rapidly spinning BH, $a = 0.99$. The large size of the torus allows us to insert a large poloidal magnetic field loop into it, with its center, or the O-point (pronounced "oh point"), located at $r_{\rm O} \simeq 600r_g$ (Fig. 3.7b). The entire magnetic flux contained between the inner edge of the torus and the O-point is available for saturating the BH.

As we will see below, the initial condition (IC) described above and shown in Fig. 3.7a–d does contain a sufficient magnetic flux to saturate the BH with magnetic

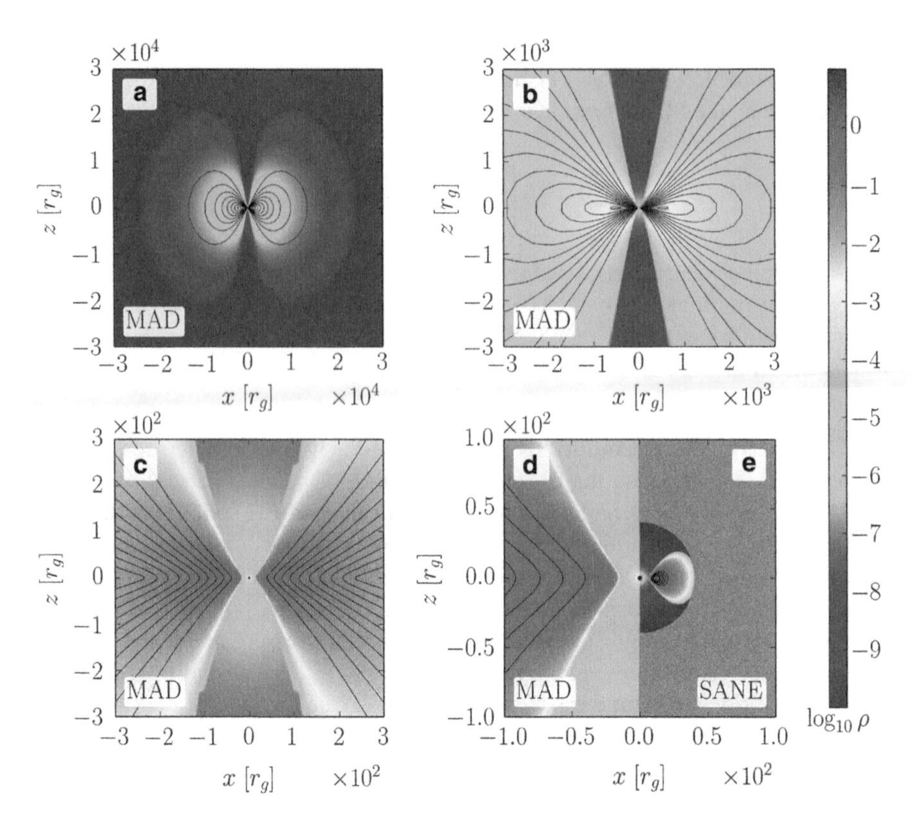

Fig. 3.7 A series of zooms into vertical slices through the simulation initial conditions (ICs). Color shows fluid-frame mass density ρ (*red* shows high and *blue* low values; see the color bar), *solid black lines* show poloidal magnetic field lines, and the *black circle* shows the BH. Panel (**a**): It is standard to initialize the simulations of BH accretion with a gas torus. This IC is similar, but the torus is chosen to be particularly large, allowing us to insert into it a magnetic flux large enough to readily flood the BH with magnetic flux and lead to a MAD. Hence, we call these "MAD" ICs. The computational domain extends out to Rout = $10^5 r_g$, i.e., to scales somewhat larger than the extent of the image. The magnetic field is weak, with $\beta = p_{\rm gas}/p_{\rm mag} \geq 100$, so it does not disturb the torus. Panel (**b**): A zoom-in on the MAD ICs. The O-point of the magnetic flux distribution is located at $r \simeq 600 r_g$, with all of the magnetic flux inside the O-point available for flooding the BH. Panel (**c**): A further zoom-in on the MAD ICs shows the large-scale magnetic flux of the same sign extends radially for more than an order of magnitude. Panel (**d**): This zoom-in on the MAD ICs shows the peak of the density distribution, which is located at $r_{\rm max} = 34 r_g$, and the inner edge of the torus, which is located at $r_{\rm in} = 15 r_g$. Panel (**e**): The standard ICs used in most simulations of BH accretion, the "SANE" ICs (this IC was generated by an open-source code HARM2D; see text for more details). They also start with a torus of gas threaded with a loop of weak magnetic field, but both the torus and the loop are much smaller than in the MAD ICs, as is apparent from the comparison to panel (**d**). For this reason, SANE ICs do not contain enough magnetic flux to readily flood the BH with magnetic flux. However, the magnetic flux is just a factor of few short of flooding the BH, highlighting the fine-tuned nature of such ICs (see Sect. 3.6). The computational domain extends out to $R_{\rm out} = 40 r_g$, with the exterior of the computational domain shown in *gray color*

flux and lead to a MAD, and we will refer to this type of IC as the MAD IC. What is the main difference of this type of IC from the standard ICs used in general relativistic MHD simulations of BH accretion (De Villiers et al. 2003b; Gammie et al. 2003; McKinney and Gammie 2004; McKinney 2005; Hawley and Krolik 2006; Barkov and Baushev 2011)? Figure 3.7e shows an example of a standard IC, which is generated by the default setup of a freely available code HARM2D,[3] and which we will refer to as the SANE IC, which stands for "standard and normal evolution" (Narayan et al. 2012). It contains a much smaller magnetic flux than the MAD IC. The initial conditions of this type usually do not contain enough magnetic flux to reach the MAD state over the attempted duration of simulations. However, this is not always the case: for instance, Fig. 4 in Sądowski et al. (2013) shows that $\phi_{\rm BH}$ in a SANE simulation for a BH with $a = 0.7$ slowly increases over time and eventually reaches $\phi_{\rm BH} \sim 40$, i.e., the simulation enters the MAD state. In fact, any value of $\phi_{\rm BH}$ between zero and $\sim$50 is possible in SANE simulations, and the value of $\phi_{\rm BH}$ reached in any given simulation is determined by the initial distributions of large-scale magnetic flux and gas density. As we will see below, MADs reach $\phi_{\rm BH} \sim 50$, essentially independent of initial conditions.

Clearly, the main difference between MAD and SANE ICs is the amount of available large-scale poloidal magnetic flux: it is much larger in the MAD ICs. Do we expect there to be a sufficient amount of large-scale magnetic flux in the environment of a supermassive BH to "flood" the hole with magnetic flux up to the MAD limit? How likely is it for such a flux to exist in a supermassive BH vicinity? Magnetic fields at the edge of the sphere of influence of a supermassive BH, i.e., $r \sim 100\,\text{pc}$, are plausibly $B \sim \mu\text{G}$. Suppose these fields maintain their coherence over a roughly similar length scale. Does a patch of this size contain enough magnetic flux to make the central BH "go MAD" if the flux accretes down to the event horizon? Clearly, the magnetic flux contained in such a patch, $\Phi_{\rm patch} \approx 10^{35}\ \text{G cm}^2$, is much larger than the flux necessary to saturate a BH with magnetic field given by Eq. (3.14), $\Phi_{\rm MAD} \approx 10^{33.5}\,\text{G cm}^2$ (Narayan et al. 2003) (i.e., for a BH of mass $M_{\rm BH} = 10^9 M_\odot$ accreting at $L = 0.1 L_{\rm Edd}$). It is thus plausible that MADs occur around supermassive BHs. Below we will see that there are observational indications that MADs are at work in a variety of astrophysical systems (see Sects. 3.7–3.10). Therefore, MADs are not rare or unusual as their name might imply, but in fact quite the opposite.

We carry out a simulation starting with our MAD IC shown in Fig. 3.7a–d. To maximize jet power, we consider a rapidly spinning BH, with $a = 0.99$. We carry out the simulations in 3D, using a numerical code HARM (Gammie et al. 2003; McKinney and Gammie 2004; Tchekhovskoy et al. 2007, 2009; McKinney and Blandford 2009; Tchekhovskoy et al. 2011; McKinney et al. 2012), which discretizes equations of general relativistic MHD in a conservative and shock-capturing form. We use the resolution of $288 \times 128 \times 64/128$ cells in $r-$, $\theta-$, and $\varphi-$directions, respectively (at $t \approx 15{,}000 r_g/c$ we double the $\varphi-$resolution from 64

[3]You can download the code at http://rainman.astro.illinois.edu/codelib

to 128 cells). The computational grid extends radially from $0.83r_{\rm H}$ to $10^5 r_g$ and uses a logarithmically-spaced radial grid, $\Delta r \propto r$, near the BH. The θ-grid is adjusted so as to resolve both the collimating polar jets and the MRI in the equatorial disk. The φ-grid is uniform.

MAD Simulation Results The outcome of the simulation is shown in Fig. 3.8. This simulation as well as all other simulations discussed in this chapter, do not include any radiative losses or cooling, which is appropriate in geometrically-thick disks that are strongly sub-Eddington or super-Eddington (see Sect. 3.5). Figure 3.8a shows horizontal and vertical slices through the same IC as that shown in Fig. 3.7a–d. Figure 3.8e shows the rest mass energy flux into the BH, $\dot{M}(r_{\rm H})c^2$, as a function of time. Until a time $t \sim 2{,}000 r_g/c$, the MRI is slowly building up inside the torus and there is no significant accretion. After this time, $\dot{M}$ steadily grows until it saturates at $t \sim 4{,}000 r_g/c$. Beyond this time, the accretion rate remains more or less steady at approximately 10 code units until the end of the simulation at $t \sim 30{,}000 r_g/c$. The fluctuations seen in $\dot{M}$ are characteristic of turbulent accretion via the MRI.

Figure 3.8f shows the time evolution of the dimensionless magnetic flux $\phi_{\rm BH}$ at the BH horizon. Since the accreting gas continuously brings in new flux, $\phi_{\rm BH}$ continues to grow even after $\dot{M}$ saturates. However, there is a limit to how much flux the accretion disc can push into the BH. Hence, at $t \sim 6{,}000 r_g/c$, the flux on the BH saturates and after that remains roughly constant at a value around 50, with the flow near the BH being highly magnetized. Panel (b) shows that magnetic fields near the BH are so strong that they compress the inner accretion disc vertically and decrease its thickness near the BH down to $h/r \sim 0.05$; at larger distances the disk thickness settles to $h/r \approx 0.3$. The accreting gas, of course, continues to bring even more flux, but this additional flux remains outside the BH, obstructs gas inflow, and causes the disk to become a MAD (Tchekhovskoy et al. 2011; Narayan et al. 2003). Panels (c) and (d) show what happens to the excess flux. Even as the gas drags the magnetic field in, field bundles erupt outward (Igumenshchev 2008), leaving the time-average flux on the BH constant. For instance, two flux bundles are seen at $x \sim \pm 20 r_g$ in Fig. 3.8c which originate in earlier eruption events. Other bundles are similarly seen in Fig. 3.8d. During each eruption, the mass accretion rate is partially suppressed, causing a dip in $\dot{M}$ (Fig. 3.8e); there is also a corresponding temporary dip in $\phi_{\rm BH}$ (Fig. 3.8f). Note that, unlike 2D (axisymmetric) simulations (e.g., Proga and Begelman 2003), there is never a complete halt to the accretion (Igumenshchev et al. 2003) and even during flux eruptions accretion proceeds via spiral-like structures, as seen in Fig. 3.8d.

The energy outflow efficiency shows considerable fluctuations with time (Fig. 3.8g), reaching values as large as $\eta \gtrsim 200\,\%$ for prolonged periods of time, with a long-term average value, $\langle \eta \rangle = 140 \pm 15\,\%$. This may explain sources with very efficient jets (McNamara et al. 2011; Fernandes et al. 2011; Punsly 2011). This value of efficiency is much larger (by a factor of 5–10) than the maximum efficiencies seen in earlier simulations (McKinney 2005; Hawley and Krolik 2006; Barkov and Baushev 2011). The key difference is that, in our simulation, we

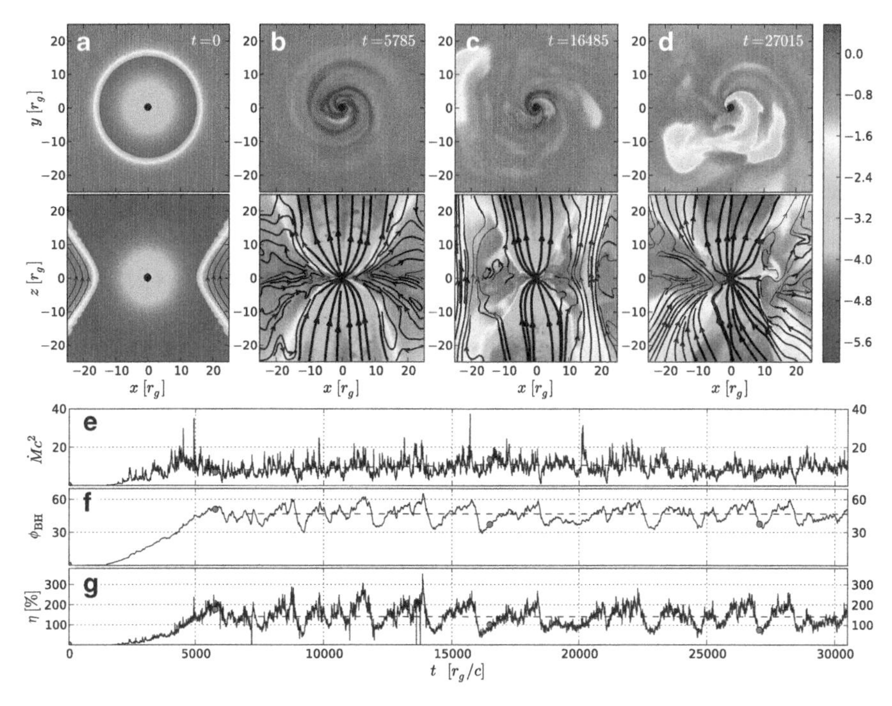

Fig. 3.8 Snapshots and time-dependence of in a simulation of a magnetically-arrested disk (MAD, taken from Tchekhovskoy et al. 2011), around a rapidly spinning BH, with $a = 0.99$. A movie is available at http://youtu.be/nRGCNaWST5Q. Panels (**a**)–(**d**): The *top* and *bottom rows* show, respectively, equatorial ($z = 0$) and meridional ($y = 0$) snapshots of the flow, at the indicated times. Colour represents the logarithm of the fluid frame mass density, $\log_{10} \rho$ (*red* shows high and *blue* low values; see colour bar), *filled black circle* shows BH horizon, and *black lines* show field lines in the image plane (the dominant, out-of-plane, magnetic field component is not shown; but see Fig. 3.9a). Panel (**e**): Time evolution of the rest-mass accretion rate, $\dot{M}c^2$. The fluctuations are due to turbulent accretion and are normal. The long-term trends, which we show with a Gaussian smoothed (with width $\tau = 1{,}500 r_g/c$) accretion rate, $\langle \dot{M} \rangle_\tau c^2$, are small (*black dashed line*). *Red dots* in the *three bottom panels* indicate the times of snapshots shown in the *top two rows* of panels. Panel (**f**): Time evolution of the large-scale magnetic flux, $\phi_{\rm BH}$, threading the BH horizon, normalized by $\langle \dot{M} \rangle_\tau$. At $t = 0$, the accretion flow contains a large amount of large-scale magnetic flux and there is zero flux on the BH. BH magnetic flux grows until $t \approx 6{,}000 r_g/c$. At this time the BH is saturated with magnetic flux. However, the accretion flow brings in even more flux, which impedes the accretion and leads to a magnetically arrested disk at $t \gtrsim 6{,}000 r_g/c$ (Panels (**c**) and (**d**) are during this period). Some of the flux escapes from the BH via magnetic interchange and flux eruptions, several of which are seen in panels (**c**) and (**d**), which frees up room for new flux. Panel (**g**) Time evolution of the energy outflow efficiency η (defined in Eq. 3.12 and here normalized to $\langle \dot{M} \rangle_\tau c^2$). During the initial stage of the simulation, $t \lesssim 6{,}000 r_g/c$, the power of outflows η grows, roughly proportional to $\phi_{\rm BH}^2$, as expected from Eq. (3.11). The strength of magnetic flux reaches saturation around $t \sim 6{,}000 r_g/c$ and the power of the outflow is maximum. During the subsequent quasi-periodic accumulation and rejection of magnetic flux by the BH, η fluctuates and its average value is $\eta \approx 140\,\%$. Since this exceeds $100\,\%$, the outflows carry more energy than the entire rest-mass energy supplied by accretion. This was the first demonstration of *net* energy extraction from a spinning BH in a realistic astrophysical setting

maximized the magnetic flux around the BH. This enables the system to produce a substantially more efficient outflow. Since $\eta > 100\,\%$, jets and winds carry more energy than the entire rest-mass supplied by the accretion. This is the first demonstration of *net* energy extraction from a BH in a realistic astrophysical scenario, a long sought result. Thicker MADs ($h/r \approx 0.6$) produce outflows at an even higher efficiency, $\eta \simeq 300\,\%$ (McKinney et al. 2012; Tchekhovskoy and McKinney 2012).

Note that Fig. 3.8 shows the magnetic field in the image plane, and the toroidal magnetic field component, B_φ, is not shown. As is clear from the 3D rendering of the accretion system shown in Fig. 3.9a, B_φ is actually the dominant component of the field in the jets. It can be seen from Fig. 3.9a that the jets extend out to much larger distances than the BH horizon radius and are collimated into a small opening angle by the disk wind.

Figure 3.8 shows that the accretion flow is highly time-variable and does not appear to resemble the idealized picture of regular magnetic field lines seen, for instance, in Fig. 3.3c. Does this mean that the simple, time-steady models are not at all applicable to the accretion simulations? To check this, let us average the accretion flow in time and in azimuth (i.e., in the φ-direction); the result is shown in Fig. 3.9b. Clearly, once the large-amplitude variability is averaged over, what is left is remarkably similar to the sketch shown in Fig. 3.3c.

Firstly, let us focus on the poloidal magnetic field lines, which are shown with black solid lines in Fig. 3.9b. They have the shape of a parabola, i.e., they curve up toward the polar regions as they move away from the BH. The group of field lines highlighted in green connects to the BH and makes up the twin polar jets. The jet field lines extract BH rotational energy and carry it away to large distances. These field lines have little to no gas attached to them and are therefore highly magnetized (since disk gas cannot cross magnetic field lines and is thus blocked from getting to the polar region, the jet field lines either drain the gas to the BH or fling the gas away by the rotation). The fact that a large amount of energy is coupled to these field lines but very little gas, allows them to accelerate efficiently to highly relativistic velocities. This acceleration occurs at distances much larger than the ones shown in Fig. 3.9b (Tchekhovskoy et al. 2008, 2009, 2010). As mentioned above, we denote the power of the jets as $P_{\rm jet}$. The rest of the field lines, highlighted in blue, connect to the disk body and make up the magnetic field bundle that produces the slow, heavy disk wind. We denote its power as $P_{\rm wind}$.

The flow of gas in Fig. 3.9b has the standard "hourglass" shape: part of the disk inflow turns around and forms the disk wind: disk wind streamlines originate in the disk body. Jet streamlines are connected to the BH. Note that in a time-average sense, disk flow streamlines cross magnetic field lines (i.e., red lines cross black lines): this would be impossible in axisymmetric ideal MHD. However, this is possible in 3D ideal MHD simulations because non-axisymmetric gas motions allow disk gas to go around magnetic field lines (e.g., via interchange instability). Note, however, that the disk wind streamlines do not cross into the jet boundary, i.e., no streamlines cross from the blue into green region: this is the reflection of suppressed turbulence and mixing in the polar regions.

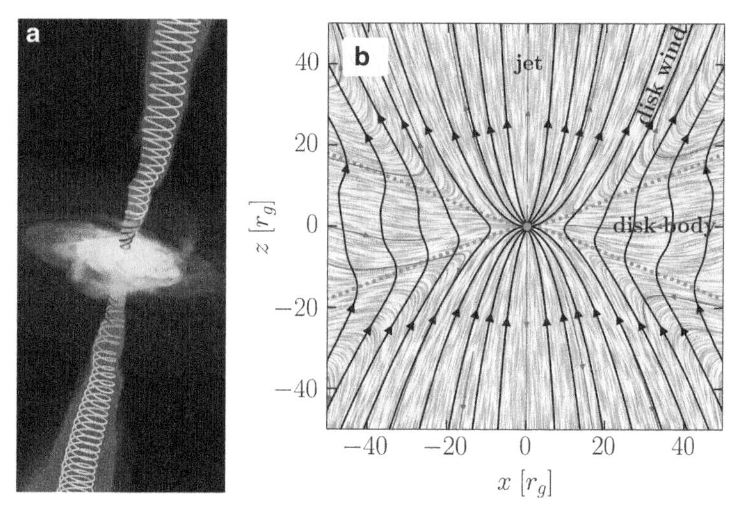

Fig. 3.9 Panel (**a**): A 3D rendering of our MAD $a = 0.99$ model at $t = 27,015 r_g/c$ (i.e., the same time as Fig. 3.8d). Dynamically-important magnetic fields are twisted by the rotation of a BH (too small to be seen in the image) at the center of an accretion disk. The azimuthal magnetic field component clearly dominates the jet structure. Density is shown with color: disk body is shown with *yellow* and jets with *cyan-blue color*; we show jet magnetic field lines with *cyan bands*. The image size is approximately $300 r_g \times 800 r_g$. Panel (**b**): Vertical slice through our MAD $a = 0.99$ model averaged in time and azimuth over the period, $25{,}000 r_g/c \leq t \leq 35{,}000 r_g/c$. Ordered, dynamically-important magnetic fields remove the angular momentum from the accreting gas even as they obstruct its infall onto a rapidly spinning BH ($a = 0.99$). *Gray filled circle* shows the BH, *black solid lines* show poloidal magnetic field lines, and *gray dashed lines* indicate density scale height of the accretion flow, $|\theta - \pi/2| = h/r$. The symmetry of the time-average magnetic flux surfaces is broken, due to long-term fluctuations in the accretion flow. This is also seen from the streamlines of velocity, u^i, which we show in two ways: with directed *thin red lines* and with colored "iron filings", which are better at indicating the fine details of the flow structure. The flow pattern is a standard hourglass shape: equatorial *disk inflow* at low latitudes, which turns around and forms a *disk wind outflow* (labeled as "disk body" and "disk wind", respectively, and highlighted in *blue*), and twin polar *jets* at high latitudes (labeled as "jet" and highlighted in *green*). We show "iron filings" using *linear integral convolution* (LIC) method, which is available as a compilable Python module at http://wiki.scipy.org/Cookbook/LineIntegralConvolution; to increase the visual contrast of the LIC, we use the technique described at http://paraview.org/Wiki/ParaView/Line_Integral_Convolution#Contrast_enhancement

Whereas field lines in Fig. 3.3c show a sharp equatorial kink, it is not present in Fig. 3.9b. The kink occurs due to the simplifying assumption that the accretion disk and the electric current it carries are of a zero thickness, $h/r = 0$. However, in the simulation the disk thickness is finite, $h/r \approx 0.3$. This converts a singular equatorial current in Fig. 3.3c into a current sheet distributed over the disk thickness in Fig. 3.9b: most of the field line curvature is concentrated within the disk body, whose boundaries are approximately indicated by gray dashed lines defined by $|\theta - \pi/2| = h/r$.

Importantly, in the MAD state η is essentially independent of the initial amount of magnetic flux in the accretion flow, i.e., η depends only on BH spin, a, and disk density angular thickness, h/r (Tchekhovskoy and McKinney 2012). This allows us to reliably study spin-dependence of various quantities, shown in Fig. 3.10. Firstly, note that both prograde ($a > 0$; BH rotating in the same sense as the disk) and retrograde BHs ($a < 0$; BH is rotating in the opposite sense to the disk) have quite similar values of magnetic flux. We will focus on prograde BHs. The dimensionless BH flux, $\phi_{\rm BH}$, varies between 40 and 60 (Fig. 3.10a), with a characteristic value,

$$\text{In a MAD:} \qquad \phi_{\rm MAD} \approx 50. \quad \text{(spin-average)} \tag{3.16}$$

Note that this is quite similar to our back of the envelope estimate of BH magnetic flux, Eq. (3.15). The spin-dependence of $\phi_{\rm BH}$ can be approximated for $a \geq 0.3$ as

$$\text{In a MAD:} \qquad \phi_{\rm MAD} \approx 70(1 - 0.38\omega_{\rm H})h_{0.3}^{1/2}, \tag{3.17}$$

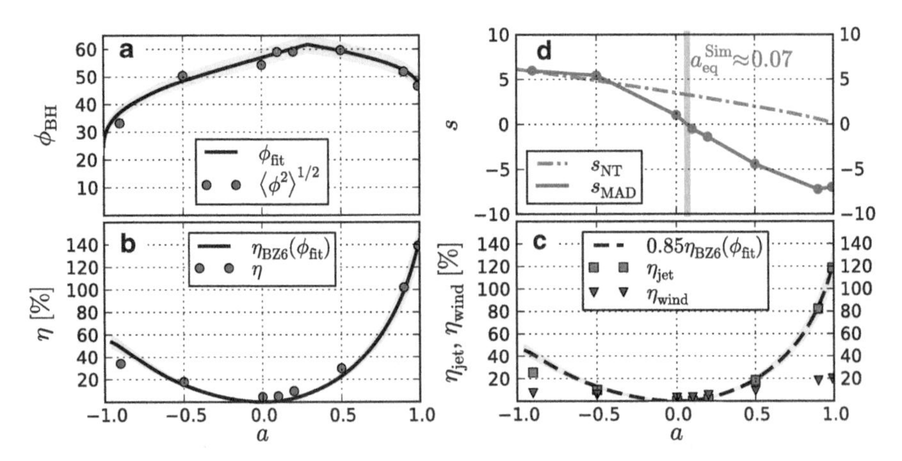

Fig. 3.10 Spin-dependence of various quantities for MADs with $h/r \approx 0.3$ (Taken from Tchekhovskoy et al. (2012)). Panel (**a**) Spin-dependence of dimensionless BH spin, $\phi_{\rm BH}$: *red dots* show simulation results, and the *black line* shows a by-eye fit, $\phi_{\rm fit}$, which is comprised of two linear segments in a $\phi_{\rm BH}$–$\Omega_{\rm H}$ plane. *Blue bands* show a 5 % uncertainty on the fit. Panel (**b**) Spin-dependence of energy outflow efficiency, η: *red dots* show simulation results, *black line* shows the BZ6 approximation for efficiency (Eq. 3.7) assuming $\phi_{\rm BH}(a) = \phi_{\rm fit}(a)$, and the *blue band* shows the 10 % uncertainty on the fit. Panel (**c**) *Green squares* show jet efficiency $\eta_{\rm jet}$ and inverted *blue triangles* wind efficiency $\eta_{\rm wind}$. *Dashed line* shows 85 % of the above BZ6 efficiency (a good estimate of jet power for prograde BHs (Tchekhovskoy and McKinney 2012)), and a *blue band* shows a 10 % uncertainty band. Panel (**d**) Connected *red dots* show spin-dependence of BH spin-up parameter in MAD simulations, $s_{\rm MAD}$ (see Eq. 3.20), and *green dash-dotted line* shows the spin-up parameter for a geometrically-thin Novikov-Thorne disk, $s_{\rm NT}$ (Novikov and Thorne 1973). Whereas for thin disks the equilibrium value of BH spin is $a_{\rm eq}^{\rm NT} = 1$, for MAD simulations it is much lower, is $a_{\rm eq}^{\rm Sim} \approx 0.07$, a value indicated by a *vertical red band*. Such a low equilibrium spin value results from a combination of two magnetic effects: (i) efficient extraction of BH spin energy by strong, dynamically-important magnetic fields threading the BH and (ii) removal of disc angular momentum by magnetized disk winds, so little angular momentum reaches the BH

where $h/r = 0.3h_{0.3}$. The corresponding BZ6 efficiency (see Eq. 3.7) is then

$$\text{In a MAD:} \qquad \eta_{\rm MAD} \approx F(\omega_{\rm H})h_{0.3} \times 100\,\% \approx 1.3h_{0.3}a^2 \times 100\,\%, \tag{3.18}$$

where we used the fact that the spin-dependent factor entering jet power, $F(\omega_{\rm H}) = 4.4\omega_{\rm H}^2(1-0.38\omega_{\rm H})^2 f(\omega_{\rm H})$, can be approximated as $F \approx 1.3a^2$ to 10 % accuracy for $a \geq 0.3$, where $f(\omega_{\rm H})$ is given by Eq. (3.8). The values of η for prograde and retrograde BHs are within a factor of a few of each other, suggesting that both of them can be responsible for producing powerful jets (Tchekhovskoy and McKinney 2012). The h/r dependence of $\phi_{\rm MAD}$ and $\eta_{\rm MAD}$, given in Eqs. (3.17) and (3.18), will be derived elsewhere. Panel (c) shows the division of total outflow efficiency into highly magnetized jet and weakly magnetized wind components, with efficiencies $\eta_{\rm jet}$ and $\eta_{\rm wind}$, respectively. Jet efficiency at $a \gtrsim 0.2$ is well-approximated by:

$$\text{In a MAD:} \qquad \eta_{\rm jet} \approx 0.65h_{0.3}a^2(1+0.85a^2) \times 100\,\%. \tag{3.19}$$

Since jets are BH spin-powered (Eq. 3.7), for $a = 0$ jet efficiency vanishes, but winds still derive their power from an accretion disk via a Blandford-Payne–type mechanism (Blandford and Payne 1982). The larger the spin, the more efficient jets and winds. However, for rapidly spinning BHs most of the energy—about 85 % for prograde BHs—is carried by relativistic jets (Tchekhovskoy and McKinney 2012). Thus, for rapidly spinning BHs, the power of BH-powered jets exceeds by a factor of 5 the power of disk-powered winds, demonstrating that BH spindown power can dominate the total power output of an accretion system, which makes it plausible to use jets as diagnostic tools of the central BHs. This important result resolves a long-standing debate on the dominant source of power behind BH outflows (Ghosh and Abramowicz 1997; Livio et al. 1999). Importantly, even rather slowly spinning BHs, with $a \sim 0.5$, produce prominent BH spin-powered jets that outshine disk-powered winds.

Do our highly efficient jets affect the spin of central BHs? Figure 3.10d shows spin-dependence of BH spin-up parameter (Gammie et al. 2004),

$$s = \frac{M_{\rm BH}}{\dot{M}} \times \frac{{\rm d}a}{{\rm d}t}. \tag{3.20}$$

If $s > 0$, the BH is spun-up by the sum of accretion and jet torques. If $s = 0$, BH is in spin equilibrium, i.e., its spin does not change in time. If $s < 0$, BH spin decreases. For standard geometrically thin accretion disks, $s > 0$, at all values of spin (see Fig. 3.10d and Novikov and Thorne 1973), and the equilibrium spin is $a_{\rm eq}^{\rm NT} = 1$ (Bardeen 1970) (we neglect photon capture by the BH, which would limit the spin to $a \approx 0.998$ Thorne 1974). Thick accretion flows in time-dependent numerical simulations (McKinney and Gammie 2004; Gammie et al. 2004; Krolik et al. 2005) typically have $a_{\rm eq} \sim 0.9$. Figure 3.10d shows that the equilibrium value of spin for our MADs (with $h/r \approx 0.3$) is much smaller, $a_{\rm eq}^{\rm Sim} \approx 0.07$, due to large BH spin-down torques by the powerful jets. For example, in an AGN accreting at $L = 0.1 L_{\rm Edd}$, the central BHs would be spun down to near-zero spin, $a \lesssim 0.1$, in $\tau \simeq 3 \times 10^8$ years. This value is interesting astrophysically because it is comparable to characteristic quasar lifetime (but could be much longer than the duration of the FRII phase of AGN). Over quasar lifetime, jets can extract a substantial fraction of the central BH spin energy and deposit it into the ambient medium. The central galaxy in the cluster MS0735.6+7421 can be one such example (McNamara et al. 2009).

How does the power of jets from MADs relate to previously reported results? Figure 3.11 compares the approximation for $\eta_{\rm jet}$ in MAD simulations (given by Eq. 3.19) to those for simulations using SANE ICs (McKinney 2005; Hawley and Krolik 2006). Clearly, MADs produce much more powerful jets than found in previous work, by a factor 5–10, depending on the value of the spin. This is not surprising because MADs achieve the maximum possible amount of magnetic flux threading the BH and thus achieve the maximum possible jet power: for fixed BH spin and $\dot{M}$, any SANE simulation will have jet power *below* that of a MAD. On the other hand, a deficit by a factor of 10 in jet power translates into a deficit by a factor of ~3 in magnetic flux. This means that the magnetic flux in SANE initial

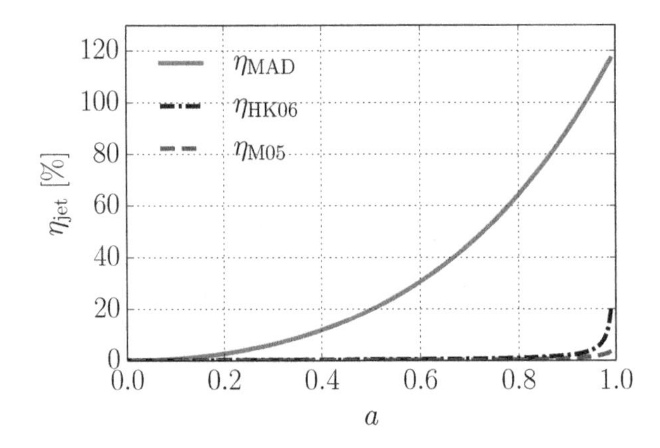

Fig. 3.11 Comparison of jet energy production efficiency obtained in MAD simulations, $\eta_{\rm jet}$ (due to Eq. 3.19), which is shown with *green solid line*, with previously reported approximations of simulated jet power: HK06, which is shown with *black dash-dotted line* (Hawley and Krolik 2006), and M05, which is shown with *brown dashed line* (McKinney 2005), plotted over the range $0 \leq a \leq 0.99$. Clearly, MADs produce much more efficient jets than in previous work: this is not surprising because for a given $\dot{M}$, MADs have the maximum possible amount of magnetic flux threading the BH and thus achieve the maximum possible jet power. Note that the shape of the spin-dependence of jet power is different in MAD simulations than in previous work, suggesting that the dependence of jet power on BH spin in SANE simulations is affected by the distribution of gas and magnetic flux in the initial conditions

conditions is *tuned* to be just a factor of few below the MAD ICs. Therefore, a mere increase of magnetic flux by $\gtrsim 3\times$ would cause SANE simulations to "go MAD". That SANE simulations did not become MAD is a consequence of the particular choice of initial torus size, magnetic flux distribution, and potentially limited run time. In fact, some SANE simulations eventually "go MAD", as we discussed when we first introduced SANE ICs.

3.7 Correlation of Jet Power and BH Spin for Stellar-Mass BHs

For a while astrophysicists have been able to measure the masses of BHs in AGN and BHBs. However, only in the past decade have they been able to reliably measure the spins of the BHs. There are two major methods of BH spin measurement. Both methods operate best when the accretion disk is close to the geometrically-thin disk state (Fig. 3.5b), which is the best understood of all BH accretion states. Both of the methods rely on the fact that disk emission cuts off (Shafee et al. 2008; Penna et al. 2010; Kulkarni et al. 2011; Zhu et al. 2012) inside the *innermost stable circular orbit*, or ISCO, whose radius has a monotonic, strong dependence on BH spin: $r_{\rm ISCO} = r_g$ for maximally spinning BHs (with $a = 1$) and $r_{\rm ISCO} = 6r_g$ for non-spinning BHs (with $a = 0$; see, e.g., Shapiro and Teukolsky 1986). By measuring the radius of the "hole in the disk", or $r_{\rm ISCO}$, one can then determine the BH spin.

In the continuum fitting method the radius of the ISCO is found via the analysis of the continuum black-body–like emission from the inner disk (see McClintock et al. 2011, 2013 for a recent review). On a qualitative level, since most of the disk emission is produced in a ring of radius $\simeq r_{\rm ISCO}$, the emergent luminosity is given by $L \sim \pi r_{\rm ISCO}^2 \sigma T^4$, where σ is Stephan-Boltzmann constant and T is the blackbody temperature of the inner disk. By measuring L from the normalization of the X-ray spectrum and T from the shape of the spectrum, one can then solve for $r_{\rm ISCO}$ and a. The iron line method relies instead on analyzing the shape of iron emission lines (as well as other "reflection" features), which are good tracers of inner disk dynamics: the red wing of the iron line profile is sensitive to the position of disk inner edge and thus its modeling allows one to measure BH spin (see Fabian et al. 2000; Reynolds 2013a for an introduction).

If it is the central BHs that power the jets, we would expect jet efficiency to correlate with BH spin. While there is no evidence for such a correlation for continuous jets in stellar-mass BHs (Fender et al. 2010), recently such a correlation was found for transient stellar-mass BH jets (Narayan and McClintock 2012). Whereas jet power cannot be measured directly, one can measure its proxy, radio emission, or more specifically, the luminosity at 5 GHz radio frequency, $L_{5\,\rm GHz,peak}$, at the peak of its emission. Note that, we are interested not in jet power but in jet energy efficiency η. In order to find it, we would need to divide jet power by mass accretion rate $\dot{M}$. However, $\dot{M}$ is difficult to measure for transient events. Instead, one can make use of the fact that state transitions often behave as "standard candles": mass accretion rate reaches an order unity fraction of Eddington luminosity at its peak during the state transition and is thus proportional to BH mass M. Thus, an observational proxy for jet efficiency is $\eta_{\rm obs} \propto L_{\rm 5GHz,peak}/M$, and we expect this quantity to correlate with BH spin. Indeed, a correlation between $\eta_{\rm obs}$ and a is observed and shown in Fig. 3.12 (but see Russell et al. 2013).

This correlation is consistent with the picture that transient jets are powered by magnetically-extracted BH spin energy. This correlation also implies that different BHs are filled with magnetic flux to the same degree, i.e., that they have similar values of the dimensionless magnetic flux $\phi_{\rm BH}$. This could be understood if the production of most transient jets is accompanied with the formation of a MAD, when $\phi_{\rm BH}$ saturates at the maximum possible value. In some cases, however, there can be an insufficient amount of magnetic flux to saturate the BH and lead to a MAD. Such sources would fall below the correlation shown in Fig. 3.12, so it is possible that the correlation gives *an upper envelope* of jet power. If so, then the correlation translates a measurement of jet power into a *lower limit* on BH spin. A robust correlation between jet power and BH spin is an extremely useful tool since

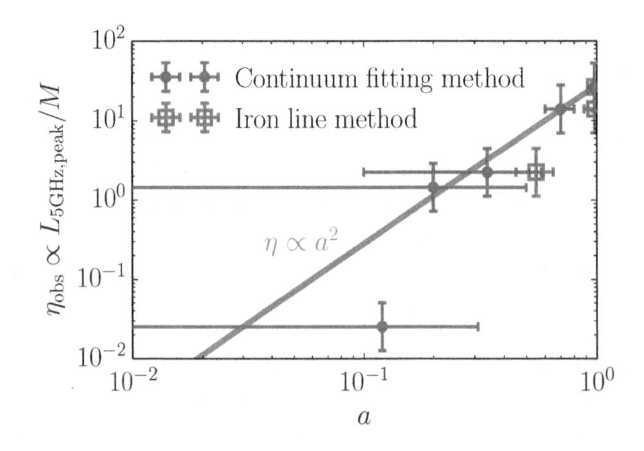

Fig. 3.12 Observational evidence for a correlation between the power of transient BHB jets and BH spin (using the updated data set from Steiner et al. 2013; see also Narayan and McClintock 2012). This figure plots spin versus a proxy for transient jet power, the 5 GHz radio luminosity $L_{\rm 5\,GHz,peak}$. Jet power has been corrected for beaming assuming a Lorentz factor $\Gamma = 2$, and normalizing by BH mass M to obtain jet efficiency (see the text for those details). Retrograde spins are not considered here, and sources with poorly constrained inclinations have likewise been omitted. BH spin was measured using the continuum fitting (shown with *red circles*) and iron line (shown with *open green squares*) methods. Jet power increases with increasing BH spin a. The *blue solid line* shows a quadratic dependence $\eta \propto a^2$ that is consistent with the data. If this curve gives the upper envelope of jet power, one expects all X-ray binaries to fall *under* this curve. The data is shown, from low-to-high $\eta_{\rm obs}$, for A0620–00, H1743–322, XTE J1550–564, GRO J1655–40, and GRS 1915+105. Error bars on the spin are 1-σ. As illustrated in the figure, the two spin measurement methods generally agree to within 1-σ error. Error bars on the jet power are taken to be a factor of 2

it allows one to convert a relatively easy measurement of jet power into a hard-to-measure value of BH spin, as was recently demonstrated (Steiner et al. 2013).

3.8 Microquasars as "Quasars for the Impatient"

Jet-producing AGN fall into two classes (Fanaroff and Riley 1974): (i) low-luminosity AGN that produce jets whose emission is centrally-dominated (so-called FRI sources, e.g., M87 galaxy) and (ii) high-luminosity AGN that produce jets whose emission is dominated by a pair of hot lobes in which the twin jets interact with the ambient medium (so-called FRII sources, e.g., Cygnus A galaxy). If an FRI jet points toward us, it appears as a *BL Lac* object, whereas if an FRII jet points at us, it appears as a *blazar* (see Urry and Padovani 1995 for a review). Both of these classes of sources are referred to as *radio-loud* AGN.

As we discussed in Sect. 3.1, it is compelling to identify a single mechanism responsible for producing jets across the entire mass range of BHs, i.e., both stellar-mass BHs in BHBs and supermassive BHs in AGN. It is appealing to think of stellar-mass BHs as scaled-down supermassive BHs, or "quasars for the impatient" (Blandford 2005).

However, when we try to draw such an analogy, we run into apparent difficulties. In BHBs, there is clear evidence that geometrically-thin disks do not produce any jets or associated radio emission (see, e.g., Russell et al. 2011). However, detections of a "big blue bump" in the spectra of some blazars (Tavecchio et al. 2011; Cowperthwaite and Reynolds 2012) seemingly indicate the presence of a geometrically-thin accretion disk and imply that—in contrast to stellar-mass BHs—supermassive BHs with geometrically-thin disks are capable of producing jets! Does this mean that there is really a fundamental difference in the physics of jet production between stellar-mass and supermassive BHs and that no useful analogies between the two BH populations can be drawn?

Let us make an attempt to sort things out. FRI jets are clear analogs of continuous BHB jets: both occur at low accretion luminosities, $L \lesssim 0.01 L_{\mathrm{Edd}}$. FRII jets are much more powerful than FRI jets, and it is less clear what their stellar-mass analogs are. There are, however, not many candidates to choose from.

Could transient stellar-mass BH jets—these are the same jets that appear during hard-to-soft accretion disk spectral state transitions and that we discussed in Sect. 3.5—be the low-mass analogs of FRII jets (Sera Markoff, private communication; see also van Velzen and Falcke (2013))? Indeed, just like FRII jets are more powerful than FRI jets, transient jets are more powerful than the continuous jets. Also, both FRII jets and transient jets appear over roughly the same luminosity range, $L \sim (0.01-1) L_{\mathrm{Edd}}$ (see, e.g., Sikora et al. 2007; Fender et al. 2004).

According to this analogy, FRII objects are in a "transient" evolutionary phase of AGN. During this phase they transition from being FRI AGN to becoming jet-less quasars. If the characteristic duration of state transitions scales linearly with central BH mass, this phase lasts 10^4–10^7 years in AGN (see Sect. 3.5). What happens to the inner regions of the accretion disk during this transition? In analogy with BHBs, mass accretion rate increases to a substantial fraction of Eddington. This increase is accompanied with the change in the state of the inner accretion disk: it switches from a radiatively-inefficient sub-Eddington accretion flow in FRI stage (Fig. 3.5c) to a radiatively-efficient geometrically-thin accretion flow in the quasar stage (Fig. 3.5b).

Realization that FRII jets and transient BHB jets are the same physical phenomenon, only occurring on different mass scales, resolves the puzzle that we started with: the detections of "big blue bump" emission in the spectra of blazars *do not* imply that their accretion disks are canonical geometrically-thin accretion disks. In contrast, these are perturbed disks in the process of changing their identity. In fact, the spectra of stellar-mass BHs undergoing the hard-to-soft spectral state transition also show a blackbody-like component whose strength increases as the transition progresses (Fender et al. 2004), reflecting the underlying change from a geometrically-thick to geometrically-thin disk. In fact, the inner edge of the geometrically-thin disk appears to *move inward* as the state transition progresses, reflecting a "refilling" geometrically-thin disk (Fender et al. 2004). This might be precisely what is observed in a blazar 3C120 (Cowperthwaite and Reynolds 2012).

Recently van Velzen and Falcke (2013) reported a tight correlation, for a sample of FRII AGN, between the radio lobe luminosity (which is expected to trace jet power) and the optical luminosity (which is expected to trace mass accretion rate). In fact, this correlation was so tight that the authors concluded that FRII jets could not be powered by BH spin: any conceivable spread in BH spin distribution would have led to scatter in jet power in excess of the observed one. However, an alternative explanation is equally plausible: that most of the BHs are near maximally spinning. To resolve this puzzle, it would be extremely useful to obtain independent measurements of BH spins of jet-producing BHs. However, these are difficult to come by because the accretion disk during the state transition is strongly perturbed away from the standard thin disk state. Indeed, most of the supermassive BH spin measurements have been performed for jet-phobic BHs that accrete via standard thin disks (Reynolds 2013b), and these measurements suggest that many jet-phobic supermassive BHs are rapidly spinning. If jet-producing AGN represent a relatively short phase of an AGN lifecycle, both jet-producing and jet-phobic supermassive BHs share the same BH spin distribution. Hence, it is conceivable that most jet-producing supermassive BHs are rapidly spinning as well, thereby resolving the puzzle. It is also possible that the optical luminosity is contaminated by jet emission and thus naturally strongly correlates with the jet power, so the detected correlation is between radio and optical emission *from the same jet* and does not tell us anything about the disk-jet connection.

3.9 MADs in Radio-Loud AGN

In Sect. 3.8 we argued that jets in blazars are supermassive analogs of transient jets in stellar-mass BHs. According to this analogy, if transient jets are powered by MADs (see Sect. 3.7), we expect the jets in blazars and FRII AGN to be powered by MADs as well. How can we test this hypothesis observationally? One way is to measure the jet power $P_{\rm jet}$ and compare it to accretion power $\dot{M}c^2$: if we robustly find $P_{\rm jet} \geq \dot{M}c^2$, this is an indication that MADs power these jets. Many studies point toward an approximate equality between jet and accretion powers (Rawlings and Saunders 1991; Ghisellini et al. 2010; Fernandes et al. 2011; McNamara et al. 2011; Punsly 2011; Martínez-Sansigre and Rawlings 2011). One of the most robust ways of constraining jet efficiency is to measure jet power $P_{\rm jet}$ from the energetics of X-ray emitting cavities inflated by the jets and divide it by mass-energy accretion rate $\dot{M}_{\rm B}c^2$ measured from the density of X-ray emitting gas within the sphere of influence of the BH called the Bondi radius, $r_{\rm B} \sim 10^5 r_g$. This gives quite a low efficiency of jet production (Allen et al. 2006). However, if one accounts for the gas expelled from the disk by its wind, we find that $\dot{M}$ near the BH is much lower than $\dot{M}_{\rm B}$. Jet efficiency then is well above 100 % for a sample of nearby low-luminosity AGN, suggesting that they are powered by MADs (Nemmen and Tchekhovskoy 2014).

We can also try and directly measure magnetic field strengths threading the central supermassive BHs. One way to do so is through Faraday rotation, which is the rotation of polarization plane of emission as it propagates through magnetized gas. For a recently discovered radio source in the vicinity of the supermassive BH at the center of our Galaxy, SgrA*, this gives a field strength of $B \sim 8\,\mathrm{mG}$ at a distance $r \sim 10^5 - 10^7 r_g$ from central BH (Eatough et al. 2013). Assuming the radial scaling $B \propto r^{-1}$, one arrives at an estimate of magnetic field at the BH horizon, $\simeq$ few $\times$ 100 G, which is close to the maximum possible magnetic field on the BH, as given by the estimate (3.14).

Such magnetic field measurements are quite difficult to come by for supermassive BHs in other galaxies, where there is a paucity of linearly polarized background radio sources close to the central nucleus that are bright enough for Faraday rotation measurement to be feasible. In order to find out if dynamically-important magnetic fields are an outlier or a norm for supermassive BHs, we need to measure field strengths near many of them. For this, we need to change our strategy. Instead of getting a handle on the magnetic field strength in accretion disks, we can measure it in the jets. One way of doing so is via a *core-shift* method. It focuses on jet's photosphere, which is the surface at which the jet becomes transparent to its own synchrotron radiation and which appears as a bright radio-emitting feature called the *radio core*. The distance from the BH to the *core* is of order $r_{\rm core} \sim 10^3 r_g$ and depends in a well-understood way on both the jet magnetic field strength *and* the observational frequency. By measuring the distance $\Delta r_{\rm core}$ between the positions of the core at two adjacent frequencies, or the *shift*, one can obtain the field strength in the frame of the jet, $B'_{\rm core} \propto \Delta r_{\rm core}^{3/4}$ and, after some manipulation, the magnetic flux through the jet, $\Phi_{\rm jet}$. The luminosity of the accretion flow gives the mass accretion rate, $\dot{M} = L_{\rm acc}/\epsilon c^2$, where one adopts a characteristic disk radiative efficiency for a rapidly spinning BH, $\epsilon \approx 0.4$. Now we are in a position to compute the prediction for the maximum possible value of magnetic flux that a BH can hold, $\Phi_{\rm MAD} = 50G(\epsilon c^5)^{-1/2} L_{\rm acc}^{1/2} M$ (see Eqs. 3.10 and 3.16), which is shown in Fig. 3.13 with the dashed line. As is clear from Fig. 3.13, the measured values of $\Phi_{\rm jet}$ agree with the MAD prediction over seven orders of magnitude in $L_{\rm acc}$. This suggests that the central BHs of these AGN are threaded with dynamically important magnetic field and their accretion disks are in the MAD regime.

Thus, the accretion disks in radio-loud AGN are *not* standard geometrically-thin accretion disks. This is a key difference from jet-less quasars, which most likely contain standard geometrically-thin accretion disks and hence are incapable of producing jets. This may resolve the apparent puzzle that most quasars, for which central BH spins have been measured, do not produce jets even though their central BHs tend to be rapidly spinning (Reynolds 2013b).

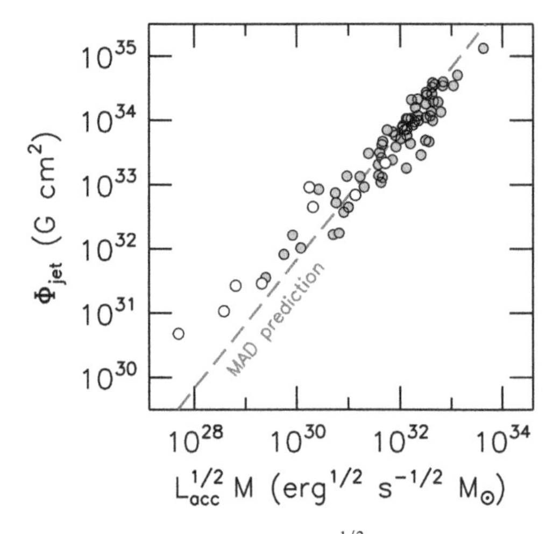

Fig. 3.13 Measured jet magnetic flux, $\Phi_{\rm jet}$, versus $L_{\rm acc}^{1/2} M$, where $L_{\rm acc}$ is the luminosity of the accretion flow and M is BH mass (Taken from Zamaninasab et al. (2014)). The observational data for $\Phi_{\rm jet}$ is shown with *open circles* for BL Lacs and *filled circles* for blazars. It is consistent with the MAD prediction for the magnetic flux on the BH, which is shown with the *dashed line* (see text for details). This suggests that the central BHs of radio-loud AGN—including blazars and BL Lacs—are threaded with dynamically-important magnetic fields and accrete surrounding gas in the MAD regime. The MAD prediction is computed assuming a radiatively efficient accretion flow with efficiency $\epsilon = 0.4$, which is appropriate for rapidly spinning BHs. As expected, BL Lac points are systematically shifted to the left of the correlation because their disks are radiatively inefficient

3.10 MADs in Tidal Disruption Events and Gamma-Ray Bursts

So far we discussed evidence for MADs in BHBs and radio-loud AGN. In fact, it is likely to find a MAD in essentially *any* BH accretion system whose mass accretion rate decreases asymptotically to zero. The reason for this is simple. Suppose an accretion flow contains a small but nonzero amount of large-scale magnetic flux. The flow drags this flux into the BH, so the strength of BH magnetic flux increases in time until nearly all of the flux ends up threading the BH. However, even as the BH magnetic flux increases, the mass accretion rate decreases in time. Therefore, it is inevitable that after some time BH magnetic flux becomes dynamically important, and a MAD forms. We will now discuss two examples of such transient systems—tidal disruption events (TDEs) and gamma-ray bursts (GRBs)—and the observational manifestations of their MADs.

MADs in Tidal Disruption Events An unfortunate star that passes too close to a supermassive BH and becomes tidally disrupted by the hole's gravity offers a unique probe of general relativity and accretion physics. As the star is torn apart by BH tidal forces, about half of its material becomes bound to the BH, forms an accretion disk and produces an optical, UV, and X-ray flare lasting for months to years (Ulmer 1999). Recently, two X-ray/soft gamma-ray events detected by the *Swift* observatory, Sw J1644+57 (Bloom et al. 2011; Burrows et al. 2011; Zauderer et al. 2011) and Sw J2058+05 (Cenko et al. 2012), have been associated with such TDEs. The association is based on the close proximity of the flare to the center of the host galaxy (and thus the central supermassive BH) and on the X-ray light curve time-dependence, $L_X \propto t^{-5/3}$, as seen in Fig. 3.14 and theoretically expected during a TDE for the mass fallback rate $\dot{M}_{fb}$ (Rees 1988) and BH mass accretion rate, $\dot{M}$, which is plausibly a fraction of $\dot{M}_{fb}$. However, unlike the usual TDEs, here the observed X-ray emission is believed to be produced by a relativistic collimated magnetized jet with an opening angle $\theta_j \sim 0.1$ radians. This is based primarily on the highly super-Eddington nature of the event, $L_X/L_{Edd} \sim 100$, as shown in Fig. 3.14 on the right $y-$axis,[4] and the detection of a radio afterglow due to a collimated jet running into the ambient medium (Giannios and Metzger 2011; Zauderer et al. 2011, 2013; Metzger et al. 2012; Berger et al. 2012; Wiersema et al. 2012).

It is highly likely that a stellar orbit was mis-aligned relative to BH midplane, since a star approaching the BH toward its death from a large distance did not know about BH's spin direction. However, there is no observational indication of such a misalignment: if the stellar orbit were misaligned, we would expect the resulting accretion disk to be misaligned as well. Such disks undergo precession around the BH (see, e.g., Fragile et al. 2007). If the direction of the jets follows the disk, the jets

[4]We do not expect disk luminosity to exceed L_{Edd} by more than a factor of a few, but there is no such constraint on jet luminosity.

would precess as well. However, this is ruled out: after the initial flaring ended, the X-ray emitting jet appeared to point steady at us, with no sign of strong precession (Stone and Loeb 2012). Does this mean that we were extremely lucky and the stellar orbit was nearly perfectly aligned with BH midplane? Not necessarily! In tilted disk-jet systems, the formation of a MAD causes the jets to reorient along BH spin axis (McKinney et al. 2013); thus, dynamically-important magnetic fields can provide the required stability of jet orientation in Sw J1644+57 without fine-tuning.

One of the most surprising features of Sw J1644 light curve, is the extremely strong early-time variability, or "flaring", seen in Fig. 3.14 during the first ~10 days. If BH spin axis points at us and the stellar orbit is tilted, the jets reorient along the BH spin axis only after the MAD forms. However, this reorientation is not instantaneous nor is it clean: as the jets work to reorient themselves, they punch holes through the disk and undergo a period of intense wobbling (McKinney et al. 2013). During this period, every time a jet passes in front of us, its emission beams into our line of sight, and we see a flare. This jet rearrangement process can naturally explain the initial period of strong flaring in the X-ray light curves of Sw J1644+57 (Tchekhovskoy et al. 2014). Moreover, prior to the MAD formation, the jets are misaligned, and their X-ray emission is beamed away from us, suggesting that the

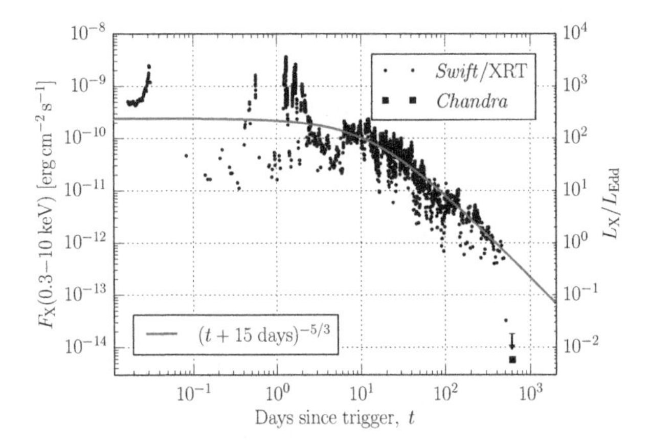

Fig. 3.14 X-ray lightcurve of the jetted tidal disruption event, Sw 1644+57, vs the time since *Swift*'s gamma-ray trigger (Adapted from Tchekhovskoy et al. (2014)). *Black dots* and the upper limit show data from the XRT telescope on board the *Swift* satellite, and the *black square* shows data from the *Chandra* X-ray observatory. The light curve shows several features: (i) the early-time "plateau" stage at $t \lesssim 10$ days, during which the X-ray luminosity is strongly variable but is on average constant, (ii) power-law decline stage, in which the light curve is consistent with the standard scaling expected for mass fallback rate in a TDE, $L_X \propto t^{-5/3}$, (iii) abrupt jet shutoff at $t \sim 500$ days. If we associate the jet shutoff with the spectral state transition ($a \rightarrow c$ in Fig. 3.5) from super-Eddington jet-producing accretion disk to jet-phobic standard geometrically-thin disk, which occurs at $\dot{M} \lesssim \dot{M}_{\rm edd}$, we can infer the mass accretion rate shown on the right y−axis in Eddington units. We see that the accretion flow starts out highly super-Eddington. Apart from the variability, the light curve is consistent with a single power-law dependence, whose zero-point by $\sim 15^{+15}_{-7}$ days precedes the trigger, $L_X \propto (t + 15\text{ days})^{-5/3}$ (see text for more details)

central BH was active for some time before the trigger. Indeed, an entire light curve, apart from the flaring, can be fit by a single power-law that starts 15^{+15}_{-7} days before the trigger, e.g., by $L_{\rm jet} \propto (t + 15\ {\rm days})^{-5/3}$, as shown with the solid red line in Fig. 3.14 (see Tchekhovskoy et al. 2014 for details).

The presence of a MAD also explains why the jet luminosity $L_{\rm jet}$ (which we assume makes up a fixed fraction of $P_{\rm jet}$) follows the power-law scaling that is expected for a mass accretion rate $\dot{M} \propto \dot{M}_{\rm fb} \propto t^{-5/3}$. This is because MADs naturally give $P_{\rm jet} \propto \dot{M}$ (see Figs. 3.8 and 3.10). Absent a MAD, obtaining such a scaling is non-trivial: $P_{\rm jet}$ is set by the large-scale magnetic flux threading the BH (see Eq. 3.7), which in general is independent of $\dot{M}$ (Tchekhovskoy et al. 2014), and below we consider an example of such a system.

MADs in Gamma-ray Bursts What other astrophysical systems have decreasing $\dot{M}$? Essentially, any transient system. Note that nearly all systems that we have discussed so far in the context of MADs are transient systems: (i) outbursts of BHBs accompanied by transient jets, (ii) outbursts of AGN accompanied by FRII jets, (iii) outbursts of AGN powered by TDEs. Another important system of this type is a core-collapse gamma-ray burst (GRB), in which a massive star ends its life, the stellar core collapses into a BH,[5] and the rest of the star accretes on to the BH with a steep dependence of mass-accretion rate on time, $\dot{M} \propto t^{-3}$–t^{-20} (Woosley 1993; MacFadyen and Woosley 1999; Popham et al. 1999). The BH powers a pair of ultra-relativistic jets that emerge out of the star and produce the GRB, an energetic burst of gamma-rays that is detectable if one of the jets happens to point at us. However, here lies a puzzle: if a jet is characterized by a constant energy efficiency, its power would decrease proportionally to $\dot{M}$: $P_{\rm jet} \propto \dot{M}$. But that is not what is observed: gamma-ray luminosity, or jet power, apart from random fluctuations, remains constant for most of the burst, 10–100 s, and then abruptly declines. This sudden drop at the end of the GRB is called the *steep power law decline*.

What causes such a rapid change in jet power? At the start of the burst, magnetic fields are not dynamically-important, or else they would lead to GRB luminosity far in excess of what is observed ($B_{\rm MAD} \sim 10^{17}$ G, whereas observationally $B_{\rm BH} \lesssim 10^{15}$ G). The power of the jets is limited by the stellar magnetic flux collected on the BH by accretion, i.e., at the early time the jet power is roughly constant. However, as $\dot{M}$ decreases, MAD eventually forms: BH magnetic flux becomes dynamically-important and leaves the BH, with jet power dropping precipitously, $P_{\rm jet} \propto \dot{M} \propto t^{-3}$–$t^{-20}$. This provides a natural explanation for the origin of the steep power law decline of GRB light curves (Tchekhovskoy and Giannios 2014).

[5] It is also possible that instead of the BH, a NS with strong magnetic fields, or a magnetar, powers the GRB.

Discussion and Conclusions

We started this chapter with describing a Blandford-Znajek process by which a large-scale poloidal magnetic field extracts and carries away BH rotational energy. The Blandford-Znajek mechanism is a very attractive source of power for jets as it works equally well for BHs of all sizes, stellar-mass and supermassive BHs. Over most of the range of BH spin ($a \lesssim 0.95$), the expression for power of Blandford-Znajek powered jets is very simple: $P_{\rm jet} \propto \Omega_{\rm H}^2 \Phi_{\rm BH}^2$ (Eq. 3.6), where $\Omega_{\rm H}$ is BH angular frequency and $\Phi_{\rm BH}$ is the poloidal magnetic flux threading the BH.

Because of the freedom in $\Phi_{\rm BH}$, for a fixed value of BH spin a and mass accretion rate $\dot{M}$ we expect a range of jet powers from zero (no jet) up to a maximum value of power at which the BH magnetic flux is so strong that it stops gas infall: this value of flux is approximately $\Phi_{\rm MAD} \approx 50(\dot{M} r_g^2 c)^{1/2}$ (see Eqs. 3.10 and 3.16). Such a strong magnetic flux leads to the formation of a *magnetically-arrested disk*, or a MAD. Since the BH magnetic flux is maximum in the MAD state, MADs achieve the maximum possible efficiency of jet production, $\eta_{\rm jet} = (P_{\rm jet}/\dot{M}c^2) \times 100\,\%$ that exceeds 100 % for rapidly spinning BHs (see Sect. 3.6 for details). Thus, MADs around rapidly spinning BHs perform *net* energy extraction from a rotating BH in a realistic astrophysical scenario, a long-sought result. Hence, it is not surprising that jet efficiency of MADs is much higher than the previously reported simulation results, $\eta_{\rm jet} \lesssim 20\,\%$ (e.g., McKinney 2005; De Villiers et al. 2005; Hawley and Krolik 2006; Barkov and Baushev 2011).

MADs produce powerful jets that extract BH spin energy so efficiently that a BH accreting at 10 % of the Eddington rate spins down to near zero spin in $\simeq 3 \times 10^8$ years, a time scale comparable to the quasar life time. This suggests that jets can extract a substantial fraction of BH spin energy, provide substantial feedback on the environment of the supermassive BH, and affect galaxy evolution.

However, not all accreting BHs produce jets. We reviewed different spectral states of BH accretion disks and their ability to launch relativistic jets in Sect. 3.5. We saw that continuous jets are launched at high, $L \gtrsim L_{\rm Edd}$, and low, $L \lesssim 0.01 L_{\rm Edd}$, accretion luminosities, both by stellar-mass and supermassive BHs. Transient jets from stellar-mass BHs are launched in the intermediate luminosity range, $L \sim (0.01{-}1) L_{\rm Edd}$. This range is similar to the one for FRII AGN jets, and we suggest that FRII jets are the supermassive analogs of transient jets in stellar-mass BHs. Following this analogy, we argued that the disks in FRII AGN are perturbed, *transient* disks in the process of changing their identity, similar to the transient disks in stellar-mass BHs during spectral state transitions (Sect. 3.8).

(continued)

Standard geometrically-thin disks exist in the same luminosity range as the transient stellar-mass BH disks, $L \sim (0.01 - 1)L_{\rm Edd}$, but do not produce jets or associated radio emission. What makes transient disks and standard thin disks so different from each other if they have the same value of $\dot{M}$? Recently, magnetic field strength threading supermassive BHs was measured, and it was found that the central BHs of many radio-loud AGN are accreting in the MAD regime (Zamaninasab et al. 2014). It is thus plausible that transient disks contain much stronger large-scale magnetic flux than the standard geometrically-thin disks (Sects. 3.7 and 3.9). This strong magnetic flux obstructs gas infall, modifies the nature of accretion, and leads to strong jets. MADs have also been inferred in tidal disruption events and core-collapse gamma-ray bursts (Sect. 3.10).

If BH accretion in the MAD regime is common, in many astrophysical systems the strength of BH magnetic flux is constrained by the dynamics of the accretion flow (see, e.g., Eq. 3.16) and is not a free parameter. This, for the first time, opens the possibility of modeling accreting BHs with 0 (*zero*) variables that parametrize our ignorance of physics. Thus, specifying just three physical parameters—$M_{\rm BH}$, a, and $\dot{M}$—is sufficient to completely define a simulated MAD system.[6] With the advent of global simulation methods that self-consistently account for radiation feedback on the structure and dynamics of accretion flow (Sądowski et al. 2014; McKinney et al. 2014) and give us the ability to compute the emergent spectra, we will be in a position to make quantitative predictions for spectra and variability of accreting BHs. This will allow us to carry out precision experiments that will test theoretical models of BH accretion and jets on a new quantitative level and give us the ability to measure masses and spins for a large number of BHs.

Acknowledgements AT thanks James Steiner for providing data for Fig. 3.12 and the detailed comments on the manuscript that substantially improved its clarity, Denise Gabuzda for her encouragement that made this work possible, Alexander Philippov for helpful suggestions. AT was supported by NASA through Einstein Postdoctoral Fellowship grant number PF3-140115 awarded by the Chandra X-ray Center, which is operated by the Smithsonian Astrophysical Observatory for NASA under contract NAS8-03060, and NASA via High-End Computing (HEC) Program through the NASA Advanced Supercomputing (NAS) Division at Ames Research Center that provided access to the Pleiades supercomputer, as well as NSF through an XSEDE computational time allocation TG-AST100040 on NICS Kraken, Nautilus, TACC Stampede, Maverick, and Ranch. We used Enthought Canopy Python distribution to generate some of the figures for this work.

[6]With radiation feedback on the structure of the accretion flow accounted for, the ratio $\dot{M}/\dot{M}_{\rm Edd}$ is expected to determine the accretion disk angular thickness, h/r, which controls the strength of BH magnetic flux and jet power in the MAD state.

References

Allen, S.W., Dunn, R.J.H., Fabian, A.C., Taylor, G.B., Reynolds, C.S.: MNRAS **372**, 21 (2006)
Balbus, S.A., Hawley, J.F.: ApJ **376**, 214 (1991)
Bardeen, J.M.: Nature **226**, 64 (1970)
Barkov, M.V., Baushev, A.N.: New Astron. **16**, 46 (2011)
Beckwith, K., Hawley, J.F., Krolik, J.H.: ApJ **678**, 1180 (2008)
Begelman, M.C.: Accretion disks in active galactic nuclei. In: Miller, J.S. (ed.) Astrophysics of Active Galaxies and Quasi-Stellar Objects, pp. 411–452. University Science Books, Mill Valley (1985)
Begelman, M.C., Armitage, P.J.: ApJ **782**, L18 (2014)
Berger, E., Zauderer, A., Pooley, G.G., et al.: ApJ **748**, 36 (2012)
Bisnovatyi-Kogan, G.S., Ruzmaikin, A.A.: Ap&SS **28**, 45 (1974)
Bisnovatyi-Kogan, G.S., Ruzmaikin, A.A.: Ap&SS **42**, 401 (1976)
Blandford, R.D.: Beyond the fringe. In: Romney, J., Reid, M. (eds.) Future Directions in High Resolution Astronomy. Astronomical Society of the Pacific Conference Series, vol. 340, p. 3. Astronomical Society of the Pacific, San Francisco (2005)
Blandford, R.D., Payne, D.G.: MNRAS **199**, 883 (1982)
Blandford, R.D., Znajek, R.L.: MNRAS **179**, 433 (1977)
Bloom, J.S., et al.: Science **333**, 203 (2011)
Burrows, D.N., et al.: Nature **476**, 421 (2011)
Cenko, S.B., et al.: ApJ **753**, 77 (2012)
Chakrabarti, S.K.: ApJ **288**, 1 (1985)
Chiueh, T., Li, Z.Y., Begelman, M.C.: ApJ **377**, 462 (1991)
Cowperthwaite, P.S., Reynolds, C.S.: ApJ **752**, L21 (2012)
Davis, S.W., Narayan, R., Zhu, Y., Barret, D., Farrell, S.A., Godet, O., Servillat, M., Webb, N.A., ApJ **734**, 111 (2011)
De Villiers, J.P., Hawley, J.F.: ApJ **589**, 458 (2003)
De Villiers, J.P., Hawley, J.F., Krolik, J.H.: ApJ **599**, 1238 (2003)
De Villiers, J.P., Hawley, J.F., Krolik, J.H., Hirose, S.: ApJ **620**, 878 (2005)
Dexter, J., McKinney, J.C., Markoff, S., Tchekhovskoy, A.: MNRAS **440**, 2185 (2014)
Eatough, R.P., Falcke, H., Karuppusamy, R., Lee, K.J., Champion, D.J., Keane, E.F., Desvignes, G., Schnitzeler, D.H.F.M., Spitler, L.G., Kramer, M., Klein, B., Bassa, C., Bower, G.C., Brunthaler, A., Cognard, I., Deller, A.T., Demorest, P.B., Freire, P.C.C., Kraus, A., Lyne, A.G., Noutsos, A., Stappers, B., Wex, N.: Nature **501**, 391 (2013)
Esin, A.A., McClintock, J.E., Narayan, R.: ApJ **489**, 865 (1997)
Esin, A.A., Narayan, R., Cui, W., Grove, J.E., Zhang, S.N.: ApJ **505**, 854 (1998)
Fabian, A., Iwasawa, K., Reynolds, C., Young, A.: Publ. Astron. Soc. Pac. **112**(775), 1145 (2000)
Fanaroff, B.L., Riley, J.M.: MNRAS **167**, 31P (1974)
Farrell, S.A., Webb, N.A., Barret, D., Godet, O., Rodrigues, J.M.: Nature **460**(7251), 73 (2009)
Fender, R.P., Belloni, T.M., Gallo, E.: MNRAS **355**, 1105 (2004)
Fender, R.P., Gallo, E., Russell, D.: MNRAS **406**, 1425. ArXiv:1003.5516 (2010)
Fernandes, C.A.C. et al.: MNRAS **411**, 1909 (2011)
Fragile, P.C., Blaes, O.M., Anninos, P., Salmonson, J.D.: ApJ **668**, 417 (2007)
Frank, J., King, A., Raine, D.J.: Accretion Power in Astrophysics, 3rd edn. Cambridge University Press, Cambridge (2002)
Gammie, C.F., Narayan, R., Blandford, R.: ApJ **516**, 177 (1999)
Gammie, C.F., McKinney, J.C., Tóth, G.: ApJ **589**, 444 (2003)
Gammie, C.F., Shapiro, S.L., McKinney, J.C.: ApJ **602**, 312 (2004)
Ghisellini, G., et al.: MNRAS **402**, 497 (2010)
Ghosh, P., Abramowicz, M.A.: MNRAS **292**, 887 (1997)
Giannios, D., Metzger, B.D.: MNRAS **416**, 2102 (2011)
Guilet, J., Ogilvie, G.I.: MNRAS **424**, 2097 (2012)

Guilet, J., Ogilvie, G.I.: MNRAS **430**, 822 (2013)
Hawley, J.F., Krolik, J.H.: ApJ **641**, 103 (2006)
Heinz, S., Sunyaev, R.A.: MNRAS **343**, L59 (2003)
Hui, Y., Krolik, J.H.: ApJ **679**, 1405 (2008)
Igumenshchev, I.V.: ApJ **677**, 317 (2008)
Igumenshchev, I.V.: ApJ **702**, L72 (2009)
Igumenshchev, I., Narayan, R., Abramowicz, M.: ApJ **592**, 1042 (2003)
Komissarov, S.S.: MNRAS **326**, L41 (2001)
Komissarov, S.S., Barkov, M.V.: MNRAS **397**, 1153 (2009)
Krolik, J.H., Hawley, J.F., Hirose, S.: ApJ **622**, 1008 (2005)
Kulkarni, A.K., Penna, R.F., Shcherbakov, R.V., Steiner, J.F., Narayan, R., Sä Dowski, A., Zhu, Y., McClintock, J.E., Davis, S.W., McKinney, J.C.: MNRAS **414**, 1183 (2011)
Livio, M., Ogilvie, G.I., Pringle, J.E.: ApJ **512**, 100 (1999). doi:10.1086/306777
Lubow, S.H., Papaloizou, J.C.B., Pringle, J.E.: MNRAS **267**, 235 (1994)
Maccarone, T.J.: A&A **409**, 697 (2003)
MacFadyen, A.I., Woosley, S.E.: ApJ **524**, 262 (1999)
Martínez-Sansigre, A., Rawlings, S.: MNRAS **414**, 1937 (2011)
McClintock, J.E., Narayan, R., Davis, S.W., Gou, L., Kulkarni, A., Orosz, J.A., Penna, R.F., Remillard, R.A., Steiner, J.F.: Class. Quantum Gravity **28**(11), 114009 (2011)
McClintock, J.E., Narayan, R., Steiner, J.F.: Space Sci. Rev., **183**, 295–322 (2013)
McKinney, J.C.: ApJ **630**, L5 (2005)
McKinney, J.C., Blandford, R.D.: MNRAS **394**, L126 (2009)
McKinney, J.C., Gammie, C.F.: ApJ **611**, 977 (2004)
McKinney, J.C., Tchekhovskoy, A., Blandford, R.D.: MNRAS **423**, 3083 (2012)
McKinney, J.C., Tchekhovskoy, A., Blandford, R.D.: Science **339**, 49 (2013)
McKinney, J.C., Tchekhovskoy, A., Sadowski, A., Narayan, R.: MNRAS **441**, 3177 (2014)
McNamara, B.R., Kazemzadeh, F., Rafferty, D.A., Bîrzan, L., Nulsen, P.E.J., Kirkpatrick, C.C., Wise, M.W.: ApJ **698**, 594 (2009)
McNamara, B.R., Rohanizadegan, M., Nulsen, P.E.J.: ApJ **727**, 39 (2011)
Metzger, B.D., Giannios, D., Mimica, P.: MNRAS **420**, 3528 (2012)
Narayan, R., McClintock, J.E.: MNRAS **419**, L69 (2012)
Narayan, R., Igumenshchev, I.V., Abramowicz, M.A.: PASJ **55**, L69 (2003)
Narayan, R., Sądowski, A., Penna, R.F., Kulkarni, A.K.: MNRAS **426**, 3241 (2012)
Nemmen, R.S., Tchekhovskoy, A.: MNRAS (2014, submitted). ArXiv:1406.7420
Novikov, I.D., Thorne, K.S.: Astrophysics of black holes. In: De Witt, C., De Witt, B.S. (eds.) Black Holes-Les Astres Occlus. Gordon & Breach, New York (1973)
Penna, R.F., et al.: MNRAS **408**, 752 (2010)
Popham, R., Woosley, S.E., Fryer, C.: ApJ **518**, 356 (1999)
Potter, W.J., Balbus, S.A.: MNRAS **441**, 681 (2014)
Proga, D., Begelman, M.C.: ApJ **592**, 767 (2003)
Punsly, B.: ApJ **728**, L17 (2011)
Rawlings, S., Saunders, R.: Nature **349**, 138 (1991)
Rees, M.J.: Nature **333**, 523 (1988)
Remillard, R.A., McClintock, J.E.: ARA&A **44**, 49 (2006)
Reynolds, C.S.: Space Sci. Rev., **183**, 277–294 (2013a)
Reynolds, C.S.: Class. Quantum Gravity **30**(24), 244004 (2013b)
Rothstein, D.M., Lovelace, R.V.E.: ApJ **677**, 1221 (2008)
Russell, D.M., Miller-Jones, J.C.A., Maccarone, T.J., Yang, Y.J., Fender, R.P., Lewis, F.: ApJ **739**, L19 (2011)
Russell, D.M., Gallo, E., Fender, R.P.: MNRAS **431**, 405 (2013)
Sądowski, A., Narayan, R., Penna, R., Zhu, Y.: MNRAS **436**, 3856 (2013)
Sądowski, A., Narayan, R., McKinney, J.C., Tchekhovskoy, A.: MNRAS **439**, 503 (2014)
Shafee, R., McKinney, J.C., Narayan, R., Tchekhovskoy, A., et al.: ApJ **687**, L25 (2008)
Shakura, N.I., Sunyaev, R.A.: A&A **24**, 337 (1973)

Shapiro, S.L., Teukolsky, S.A. (eds.): Black Holes, White Dwarfs and Neutron Stars: The Physics of Compact Objects, pp. 672. Wiley-VCH, Weinheim (1986). ISBN:0-471-87316-0
Sikora, M., Stawarz, Ł., Lasota, J.P.: ApJ **658**, 815 (2007)
Steiner, J.F., McClintock, J.E., Narayan, R.: ApJ **762**, 104 (2013)
Stone, N., Loeb, A.: PRL **108**(6), 061302 (2012)
Straub, O., Godet, O., Webb, N., Servillat, M., Barret, D.: ArXiv e-prints (2014)
Tanabe, K., Nagataki, S.: Phys. Rev. D **78**(2), 024004 (2008)
Tavecchio, F., Becerra-Gonzalez, J., Ghisellini, G., Stamerra, A., Bonnoli, G., Foschini, L., Maraschi, L.: A&A **534**, A86 (2011)
Tchekhovskoy, A., Giannios, D.: MNRAS (2014, submitted). ArXiv:1409.4414
Tchekhovskoy, A., McKinney, J.C.: MNRAS **423**, L55 (2012)
Tchekhovskoy, A., McKinney, J.C., Narayan, R.: MNRAS **379**, 469 (2007)
Tchekhovskoy, A., McKinney, J.C., Narayan, R.: MNRAS **388**, 551 (2008)
Tchekhovskoy, A., McKinney, J.C., Narayan, R.: ApJ **699**, 1789 (2009)
Tchekhovskoy, A., Narayan, R., McKinney, J.C.: ApJ **711**, 50 (2010)
Tchekhovskoy, A., Narayan, R., McKinney, J.C.: New Astron. **15**, 749 (2010)
Tchekhovskoy, A., Narayan, R., McKinney, J.C.: MNRAS **418**, L79 (2011)
Tchekhovskoy, A., McKinney, J.C., Narayan, R.: J. Phys. Conf. Ser. **372**(1), 012040 (2012)
Tchekhovskoy, A., Metzger, B.D., Giannios, D., Kelley, L.Z.: MNRAS **437**, 2744 (2014)
Thorne, K.S.: ApJ **191**, 507 (1974)
Ulmer, A.: ApJ **514**, 180 (1999)
Urry, C.M., Padovani, P.: PASP **107**, 803 (1995)
van Velzen, S., Falcke, H.: A&A **557**, L7 (2013)
Webb, N., Cseh, D., Lenc, E., Godet, O., Barret, D., Corbel, S., Farrell, S., Fender, R., Gehrels, N., Heywood, I.: Science **337**, 554 (2012)
Wiersema, K., van der Horst, A.J., Levan, A.J., et al.: MNRAS **421**, 1942 (2012)
Woosley, S.E.: ApJ **405**, 273 (1993)
Yuan, F., Narayan, R.: ARA&A, **52**, 529. ArXiv e-prints (2014)
Yuan, F., Quataert, E., Narayan, R.: ApJ **598**, 301 (2003)
Zamaninasab, M., Clausen-Brown, E., Savolainen, T., Tchekhovskoy, A.: Nature **510**(7503), 126 (2014)
Zauderer, B.A., Berger, E., Soderberg, A.M., et al.: Nature **476**, 425 (2011)
Zauderer, B.A., Berger, E., Margutti, R., Others.: ApJ **767**, 152 (2013)
Zhu, Y., Davis, S.W., Narayan, R., Kulkarni, A.K., Penna, R.F., McClintock J.E.: MNRAS **424**, 2504 (2012)

Chapter 4
Kiloparsec-Scale AGN Jets

Martin Hardcastle

Abstract In this chapter we discuss kiloparsec-scale AGN jets and their interactions with the environment, which lead to observed structures such as radio lobes and hotspots. After a brief summary of the history of the topic, we outline the basic physics of the emission processes, the sources' key observational features (focussing on source physics) and their interpretation in terms of particle and field content, particle acceleration, and overall source dynamics.

4.1 Historical Background

The first observations of radio-loud AGN were made using low-resolution, low-frequency survey instruments (Reber 1944; Ryle et al. 1950), which are sensitive to the emission from structures on all scales. It was realised very early on (Jennison and Das Gupta 1953) that many of these sources were resolved, on scales which, when they were identified with galaxies (Baade and Minkowski 1954) implied physical sizes of hundreds of kpc or more. The spectrum of the observed radio emission, together with the early detection of polarized optical counterparts (Baade 1956) implied that the emission was synchrotron radiation (Burbidge 1956), and allowed the first inferences of the contents of the radio-emitting structures (electrons and magnetic fields) and the realization that very large amounts of energy were required to power such objects. The basic observational constraints on the nature of these objects were in place by the end of the 1950s.

The crucial development in the following decades was the development of radio interferometers with the capability both of surveying the sky at adequate resolution to allow the identification of radio sources with galaxies and quasars (e.g. Edge et al. 1959) and of making detailed images of the structures of individual objects (e.g. Macdonald et al. 1968) of the large-scale radio structure. By the 1970s, there were enough high-resolution images, and enough source identifications

M. Hardcastle (✉)
School of Physics, Astronomy and Mathematics, University of Hertfordshire, College Lane, Hatfield AL10 9AB, UK
e-mail: mjh@extragalactic.info

I. Contopoulos et al. (eds.), *The Formation and Disruption of Black Hole Jets*, Astrophysics and Space Science Library 414,
DOI 10.1007/978-3-319-10356-3_4

to allow Fanaroff and Riley (1974) to make their well-known morphological classification of radio sources as centre-brightened (class I, FRI) or edge-brightened (FRII) and enough galaxy identifications for them to be able to show that the morphological classification correlated with radio luminosity, in the sense that FRII objects tended to have luminosities above a threshold value (which, in a modern cosmology, corresponds to about 10^{26} W Hz^{-1} at 178 MHz) while FRIs lay below it. This, together with the detection of elongated kpc-scale structures connecting the large-scale emission to the nucleus, which became known as jets (e.g. Northover 1973), gave the first indications of relationships between the radio structure and the underlying physics, and motivated the adoption of the now standard 'beam model' in which outflows from the active nucleus drive the large-scale structure (Scheuer 1974; Blandford and Rees 1974), with the observed jets being the physical manifestations of these outflows. The importance of the interaction between the radio structures and the newly discovered X-ray-emitting intracluster medium was also first realised at this time (Gull and Northover 1973).

The principal observational improvements for studies of kpc-scale radio AGN physics between the 1970s and the present day were (1) the construction of the NRAO Very Large Array (VLA), whose design allowed imaging of both compact and large-scale structure at the same time, permitting detailed, high-fidelity morphological studies of individual objects (e.g. Carilli et al. 1991; Laing and Bridle 2002) and of large samples (e.g. Black et al. 1992; Fernini et al. 1993, 1997; Bridle et al. 1994; Leahy et al. 1997; Hardcastle et al. 1997a; Gilbert et al. 2004; Mullin et al. 2006); (2) the extension of radio synchrotron studies to the optical; and (3) the development of X-ray telescopes with enough sensitivity to routinely image both synchrotron and inverse-Compton emission from the radio-emitting structures and their hot-gas environments (e.g. Böhringer et al. 1993; Harris et al. 1994, 2000; Hardcastle et al. 1998c, 2001; Wilson et al. 2001; Croston et al. 2003). These developments will be discussed in the following sections.

4.2 Physics of Emission Processes

In this section we briefly summarize the important features of the relevant emission processes. For details of the derivations of these results the reader is referred to e.g. Longair (2010) or Rybicki and Lightman (1979).

4.2.1 Synchrotron Emission

Synchrotron emission is produced by relativistic electrons (and/or positrons; hereafter, for simplicity, we refer to the leptonic radiating particles as electrons) moving in magnetic fields. An individual electron with a Lorentz factor γ has a synchrotron spectrum which is strongly peaked at frequencies around

$$\nu \approx \gamma^2 \frac{eB}{2\pi m} \tag{4.1}$$

where m is the mass of the electron, e is its charge and B is the magnetic field strength. The energy loss rate for this single electron via synchrotron radiation can be shown to be

$$\frac{\mathrm{d}E}{\mathrm{d}t} = \frac{4}{3}\sigma_{\mathrm{T}} c U_B \gamma^2 \tag{4.2}$$

where σ_{T} is the Thomson cross-section and U_B is the energy density in the magnetic field, which leads to the important result that the timescale for energy loss is given by

$$\tau = \frac{E}{\mathrm{d}E/\mathrm{d}t} = \frac{3mc}{4\sigma_{\mathrm{T}} U_B \gamma} \tag{4.3}$$

so that electrons with higher energies radiate their energy away more rapidly. Combining equations 4.1 and 4.3, we can infer maximum ages since particle acceleration from observations of radiation of a given frequency, if the magnetic field strength is known.

In reality of course we observe the radiation from a population of electrons. The volume synchrotron emissivity may be written

$$J(\nu) = \frac{\sqrt{3}Be^3 \sin\theta}{4\pi\epsilon_0 cm} \int_{E_{min}}^{E_{max}} F(x)N(E)\mathrm{d}E \tag{4.4}$$

where ϵ_0 is the permittivity of free space. θ is the pitch angle of the electrons with respect to the magnetic field direction and x is defined by

$$x = \frac{4\pi m}{3e} \frac{\nu}{\gamma^2 B \sin\theta} \tag{4.5}$$

with $E = \gamma mc^2$. $F(x)$ gives the shape of the synchrotron spectrum of an individual electron,

$$F(x) = x \int_x^{\infty} K_{5/3}(z)\mathrm{d}z \tag{4.6}$$

where $K_{5/3}$ is the modified Bessel function of order 5/3. Because $F(x)$ falls off relatively sharply above $x \approx 1$, the key factor determining the observed synchrotron spectrum is the electron energy distribution $N(E)$. Observations of power-law synchrotron spectra turn out to imply power-law electron energy distributions: if $N(E) = N_0 E^{-p}$, then we can integrate equation 4.4 over electron energy and pitch angle (assumed isotropic) to obtain

$$J(\nu) = CN_0\nu^{-\frac{(p-1)}{2}}B^{\frac{(p+1)}{2}} \tag{4.7}$$

where

$$C = c(p)\frac{e^3}{\epsilon_0 c m_e}\left(\frac{m_e^3 c^4}{e}\right)^{-(p-1)/2} \tag{4.8}$$

$c(p)$ is a combination of gamma functions which numerically is of order 0.05 and depends only weakly on p. We see from Eq. 4.7 that the electron energy spectral index can be inferred from the dependence of the synchrotron luminosity on frequency.

Power-law electron energy spectra are expected to be set up by some particle acceleration processes (e.g. Bell 1978) but we saw above (Eq. 4.3) that the energy loss timescale depends on electron energy: so an initially power-law electron spectrum will start to deviate from a power law at late times. If we write the energy loss equation 4.2 as

$$\frac{\mathrm{d}E}{\mathrm{d}t} = -aE^2 \tag{4.9}$$

then it can be shown (Pacholczyk 1970) that after some time t an initial power-law energy spectrum becomes

$$N(E) = \begin{cases} N_0 E^{-p}(1 - Eat)^{p-2} & E < 1/at \\ 0 & E \geq 1/at \end{cases} \tag{4.10}$$

Thus in the absence of any continuing particle acceleration we get an electron energy spectrum with a cutoff at high energies and significant steepening below the cutoff. Putting such an electron energy distribution into Eq. 4.4, we can derive an 'aged' synchrotron spectrum that can in principle be fitted to observations to estimate t, though the details depend on assumptions about pitch angle scattering and magnetic field distribution (Kardashev 1962; Jaffe and Perola 1973; Tribble 1993; Hardcastle 2013). Again, note that the loss time depends on the magnetic field strength, so that we need to know B to apply this technique to determine absolute ages.

Determination of magnetic field strengths is a long-standing problem: from Eq. 4.4 we see that the synchrotron emissivity depends both on the normalization of the electron energy distribution and the magnetic field strength, so that we cannot infer these two quantities separately from synchrotron observations alone. This means that we cannot infer the total energy density, which we may write

$$U_{\mathrm{tot}} = \frac{B^2}{2\mu_0} + (1+\kappa)\int_{E_{\min}}^{E_{\max}} EN(E)\mathrm{d}E \tag{4.11}$$

where (as is conventional) the factor $1 + \kappa$ accounts for any energy density in non-radiating particles. However, if we assume a shape for the electron energy spectrum, such as a power law, then the integral in Eq. 4.11 can be done up to the normalization of the power law N_0, and we also have a relation between emissivity, normalization and magnetic field from Eq. 4.7, which we can use to eliminate either B or N_0 from Eq. 4.11. Burbidge (1956) was the first to show that this implies a *minimum* energy density for a region of synchrotron-emitting plasma, with a field strength that roughly corresponds to *equipartition* of energy between field and electrons (i.e. $U_B \approx U_E$) if we assume $\kappa = 0$. These two assumptions, minimum energy or equipartition, form the basis of many calculations in the literature of the total kinetic power of radio-loud AGN, but it is important to bear in mind that neither is anything but an assumption if only synchrotron emission has been observed.

Finally we note that synchrotron emission is strongly polarized. The fractional polarization Π can be found by integrating the polarized emissivity function for a single electron, $G(x) = xK_{2/3}$, over electron energy and dividing by the total emissivity, i.e.

$$\Pi = \frac{\int G(x)N(E)\mathrm{d}E}{\int F(x)N(E)\mathrm{d}E} \tag{4.12}$$

It can be seen that the fractional polarization (where we are assuming a region with a single, uniform magnetic field) does not depend on field strength but only on the properties of the electron energy spectrum: for $N(E) = N_0 E^{-p}$ it can be shown that $\Pi = (p+1)/(p+\frac{7}{3})$, or about 70 % for $p = 2$. The theoretical polarization can approach 100 % for very steep spectra (e.g. in the exponential cutoff of an aged spectrum as described above). For a region of uniform field, the direction of the polarization E-vector is perpendicular to the magnetic field direction: it is therefore common for observers to plot vectors perpendicular to the E-vectors and call them the 'magnetic field direction', but of course it must be borne in mind that the observed polarization direction comes from the emissivity-weighted sum of the polarization directions of electrons which are potentially in many different field configurations. In addition, Faraday rotation, in which the propagation of polarized radio waves through a magnetized medium of thermal electrons rotates the plane of polarization, may affect the apparent polarization of synchrotron emission. In the simplest case, where the thermal electrons are all external to the radio source and the screen is well resolved, we expect to find that the rotation angle ϕ is given by

$$\phi = \frac{c^2}{\nu^2} K \int_0^s n_{th} \mathbf{B}.\mathrm{d}\mathbf{s} \tag{4.13}$$

where K is a constant and the integral is along the line of sight to the source: thus there is a simple λ^2 dependence of the rotation angle. This equation is an adequate description for Faraday rotation of extragalactic radio sources by plasma in our own Galaxy, for example. If the Faraday screen is not resolved or not external to the radio galaxy, we will see depolarization as well as rotation, and a λ^2 dependence will in general not be present: for details see e.g. Burn (1966) and Laing (1984).

4.2.2 Inverse-Compton Emission

The inverse-Compton process is the scattering of a low-energy photon to high energies by a relativistic electron, which involves an energy gain of the scattered photon of roughly γ^2 where γ is the Lorentz factor of the electron. As seed photons from the CMB are always available, and relativistic electrons are present wherever synchrotron emission is observed, inverse-Compton is a required process in radio galaxies. If the interaction is far from the Klein-Nishina regime, i.e. if $h\nu \ll mc^2$ in the rest frame of the electron, the relevant physics is quite simple, and indeed closely mimics synchrotron emission in many ways (Rybicki and Lightman 1979): in particular, the energy loss rate to inverse-Compton in the ultrarelativistic regime is

$$\frac{\mathrm{d}E}{\mathrm{d}t} = \frac{4}{3}\sigma_{\mathrm{T}} c U_{ph} \gamma^2 \tag{4.14}$$

which is identical in form to Eq. 4.2 except that the energy density in photons appears instead of that of the magnetic field. For purposes of spectral ageing calculations as discussed above, the photon field acts like another contribution to the magnetic field energy density.

If the photon field is isotropic (as with the CMB) then the inverse-Compton emissivity can be written in a simple form which again has similarities to the synchrotron formulae, but requires an additional integral over the photon field (Hardcastle et al. 1998b):

$$J_{\mathrm{ic}}(\nu_1) = m_e^2 c^4 \nu_1 \sigma_T \int_{E_{\min}}^{E_{\max}} \frac{N(E)}{E^2} \int_{\nu_{\min}}^{\nu_{\max}} \frac{3c}{4\nu_0^2} U_{\nu_0} f(x) \mathrm{d}\nu_0 \mathrm{d}E \tag{4.15}$$

where U_{ν_0} is the spectral energy density of the photon field, x is essentially the electron energy scaled by the ratio of photon frequencies,

$$x = \frac{\nu_1}{4\gamma^2 \nu_0} \tag{4.16}$$

and $f(x)$ gives the energy-dependence of the scattering cross-section, e.g. Rybicki and Lightman (1979)

$$f(x) = 2x \ln x + x + 1 - 2x^2 \tag{4.17}$$

Thus the emissivity can be calculated if the photon spectrum is known (e.g. a black body for the CMB, or a power law for synchrotron self-Compton (SSC) emission in which the synchrotron photons themselves are scattered to high energies). For anisotropic photon fields, photon energies in the Klein-Nishina limit, or both, it is necessary to apply more complex formulae with an explicit integration over the angle between the electrons and the photons: see e.g. Brunetti (2000) for applicable methods in this case.

A very important feature of inverse-Compton emission is that the emissivity is independent of the magnetic field strength: if the input photon field is known, as it is for CMB scattering or SSC, inverse-Compton tells us about the number density of electrons with energies capable of scattering those photons into the observed band. Observations of inverse-Compton emission thus break the degeneracy between electron spectrum normalization and B discussed in the previous section. If we have synchrotron and inverse-Compton emission from the same region, we can measure the magnetic field strength B, subject only to assumptions about the electron spectrum $N(E)$ (since we often do not have observations of synchrotron emission from the *same* population of electrons that do the inverse-Compton scattering).

4.2.3 Thermal Bremsstrahlung

The dominant emission process from the hot phase of the intergalactic medium, which as the highest-pressure gas in these regions is primarily responsible for the confinement of radio lobes, is thermal bremsstrahlung, which for the temperatures of the hot phase ($\sim 10^7$–10^8 K) appears in the X-ray band. The reader is referred to reviews of X-ray emission from clusters of galaxies (e.g. Sarazin 1986) for details: the important point here is simply that the emissivity from this process goes roughly as $T^{1/2}n^2$, where n is the number density of particles. At lower temperatures line cooling of the medium is also important, which gives rise to a different temperature dependence but the same dependence on density. Thus images of the X-ray emission from groups and clusters are essentially tracing the line-of-sight integral of n^2. Thermal bremsstrahlung (together with line emission where present) have a distinctive spectrum which is quite different from the power laws produced by synchrotron or inverse-Compton emission, and so the two components can be distinguished in X-ray spectra with good signal to noise. If a *combination* of power-law inverse-Compton and thermal bremsstrahlung is expected to be present, it is rather harder to put good constraints on the normalizations of the two components (Hardcastle and Croston 2010).

4.3 Observations of Jets and Lobes

We now turn to the details of the observations of kpc-scale structures in radio-loud AGN.

4.3.1 *Total-Intensity Radio Emission*

4.3.1.1 Classical Doubles

High-resolution radio observations show that most FRII ('classical double') objects share a set of basic characteristics, including roughly symmetrical double lobes. a central sub-arcsec radio core, bright compact structures known as hotspots embedded in more diffuse structure at the ends of the lobes, and, often but not always, a jet or jets in between the core and the hotspots (Fig. 4.1), where a jet is classically defined as an elongated region of surface brightness excess (Bridle and Perley 1984). Many lobes have more than one hotspot (Laing 1989). The fact that the jet is often one-sided (e.g. Bridle et al. 1994), combined with direct evidence that the brighter jet usually points towards us (Laing 1988; Garrington et al. 1988) shows that the jet in these objects is likely to be relativistically beamed, with effective beaming speeds corresponding to $\Gamma \sim 2$ (Wardle and Aaron 1997; Hardcastle et al. 1998a; Mullin and Hardcastle 2009): the apparent luminosity is increased by a factor $[\Gamma(1-\beta\cos\theta)]^{2+\alpha}$ (Ryle and Longair 1967; Scheuer and Readhead 1979), where $\beta = v/c$, Γ is the bulk Lorentz factor $(1-\beta^2)^{-1/2}$, θ is the angle made by the jet to the line of sight, and α is the spectral index, defined in this chapter in the sense that

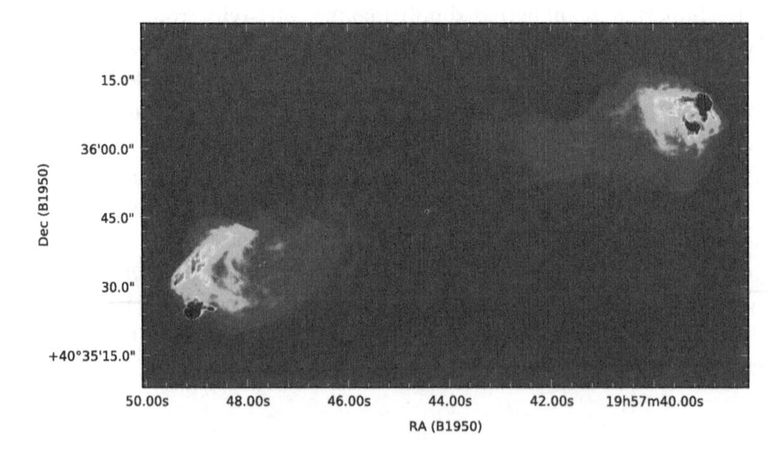

Fig. 4.1 5-GHz radio image of the classical double (FRII) radio galaxy Cygnus A (3C 405). Note the central 'core', the bright multiple hotspots in each lobe, and the faint jet and counterjet (Data from Perley et al. 1984)

$S \propto \nu^{-\alpha}$. Relativistic beaming also affects the observed brightness of the central core emission, which is unsurprising as this comes from the unresolved parsec-scale jet. In objects where the jet is very closely aligned with the line of sight this beaming gives rise to core-dominated or core-jet sources with very one-sided jets.

Broad-band spectral mapping of FRIIs shows that the spectrum is flat ($\alpha \sim 0.5$–0.7) in the hotspots and jets, where present, and tends to steepen systematically away from the hotspots towards the central regions of the lobes (e.g. Myers and Spangler 1985; Alexander and Leahy 1987; Harwood et al. 2013): this is usually interpreted in terms of radiative losses as described in Sect. 4.2.1.

A small fraction of objects with FRII luminosity and morphology differ from this standard picture. Among them are the so-called 'double-double' sources in which a pair of smaller lobes appears to be embedded in the larger-scale lobe emission (e.g. Schoenmakers et al. 2000); care is needed to distinguish these structures from a two-sided jet. 'Fat double' sources have weak or no hotspots, and may be sources with unusually dissipative jets or relics where nuclear activity has ceased.

4.3.1.2 ... and the Rest

Objects with FRI morphology, or with luminosities around the FRI/FRII break even if formally classed as FRII, exhibit a wider range of structures. On the largest scales FRIs are often assimilated to the 'classical twin-jet' behaviour shown in Fig. 4.2, where a large-scale jet/plume broadens only very gradually into structures that might be called lobes, but in fact it is more normal for the jets to be embedded in lobes similar to those of FRIIs (Parma et al. 1987). Canonically FRI jets are two-sided, or become two-sided, with a wide opening angle at the base, on scales of a few kpc; the transition from a one-sided to a two-sided appearance is thought to indicate deceleration from a relativistic to a sub-relativistic regime (Bridle and Perley 1984; Hardcastle et al. 1997b; Laing and Bridle 2002) and occurs on larger scales for more luminous sources (Parma et al. 1987). In a small minority of powerful FRIs, the so-called 'wide-angle tails', one-sided, well-collimated jets extend to distances of tens of kpc before transitioning into a plume, often with a weak compact feature at the jet termination analogous to the FRII hotspots (Hardcastle and Sakelliou 2004). The plumes in the wide-angle tails may be, but are not always, bent; such sources should not be confused with 'narrow-angle tails' which appear to be classical wide-opening-angle twin-jet sources where the jets bend through a large angle. Finally, some sources classified as FRIs are actually core-halo objects where the core may well be a steep-spectrum double source; such objects are often found in the centres of rich clusters and the canonical example is the radio galaxy 3C 84 (NGC 1275 in the Perseus cluster). Where jets and plumes are seen, spectral gradients are broadly in the sense that the spectrum is flat ($\alpha \sim 0.6$) in the inner jet and then steepens with distance away from this location.

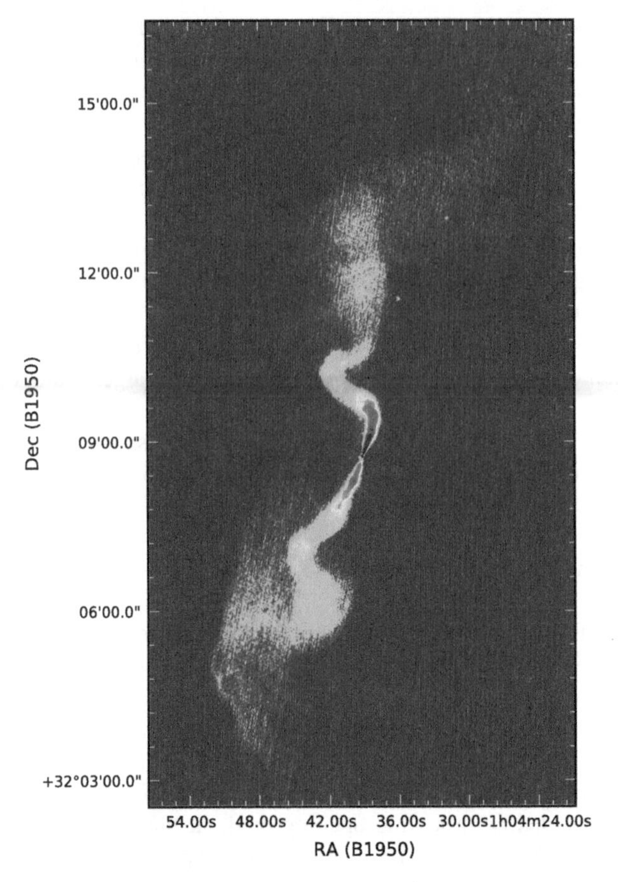

Fig. 4.2 1.6-GHz radio image of the FRI radio galaxy 3C 31. Note the bright inner jets (the northern jet is slightly brighter due to beaming) and the slow broadening into low-surface-brightness plumes, which in this image fade into the noise (Data from Laing et al. 2008)

4.3.2 Polarization in the Radio

The synchrotron emission observed from well-resolved regions of typical radio galaxies is strongly polarized at GHz frequencies, with mean fractional polarization in the range 20–40 % and regions where polarization appears comparable to the theoretical maximum of around 70 %. This need not be indicative of large-scale uniformity in the field: even a Gaussian random field with a Kolmogorov power spectrum can give rise to a mean fractional polarization of order 30 % if the largest scale of structure is significantly larger than the resolution element and, as pointed out by Laing (1980), compressing an initially random field can give rise to polarization approaching the theoretical maximum. Observed apparent magnetic field directions (see Sect. 4.2.1) in lobes, far away from the jet termination, are almost always transverse to the jet direction, except at the edges of the lobes where the field direction is aligned with the lobe boundary: the fractional polarization is generally a maximum at the lobe edges (see Fig. 4.3). The perpendicular fields in the bulk of the lobes are consistent with the idea that the toroidal component of the field dominates on large scales, as would be expected if an initially disordered field expands into the lobes, but any model in which the field is confined to planes perpendicular to the jets will give the same result (Laing 1980).

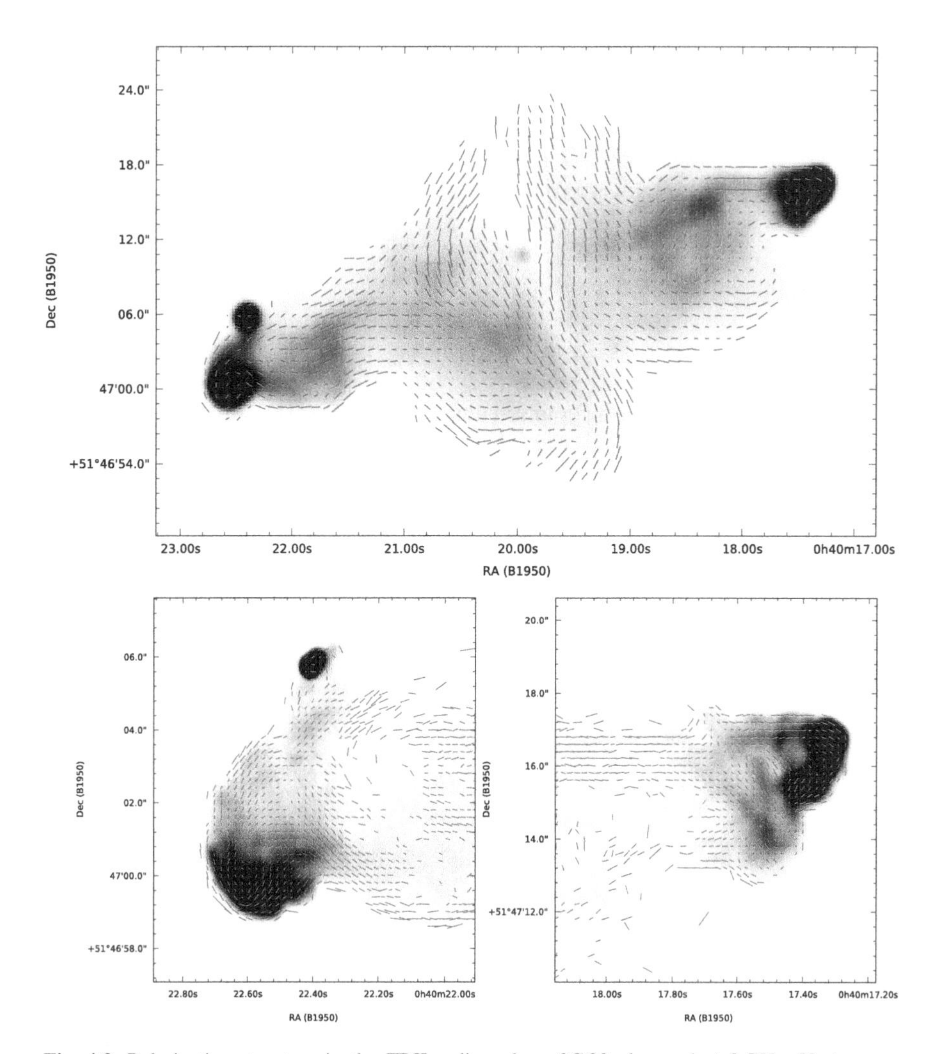

Fig. 4.3 Polarization structure in the FRII radio galaxy 3C 20 observed at 8 GHz. Vectors are plotted perpendicular to the E-vector and their magnitude shows the fractional polarization; *greyscale* shows total intensity. The overall source structure is shown in the *upper panel*, while the *lower panels* show details of the hotspots (Data from Hardcastle et al. 1997a)

Turning to the more compact components, the apparent magnetic field directions in FRII jets tends to be aligned with the jet direction, though with some transverse polarization associated with bright 'knots' (Bridle and Perley 1984). The same is true for the jets in wide-angle-tail FRIs (Hardcastle and Sakelliou 2004). The most compact hotspots in FRIIs, those associated with the jet termination, have magnetic field directions perpendicular to the jet direction (Laing 1989; Hardcastle et al. 1998a): Fig. 4.3. The wider jets and plumes in normal FRIs jets invariably show a transverse field direction on large scales, but may show a transition from a parallel to a transverse field configuration in the inner few kpc, and may around this point show a parallel field in the jet edges with perpendicular field in the centre (Hardcastle et al. 1996; Laing and Bridle 2014). Purely helical field configurations throughout any kpc-scale jets are ruled out by observations (Laing 1981).

At frequencies below a few GHz, and sometimes even at GHz frequencies, Faraday depolarization (Sect. 4.2.1) becomes an important effect. In many of the best-studied cases depolarization appears to be due to an external Faraday-rotating screen that is largely resolved at the best available resolutions (e.g. Dreher et al. 1987; Taylor and Perley 1993; Laing et al. 2008): a pure λ^2 dependence of the position angle with frequency is observed, which puts strong limits on the Faraday rotation due to material mixed with the synchrotron-emitting plasma, and allows estimates of the magnetic field in the group or cluster environment (Carilli and Taylor 2002). It has been suggested that depolarization observed in the giant lobes of Centaurus A is associated with internal thermal material (O'Sullivan et al. 2013), but the densities required are very large compared to the known limits in other sources; more work is needed to confirm this result. On small scales, depolarization may be associated with interactions between the radio source and dense, cold gas (Hardcastle et al. 2010, 2012) and in this case may be used to constrain the magnetic field in the interaction region. The tendency in large samples of quasars for the jetted lobe to be less depolarized, and hence presumably closer to us so that it is seen through less of the IGM of its host environment, has already been mentioned (Laing 1988; Garrington et al. 1988).

4.3.3 Non-thermal Emission at Other Wavelengths

Although radio-loud AGN are naturally best known for their emission at radio wavelengths, non-thermal emission is seen at all other wavebands through to high-energy γ-rays. The most striking example of this is of course the strongly beamed synchrotron and inverse-Compton emission from the pc-scale jet that dominates the entire continuum spectrum of BL Lac objects. However, on kpc scales, FRII hotspots (Saslaw et al. 1978) and FRI jets (Felten 1968; Butcher et al. 1980) have long been known to be optical synchrotron emitters; as we will discuss later, this provides important constraints on the possible locations of particle acceleration. Optical polarization can provide high-resolution information on magnetic field structure in jets and hotspots (e.g. Perlman et al. 2011).

X-ray emission from jets has been known almost as long (Schreier et al. 1979) (Fig. 4.4). Synchrotron emission is believed to be the explanation of X-ray emission from all FRI jets (Hardcastle et al. 2001) and from some FRII hotspots (Hardcastle et al. 2004). In hotspots the situation is complicated by the presence of synchrotron self-Compton emission in the X-ray from some of the radio-brightest objects (Harris et al. 1994) and inverse-Compton models, with the rest-frame CMB photon energy density enhanced by relativistic beaming, may also be able to explain the X-ray emission seen in many kpc-scale quasar jets (Tavecchio et al. 2000; Marshall et al. 2011). Diffuse X-ray emission from lobes is widely modelled as inverse-Compton scattering of CMB photons (Feigelson et al. 1995; Hardcastle et al. 2002; Kataoka and Stawarz 2005; Croston et al. 2005) and this process appears to continue to operate up to the γ-ray (Abdo et al. 2010), with implications for the magnetic field strengths in the lobes that will be discussed in Sect. 4.6.

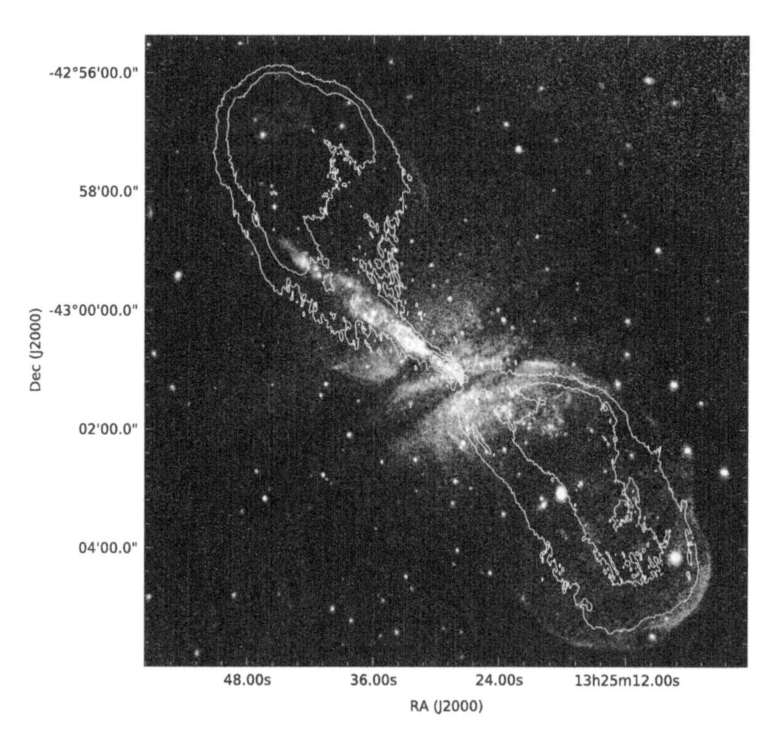

Fig. 4.4 False-colour *Chandra* X-ray image of the environment, shocks and jets of Centaurus A, with 1.4-GHz radio contours overlaid. *Red*, *green* and *blue* correspond to soft, medium and hard X-rays in the range 0.4–2.5 keV. Note the very prominent X-ray jet to the NE and the shock around the lobe to the SW (Data from Hardcastle et al. 2007a)

4.4 Hosts, Environments and Environmental Interactions

The host galaxies of most powerful radio sources have been known for a long time either to be optically classified as quasars or to have the overall appearance and colour of elliptical galaxies (Lilly and Longair 1982). A very few powerful radio sources are found in late-type galaxies, e.g. Ledlow et al. (1998), although the low-power double-lobed radio sources found associated with many nearby Seyferts have much in common physically with radio galaxies (Hota and Saikia 2006). The probability of a powerful radio source being found in a galaxy is a strongly increasing function of stellar mass (Best et al. 2005), which may provide much of the explanation of the strong preference of radio sources for elliptical hosts. There is some evidence that hosts of the most powerful radio galaxies tend to be disturbed, with more signatures of interactions and mergers (Heckman et al. 1986; Ramos Almeida et al. 2012).

Similarly, it has been known for a long while that the typical environment of a radio galaxy is a group or cluster (Longair and Seldner 1979; Prestage and Peacock 1988), which is unsurprising as their elliptical hosts also show a strong preference for such environments (Dressler 1980). The minimum pressures of the lobes of radio galaxies are comparable to the thermal pressure of the hot phase of the intragroup/intracluster medium; it is this hot, X-ray emitting phase with which the lobes predominantly interact. In detail, it is not clear what the exact observational relationship is between the intrinsic and environmental properties of radio galaxies, and how this depends on redshift (e.g. Prestage and Peacock 1988; Yates et al. 1989; Hill and Lilly 1991; Ineson et al. 2013). (An additional complication is a dependence on the AGN type of the radio galaxy, which we do not consider here.) All other things being equal, we would expect richer environments to produce more luminous radio galaxies, as the higher external pressure would give rise to a higher internal pressure and hence synchrotron emissivity for a given lobe volume (Barthel and Arnaud 1996; Hardcastle and Krause 2013) but there is at best weak evidence that this is the case.

There is, however, very strong evidence for interactions between the radio lobes and their surrounding intracluster medium. A few well-studied objects show evidence for shocks driven into the external medium by the lobes (Smith et al. 2002; Nulsen et al. 2005; Croston et al. 2009, 2011) (Fig. 4.4). Many more, particularly low-power (FRI) radio galaxies, show 'cavities', deficits of thermal X-ray emission, associated with the lobes. These were discovered some time ago (Böhringer et al. 1993; Hardcastle et al. 1998a) but have been particularly noticeable with the sensitivity and resolution of *Chandra* (for overviews of large samples see e.g. Bîrzan et al. 2004; Rafferty et al. 2006). From the point of view of AGN physics, the cavities tell us that the lobes have a much lower thermal X-ray emissivity than their surroundings, and therefore that they cannot be filled with thermal material with $T \approx T_{\rm ext}$; they also require that an amount of work of order $p_{\rm ext} V_{\rm cavity}$ must have been done on the external medium, and this can be used to calibrate empirical relationships between radio luminosity and jet power (Cavagnolo et al. 2010). Both shocks and cavities, and indeed any measurement of the external pressure, give a lower limit on the internal pressure of the lobes which affects estimates of magnetic field strength and particle content (Sect. 4.6).

There is also considerable evidence that jets affect other phases of gas in their vicinity. The earliest hints of this come from observations of extended emission-line regions (EELR), regions of ionized gas extending for tens to hundreds of kpc from the nucleus and often well aligned with the lobes (McCarthy et al. 1991, 1996). There is at least some evidence that in the most powerful objects the emission-line gas is either ionized or at least strongly affected by jet-driven shocks (Best et al. 2000; Nesvadba et al. 2008), rather than being photoionized by the AGN. More recently it has been shown (Morganti et al. 2005) that jets also appear to drive large-scale outflows of neutral hydrogen. Hardcastle et al. (2010, 2012) have argued that the neutral and ionized phases are part of a multi-phase outflow which also includes a dominant component of hot, X-ray-emitting gas, and that the overall powers of these outflows can be very considerable. This may be relevant to models of 'AGN feedback' in the high-redshift universe, since the jets are removing cold, dense gas which might otherwise fuel star formation.

In contrast to this is the notion that jets might actually cause new stars to be formed ('jet-induced star formation': Rees 1989). Observational evidence for this process was first provided by studies of individual objects (van Breugel et al. 1985; Brodie et al. 1985) and it was later argued to be a possible cause of the apparent alignment between radio lobes and the optical major axis of the host galaxy in high-redshift objects (McCarthy et al. 1987), although other explanations of this effect exist (Dey and Spinrad 1996). In view of the fact that jets do appear to be able to drive dense material out of the host galaxy, as noted above, it is not clear that the originally proposed mechanism for jet-induced star formation, in which shocks driven by the radio lobes trigger dense clouds to collapse while leaving them in situ in the galaxy, can really be operating. The most recent constraints on this process come from large-scale statistical studies with *Herschel*, which remove some of the ambiguity in studies based on rest-frame ultraviolet: (Virdee et al. 2013) show that physically smaller radio sources do appear to have more star-formation activity than their larger counterparts at the same redshift, but it is not clear whether this implies a direct role for the jet or whether it could be due to some event, such as a merger, triggering both jet and star-formation activity.

4.5 Jet Dynamics and Particle Acceleration

The basic beam model for radio-loud AGN (Scheuer 1974; Blandford and Rees 1974) is now universally accepted, as a result of observations of continuous jets on scales from the parsec to 100-kpc scales, and alternative models will not be considered here. The observed source structure is then understood to be the result of interactions between twin relativistic jets and their environment, with the radio lobes being bubbles inflated by material that has passed along the jets (dynamics of the lobes will be discussed in Sect. 4.7). Although it is still commonly proposed that FRI and FRII radio sources have intrinsically different jets, the most parsimonious models suggest that they are initially identical and that the observed differences on kpc scales are the result of deceleration of the FRI jets due to entrainment either of external material or of mass lost from stellar winds within the jets (Bowman et al. 1996): in this model, FRII objects have more powerful jets which can propagate out of the environment of the host galaxy without decelerating below their own internal sound speeds, while FRIs decelerate to sub-sonic or trans-sonic speeds (Bicknell 1995). This model is supported by the observations of jet speeds discussed above (Sects. 4.3.1.1 and 4.3.1.2): the maximum sound speed for a relativistic plasma is $c/\sqrt{3}$, which is clearly above the typical beaming speeds inferred for FRIs after their initial deceleration, but below or comparable to those inferred for FRIIs.

A jet which is supersonic with respect to its own internal sound speed must necessarily terminate in a shock. These shocks are generally identified with the hotspots seen at the ends of the lobes in FRIIs. There is little *direct* evidence that these structures really do represent shocks: the best indirect evidence is the agreement of radio through to optical hotspot synchrotron spectra, in many cases, with the predictions of simple models of first-order Fermi acceleration and downstream loss (Meisenheimer et al. 1989, 1997), although high-resolution observations in the optical and X-ray generally paint a more complicated picture (Perley et al. 1997; Hardcastle et al. 2007a). The idea that hotspots are the predominant sites of particle acceleration in FRIIs is consistent with the fact that radio spectra appear to steepen with distance away from the hotspot (Sect. 4.3.1.1). The reason for the frequent observations of multiple hotspots at the ends of lobes is not clear in this picture: some may be relics left behind when the jet termination point moves within the lobe (so-called 'dentist's drill' behaviour, Scheuer 1982) while others may be connected to continued outflow (Cox et al. 1991). The detection of X-ray synchrotron emission associated with some of these 'secondary' hotspots, which requires particle acceleration, is evidence that at least some of them are not relics but are still connected to the energy supply. It is not clear what particle acceleration processes, if any, operate in the kpc-scale jets of FRIIs, although the fact that they are visible at all in the radio and sometimes at higher frequencies in synchrotron radiation (Wilson et al. 2001) implies that such processes are present.

In FRI radio galaxies, observations of optical and X-ray synchrotron radiation require an in situ particle acceleration process, but it is not clear what that process is or how it is related to the jet dynamics. The regions of X-ray emission seem mostly to occur before the strongest jet deceleration (Laing and Bridle 2014), but it may be that particle acceleration persists out to larger scales with a lower maximum energy, and in any case some FRI jets appear to show X-ray synchrotron emission on scales of hundreds of kpc (Mack et al. 1997; Evans et al. 2005). There is some dynamical evidence associating compact X-ray features with shocks, or at any rate with features moving at slower speeds than the bulk flow (Hardcastle et al. 2003; Goodger et al. 2010) but there is also fairly strong evidence for a diffuse component of the X-ray emission which is not associated with discrete shocks, with a spectral index and radio/X-ray ratio which depends on position in the jet (Hardcastle et al. 2007b) and whose overall radio/X-ray ratio is related to jet power (Harwood and Hardcastle 2012). This diffuse component might be created by first-order Fermi acceleration from a large population of unresolved shocks or might be due to some more exotic process such as second-order Fermi at turbulence or magnetic field reconnection. The fact that the radio spectra of FRI jets steepen on scales >10 kpc suggests that these inner parts of the jet are (normally) the primary sites of particle acceleration in the source.

4.6 Magnetic Fields and Particle Content

Observations of synchrotron radiation tell us that the structures in radio-loud AGN contain relativistic electrons and magnetic field, but give us no information about either the presence of non-radiating particles (such as thermal electrons and/or protons) or the relative energy densities in electrons, field or non-radiating particles. It was recognised very early on (Burbidge 1956) that there is a minimum energy density in the components that must be present (relativistic electrons and field) to produce a given observed synchrotron emissivity, subject only to assumptions about the magnetic field geometry and the electron energy spectrum; this minimum energy density corresponds roughly to equipartition of energy between the electrons and the field. However, there was no a priori reason to suppose that this minimum energy density bore any relation to the true value, both because no physical mechanism that enforces detailed equipartition of energy on all scales was known and because the contribution from non-radiating particles might be very large. Various arguments from the properties of the pc-scale jets have been advanced to constrain the particle content, but perhaps the strongest (that heavy particles such as protons are required to allow the required power to be transported from the inner regions of the source without catastrophic inverse-Compton losses, Celotti and Fabian (1993)) can be evaded if jet power is predominantly transported as Poynting flux on the smallest scales (a key role for magnetic fields is required in current jet generation models: McKinney and Blandford (2009)).

The inverse-Compton observations discussed above (Sect. 4.3.3) provide very important constraints on magnetic fields for all kpc-scale components of radio sources. These constraints can be derived because inverse-Compton emissivity, if the photon field is known, depends only on the number density of relativistic electrons with the appropriate energies to scatter photons into the observed band: the photon gain in energy is of order γ^2, where γ is the Lorentz factor, so, for example, for inverse-Compton scattering of $z = 0$ CMB photons into the X-ray by an electron

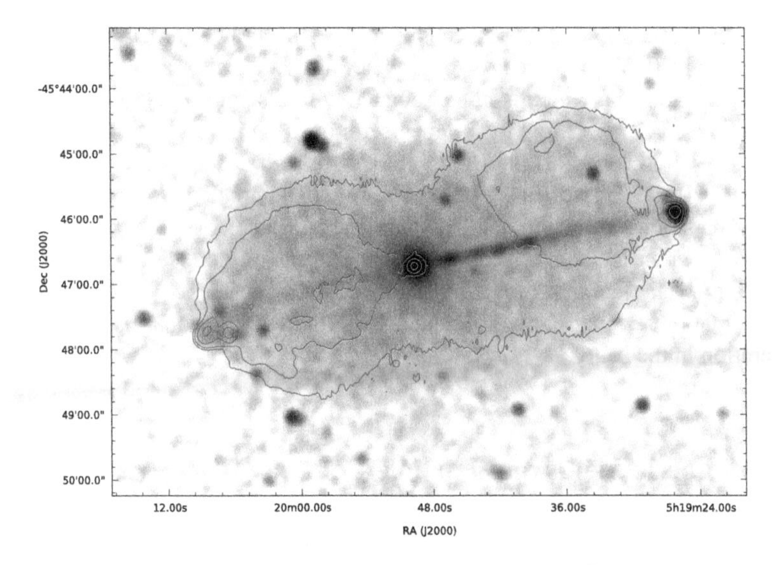

Fig. 4.5 *Chandra* 0.5–5.0 keV images of inverse-Compton emission from the lobes of the FRII radio galaxy Pictor A (Data from Marshall et al. 2010). Overlaid are 1.4-GHz radio contours (Perley et al. 1997)

population with negligible bulk speed with respect to the CMB, $\gamma \sim 1{,}000$ is required. Since the synchrotron emissivity also depends on the number density of electrons and on the energy density in magnetic field, observations of synchrotron and inverse-Compton from the same region, together with some assumption about the electron energy spectrum, are sufficient to estimate the magnetic field strength.[1]

Inverse-Compton observations of both hotspots and lobes in FRII radio galaxies (Fig. 4.5) tell a very similar story: the magnetic field strength is comparable to, but somewhat less than, the equipartition/minimum energy value, with the energy densities in the field being typically perhaps an order of magnitude less than those in the electrons (Kataoka and Stawarz 2005; Croston et al. 2005). Thus minimum-energy estimates of the jet power, if only electrons and field are considered, are not very far from the truth. Moreover, these results allow an indirect argument to be made about the particle content in non-radiating particles: if the non-radiating particles dominated the energy density, the rough agreement between the energy densities in relativistic electrons and field would have to be a coincidence, in the

[1]It is conventional to talk about 'the magnetic field strength in the lobes' even though, in practice, we expect a strongly varying point-to-point magnetic field strength. We may define a mean field, $B_0^2 = \int B^2 p(B) \mathrm{d}B$, and Hardcastle (2013) has shown in the case of a Gaussian random field with uniform electron energy density that the field estimates derived from inverse-Compton scattering of the CMB are effectively estimates of B_0. If there are also significant variations of the *mean* field strength from point to point in the lobe, as argued to be the case by e.g. Hardcastle and Croston (2005), then it is less clear what a lobe-averaged magnetic field strength means, but it seems likely that the values we measure are not far from a true lobe mean field strength.

sense that both would actually be small fractions of the true energy density and there is no obvious physical reason why those fractions should be similar. It has therefore been suggested that the inverse-Compton observations of FRIIs require the energy density in non-radiating particles to be at most comparable to that in the other components, and this is supported by the observation that the pressures estimated in FRII lobes from inverse-Compton observations are very comparable to the external thermal pressures (see below, Sect. 4.7). Hence, in a typical FRII source, we often have quite good constraints on the internal and possibly the external physical conditions on large scales. The only very uncertain region is the jet, where sidedness/jet prominence analysis and beamed inverse-Compton models require very different bulk Lorentz factors ($\sim$2 and $>$10 respectively) and it is not at all clear what the magnetic field strengths are; however, if the arguments from hotspots and lobes are to be taken seriously, we do not expect a dominant contribution to the jets' energy density from non-radiating particles. The next step would be to consider the *resolved* inverse-Compton emission with a view to looking at how the electron/field ratio varies with a function of position in the source, but this relies on high signal to noise in the inverse-Compton emission, which has so far only been available in a few sources (Isobe et al. 2002; Hardcastle and Croston 2005; Goodger et al. 2008), and is complicated by the variation of the electron energy spectrum over the source, since at GHz frequencies we are not directly observing the electrons responsible for inverse-Compton emission; there is some evidence that the magnetic field strength decreases with distance from the hotspot, but more work on this using low-frequency radio observations is necessary.

The situation is markedly different for more typical low-power radio sources (predominantly FRIs). In these objects there are few inverse-Compton measurements, with the only clear detection being that of the giant lobes of Cen A in the γ-ray, which we will discuss below. Limits on inverse-Compton emission in the TeV γ-ray shows that the inner jet of Cen A plausibly has field strengths no more than a factor of a few below equipartition (Hardcastle and Croston 2011) but this result is model-dependent and in any case may not generalize to other sources. In addition, it is quite clear from observations of FRI environments that the minimum pressures estimated from synchrotron radiation are far below the thermal pressures in the environments in which the lobes are embedded (Morganti et al. 1988; Feretti et al. 1992; Worrall et al. 1995; Hardcastle et al. 1998a; Croston et al. 2003, 2008; Dunn et al. 2005) which requires either a strong deviation from equipartition or dominance by a non-radiating particle component. Dominance by electrons can generally be ruled out from inverse-Compton upper limits (Hardcastle and Croston 2010). The fact that this discrepancy is clearly present in the inner lobes of Cen A, where the lobe pressure can be estimated from the shock driven by the lobes into the external medium (Croston et al. 2009) shows that it can set in on quite small ($\sim$10 kpc) scales, but most observations also show that the missing pressure is larger on the largest scales. Croston et al. (2008) suggest that it is related to the entrainment of thermal material known to take place in the jets of FRIs, but if so, both the expected modest entrainment rates and the fact that X-ray observations show that FRI lobes have lower X-ray emissivity than their surroundings (Sect. 4.4) require that the entrained thermal material must be heated to very high effective temperatures in order to provide pressure balance. In the specific case of Cen A, Wykes et al. (2013) have estimated this temperature to be $\sim 10^{12}$ K. However, the giant lobes of Cen A are puzzling in the context of what we know about other sources; the magnetic field appears to be in equipartition with the radiating electrons (Abdo et al. 2010), in which case, for a plausible external environment, the energy density in entrained material must exceed the energy densities in electrons and field by a significant factor. More inverse-Compton constraints on the magnetic field strengths in low-power radio galaxies are needed to make progress in this area.

4.7 Lobe Dynamics

The study of the dynamics of the lobes (of both FRIs and FRIIs) has been important since the earliest days of the beam model. Lobe expansion is driven by the momentum flux along the jets (which acts primarily in an axial direction) and the internal pressure of the lobes (which causes the lobes to expand in all directions). Scheuer (1974) identified two important limiting cases: in his 'model A', the radio source is so strongly overpressured with respect to the external medium that it expands supersonically in all directions at all times, driving an elliptical bow shock in all directions, while in his 'model C' the internal thermal pressure of the lobes becomes comparable at some point to the external pressure, causing the transverse expansion of the lobes to halt. Scheuer (1974) pointed out that the variation of the external pressure along a source's length could lead to a situation where the inner parts only are underpressured and contract, so that eventually buoyancy forces would push the radio lobes away from the galaxy, while they continued to expand both linearly and transversely at large radii. It is certainly possible to construct models of radio sources that are strongly overpressured at all times (Begelman and Cioffi 1989). Indeed, it seems very likely that this is a good description of young radio lobes. The small-scale lobes in Cen A, for example, appear to be driving a shock in all directions, though it is noticeably weaker in the inner regions of the galaxy (Croston et al. 2009). In such models the advance speed of the lobes is expected to be given by the balance between the momentum flux up the jet and the ram pressure of the external medium: for example, for a non-relativistic jet,

$$v_{\rm h} \approx \left(\frac{Q}{v_j \rho_{\rm ext} A_h}\right)^{1/2} \tag{4.18}$$

where Q is the jet power, v_j is its speed, ρ is the density of the external medium at the end of the lobes, and A_h is the area over which the jet momentum flux is distributed, taken to be of order the cross-sectional area of the lobes (Begelman and Cioffi 1989). The self-similar models used by e.g. Kaiser and Alexander (1997), which are useful because they allow many results on lobe dynamics to be derived analytically, then make the further assumption that the transverse expansion speed scales in the same way so that the axial ratio of the radio source remains constant with time, which requires that the lobes be strongly overpressured with respect to the external medium.

However, X-ray observations with the sensitivity to detect both inverse-Compton emission from the lobes and thermal bremsstrahlung from the host environment tell a rather different story (e.g. Hardcastle et al. 2002; Croston et al. 2004; Belsole et al. 2007; Shelton et al. 2011). The pressure of the lobes in 100-kpc-scale FRII radio galaxies is consistently found to be comparable to the external pressure at around the midpoint of the lobes. The X-ray emission from the host groups and clusters of radio galaxies are normally adequately described as bremsstrahlung (Sect. 4.2.3) from an environment adequately described (for our purposes) by an isothermal β model, sometimes called a King profile (Cavaliere and Fusco-Femiano 1978):

$$n = n_0 \left(1 + \frac{r^2}{r_c^2}\right)^{-3\beta/2} \tag{4.19}$$

where r is the distance from the cluster centre (normally therefore also from the AGN), r_c is known as the core radius, and β describes the limiting power-law behaviour when $r \gg r_c$. It can be seen that if the radio source size is comparable to or larger than r_c, we expect significant pressure differences along the lobes: a source in pressure balance at its midpoint may well be underpressured in the centre (so that the lobes will be forced out) but still overpressured at large radii. Although it remains possible that there is a dominant contribution to the pressure from non-radiating particles (as is believed to be the case in FRIs, see above, Sect. 4.6) there is no direct evidence for this, and numerical models of radio source growth in which the lobes are allowed to come into pressure balance with realistic group/cluster environments appear to give good agreement with observation (Hardcastle and Krause 2013). At present it is therefore necessary to be cautious when applying analytical models like those of Kaiser and Alexander (1997), particularly as these also unrealistically assume a simple power-law atmosphere (much early modelling work made either this assumption or the even worse assumption that the external pressure and density were constant). Analytic studies of this situation have been carried out (Alexander 2002).

Direct observational constraints on dynamical properties of lobes are in short supply. The sub-kpc compact symmetric objects (CSOs) show relativistic expansion speeds (An et al. 2012) and it is widely thought that some or all of these may evolve into kpc-Mpc scale radio galaxies. However, by the time these sources have expanded to scales of tens to hundreds of kpc, their expansion is no longer directly observable. The statistics of lobe length asymmetry provide some rough constraints on possible expansion speeds, giving limits on the 100-kpc scale sources of $v < 0.1c$ (Scheuer 1995), and probably as low as a few per cent of c; similar constraints were found with a more sophisticated model by Arshakian and Longair (2000). Speeds even as high as $0.1c$ would correspond to high Mach numbers, $\sim$30 for a typical cluster environment, which do not seem to be consistent with observations in the X-ray.

The other approach to determining source dynamics is 'spectral ageing', i.e. fitting models of an aged synchrotron spectrum to multi-frequency radio data as discussed in Sect. 4.2.1. If the assumptions that underlie these models are correct (i.e. in particular there is no in situ particle acceleration in the lobes, and no mixing of old and new plasma) then the oldest estimated ages should give a reasonable estimate of the source age. Speed estimates from *gradients* of spectral ages, however, measure the sum of hotspot advance speed and the reverse-directed flow from the hotspots into the lobes ('backflow') and so should be treated with caution (Alexander and Leahy 1987). In powerful sources the 'spectral age speed' has been estimated to be as high as $0.3c$ (Liu et al. 1992). However, it should be noted that equipartition field estimates are used in these analyses: since observed fields are sub-equipartition, the ages will be underestimated and speeds overestimated. The maximum spectral ages inferred for giant (Mpc-scale) radio galaxies, where inverse-Compton losses dominate and so magnetic field estimates are not so important, are of the order of 10^8 years, implying more reasonable mean lobe growth speeds of $\sim$0.015c and Mach numbers $\sim$4. Thus it seems plausible that the revised values of magnetic field strength now being found from inverse-Compton observations will reduce the discrepancy between the spectral ages/speeds and plausible dynamical values, though it is not clear that this is sufficient to solve the problem (Harwood et al. 2013). Many assumptions of the standard spectral ageing model are simplistic (e.g. the idea that a single magnetic field strength has applied throughout the lifetime of the source) and detailed numerical modelling of the spectra of sources with realistic dynamics is probably needed to make progress.

4.8 Numerical Modelling

Understanding the dynamics and evolution of radio sources through numerical modelling has been an important aid to observational and theoretical research since computer power made the first simple models possible (Norman et al. 1982; Williams and Gull 1985). Constructing a complete model of such a system presents formidable difficulties, among them the very large range of scales of interest (in principle from the jet generation scale to the 1-Mpc largest scale of the lobes), the fact that relativistic bulk motions and non-negligible magnetic fields are both expected to be present, the fact that radio sources are clearly not axisymmetric so that three-dimensional modelling is needed, the difficulty of accurately modelling particle acceleration and energy losses so as to obtain a realistic visualization of the system, and, at least in the early days of modelling, our almost complete ignorance of many of the key physical conditions in the jets and lobes. All numerical models of jet termination and dissipation (i.e. of the structures discussed in this chapter) treat the injection of the jet as a boundary condition, so that the jet-generation region need not be considered. However, they differ widely in the physics and assumptions that they do incorporate.

One class of model aims to deal with the overall dynamics of the source without being excessively concerned with the details of radio emission from the lobes or jets. Such work often consists of parameter studies where properties of the jets or environment are varied (e.g. Norman et al. 1982; Cioffi and Blondin 1992; Massaglia et al. 1996; Carvalho and O'Dea 2002; Krause 2003; O'Neill et al. 2005; Hardcastle and Krause 2013) and may include realistic environments and magnetic fields, but do not typically deal with relativistic jets. These have shown that jets must be light to inflate lobes – consistent with our understanding from observations, as discussed above, which suggests that relativistic electron/positron plasma and field alone can provide enough pressure to balance the external pressure, particularly in FRIIs. The lobes in light jets that are supersonic with respect to the lobe sound speed are formed after a jet termination shock which is expected to be analoguous to the hotspots in real radio galaxies. When realistic environments are considered, clear departures from self-similarity are seen, as expected (Hardcastle and Krause 2013). Including magnetic fields in these models (Gaibler et al. 2009) helps to suppress the large-scale Kelvin-Helmholtz instabilities that are seen in pure hydrodynamical models.

Other models aim at understanding the detailed structures seen in radio emission, e.g. the multiple hotspots (Williams and Gull 1985; Cox et al. 1991) or jet structures (Bodo et al. 1998; Rossi et al. 2008; Perucho et al. 2010). These normally require three-dimensional simulations, magnetic fields, and (for the jet studies) relativistic effects to be taken into account; they do not require large volumes or realistic environments to be modelled. These studies have shown that our simple understanding of the jet dynamics is broadly correct – for example, relativistic magnetized jets can be stable under the right conditions (Mignone et al. 2010). It is challenging to model the type of decelerating relativistic jet expected in FRIs, but this has also been attempted (Perucho and Martí 2007). Fewer detailed studies of the lobes have been carried out, but Gaibler et al. (2009) and Huarte-Espinosa et al. (2011) have considered the evolution of the field in the lobes, producing realistic field configurations from quite different initial (injection) conditions, and Huarte-Espinosa et al. (2011) has produced predictions for the polarization of lobes that may be compared to observation.

Still others have focussed on particle acceleration, which is a key ingredient in visualization of the simulated lobes (e.g. Jones et al. 1999; Tregillis et al. 2001, 2004). These show complex structures developing in the synchrotron emissivity, very much as seen in the highest-resolution radio images; although the recipes that they use for tracking particle acceleration are necessarily simple (since we do not have a good observationally derived understanding of the nature or even location of particle acceleration: Sect. 4.5) this type of work will be very important in future in modelling broad-band radio spectra (Sect. 4.7). Early work used unrealistic (uniform) external environments, but more recently it has been possible to combine realistic environments and electron transport, producing simulated radio sources that bear a strong resemblance to real objects (Mendygral et al. 2012).

Finally, a great deal of effort has been put into simulating the effects of various different classes of radio source on the external medium. As discussed above (Sect. 4.4) there is strong X-ray evidence that radio lobes have a variety of different effects on their hot-gas environment (see McNamara and Nulsen (2012) for a review), and evidence from other wavebands that cold gas is affected too. Modelling of radio source/environment interactions varies widely in the realism of the radio source models: often it is only necessary to consider the dynamics of a buoyant bubble (e.g. Churazov et al. 2001) which, however, bears little relation to the dynamics of an active radio galaxy. Where active jets are considered, there is a general consensus (e.g. Basson and Alexander 2003; Zanni et al. 2003; O'Neill et al. 2005; Gaibler et al. 2009; Hardcastle and Krause 2013) that a significant fraction of the injected energy, ~ 0.5, is put into the external medium in the form of shocks and $p\mathrm{d}V$ work by a powerful (FRII-type) source. The remaining energy in the lobes may of course be thermalized after the jet is switched off. Thus the radio galaxy can have a very significant effect on the energetics of the intracluster medium. However, it is important to note (Omma and Binney 2004) that these powerful sources do not provide an *efficient* way of solving the problem of excess cooling at the centres of clusters, since they only couple to the densest central regions with the highest cooling rates for a short time. On the other hand, modelling of powerful sources interacting with *cold* gas shows that they can have a significant effect, possibly even triggering star formation (Sutherland and Bicknell 2007; Gaibler et al. 2011, 2012) – see Fig. 4.6. Again, the regime that has not been fully explored, for both hot and cold-gas interactions, is the much more computationally complicated (but much commoner) FRI population, but hopefully this will change as available computing power continues to increase.

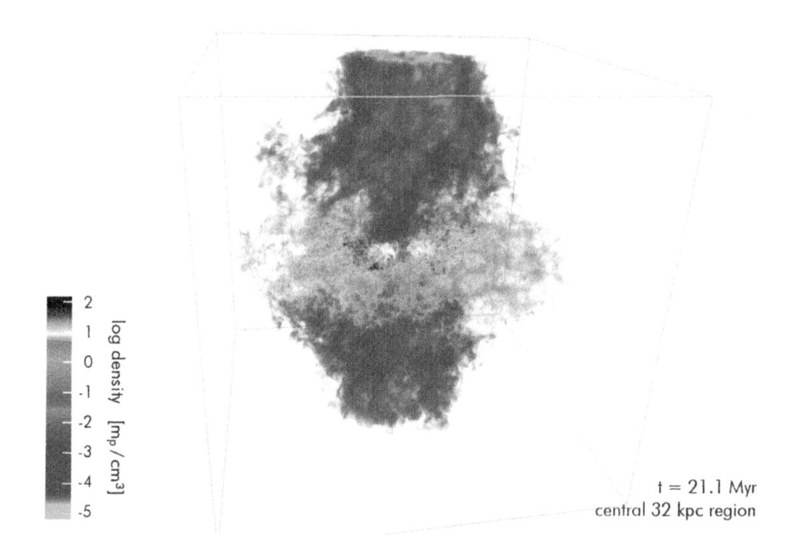

Fig. 4.6 Visualization of numerical modelling of a double radio source interacting with a massive gaseous disc (Gaibler et al. 2012). Note the complex, turbulent radio lobes and the high densities induced in the inner part of the disc by the lobes' passage. The simulations of the radio galaxy here are purely hydrodynamic, using a light, supersonic jet

Acknowledgements I am grateful to Robert Laing and Rick Perley for supplying radio images and Volker Gaibler for providing images from his numerical models. The NRAO VLA, with which all the radio images were made, is a facility of the National Science Foundation operated under cooperative agreement by Associated Universities, Inc.

References

Abdo, A.A., et al.: Fermi gamma-ray imaging of a radio galaxy. Science **328**, 725 (2010). doi:10.1126/science.1184656

Alexander, P., Leahy, J.P.: Ageing and speeds in a representative sample of 21 classical double radio sources. MNRAS **224**, 1 (1987)

Alexander, P.: On the interaction of FR II radio sources with the intracluster medium. MNRAS **335**, 610–620 (2002). doi:10.1046/j.1365-8711.2002.05638.x

An, T., Wu, F., Yang, J., Taylor, G.B., Hong, X., Baan, W.A., Liu, X., Wang, M., Zhang, H., Wang, W., Chen, X., Cui, L., Hao, L., Zhu, X.: VLBI observations of 10 compact symmetric object candidates: expansion velocities of hot spots. ApJS **198**, 5 (2012). doi:10.1088/0067-0049/198/1/5

Arshakian, T.G., Longair, M.S.: An asymmetric relativistic model for classical double radio sources. MNRAS **311**, 846 (2000)

Böhringer, H., Voges, W., Fabian, A.C., Edge, A.C., Neumann, D.M.: A ROSAT HRI study of the interaction of the X-ray emitting gas and radio lobes of NGC 1275. MNRAS **264**, L25 (1993)

Bîrzan, L., Rafferty, D.A., McNamara, B.R., Wise, M.W., Nulsen, P.E.J.: A systematic study of radio-induced X-ray cavities in clusters, groups, and galaxies. ApJ **607**, 800–809 (2004). doi:10.1086/383519

Baade, W.: Polarization in the jet of Messier 87. ApJ **123**, 550–551 (1956). doi:10.1086/146194
Baade, W., Minkowski, R.: On the identification of radio sources. ApJ **119**, 215 (1954). doi:10.1086/145813
Barthel, P.D., Arnaud, K.A.: Anomalous radio-loudness of Cygnus A and other powerful radio galaxies. MNRAS **283**, L45 (1996)
Basson, J.F., Alexander, P.: The long-term effect of radio sources on the intracluster medium. MNRAS **339**, 353 (2003)
Begelman, M.C., Cioffi, D.F.: Overpressured cocoons in extragalactic radio sources. ApJ **345**, L21 (1989)
Bell, A.R.: The acceleration of cosmic rays in shock fronts. I. MNRAS **182**, 147 (1978)
Belsole, E., Worrall, D.M., Hardcastle, M.J., Croston, J.H.: High-redshift Fanaroff-Riley type II radio sources: large-scale X-ray environment. MNRAS **381**, 1109–1126 (2007). doi:10.1111/j.1365-2966.2007.12298.x
Best, P.N., Röttgering, H.J.A., Longair, M.S.: Ionization, shocks and evolution of the emission-line gas of distant 3CR radio galaxies. MNRAS **311**, 23–36 (2000). doi:10.1046/j.1365-8711.2000.03028.x
Best, P.N., Kauffmann, G., Heckman, T.M., Brinchmann, J., Charlot, S., Ivezić, Z., White, S.D.M.: The host galaxies of radio-loud active galactic nuclei: mass dependences, gas cooling and active galactic nuclei feedback. MNRAS **362**, 25 (2005)
Bicknell, G.V.: Relativistic jets and the Fanaroff-Riley classification of radio galaxies. ApJS **101**, 29 (1995)
Black, A.R.S., Baum, S.A., Leahy, J.P., Perley, R.A., Riley, J.M., Scheuer, P.A.G.: A study of FRII radio galaxies with $z < 0.15$: I. High resolution maps of 8 sources at 3.6 cm. MNRAS **256**, 186 (1992)
Blandford, R.D., Rees, M.J.: A 'twin-exhaust' model for double radio sources. MNRAS **169**, 395 (1974)
Bodo, G., Rossi, P., Massaglia, S., Ferrari, A., Malagoli, A., Rosner, R.: Three-dimensional simulations of jets. A&A **333**, 1117–1129 (1998)
Bowman, M., Leahy, J.P., Komissarov, S.S.: The deceleration of relativistic jets by entrainment. MNRAS **279**, 899 (1996)
Bridle, A.H., Perley, R.A.: Extragalactic radio jets. ARA&A **22**, 319 (1984)
Bridle, A.H., Hough, D.H., Lonsdale, C.J., Burns, J.O., Laing, R.A.: Deep VLA imaging of twelve extended 3CR quasars. AJ **108**, 766 (1994)
Brodie, J.P., Bowyer, S., McCarthy, P.: A radio and optical study of a jet/cloud interaction in the galaxy cluster A194. ApJ **295**, L59 (1985)
Brunetti, G.: Anisotropic inverse Compton scattering from the trans-relativistic to the ultra-relativistic regime and application to the radio galaxies. Astropart. Phys. **13**, 107 (2000)
Burbidge, G.: On synchrotron radiation from Messier 87. ApJ **124**, 416 (1956)
Burn, B.J.: On the depolarization of discrete radio sources by Faraday dispersion. MNRAS **133**, 67 (1966)
Butcher, H.R., van Breugel, W., Miley, G.K.: Optical observations of radio jets. ApJ **235**, 749 (1980)
Carilli, C.L., Taylor, G.B.: Cluster magnetic fields. ARA&A **40**, 319–348 (2002). doi:10.1146/annurev.astro.40.060401.093852
Carilli, C.L., Perley, R.A., Dreher, J.W., Leahy, J.P.: Multifrequency radio observations of Cygnus A: spectral aging in powerful radio galaxies. ApJ **383**, 554 (1991)
Carvalho, J.C., O'Dea, C.P.: Evolution of global properties of powerful radio sources. II. Hydrodynamical simulations in a declining density atmosphere and source energetics. ApJS **141**, 371–414 (2002). doi:10.1086/340646
Cavagnolo, K.W., McNamara, B.R., Nulsen, P.E.J., Carilli, C.L., Jones, C., Bîrzan, L.: A relationship between AGN jet power and radio power. ApJ **720**, 1066–1072 (2010). doi:10.1088/0004-637X/720/2/1066
Cavaliere, A., Fusco-Femiano, R.: The distribution of hot gas in clusters of galaxies. A&A **70**, 677 (1978)

Celotti, A., Fabian, A.C.: The kinetic power and luminosity of parsec-scale radio jets – an argument for heavy jets. MNRAS **264**, 228 (1993)
Churazov, E., Brüggen, M., Kaiser, C.R., Böhringer, H., Forman, W.: Evolution of buoyant bubbles in M87. ApJ **554**, 261–273 (2001). doi:10.1086/321357
Cioffi, D.F., Blondin, J.M.: The evolution of cocoons surrounding light, extragalactic jets. ApJ **392**, 458–464 (1992). doi:10.1086/171445
Cox, C.I., Gull, S.F., Scheuer, P.A.G.: Three-dimensional simulations of the jets of extragalactic radio sources. MNRAS **252**, 588 (1991)
Croston, J.H., Hardcastle, M.J., Birkinshaw, M., Worrall, D.M.: *XMM-Newton* observations of the hot-gas atmospheres of 3C 66B and 3C 449. MNRAS **346**, 1041 (2003)
Croston, J.H., Birkinshaw, M., Hardcastle, M.J., Worrall, D.M.: X-ray emission from the nuclei, lobes and hot-gas environments of two FRII radio galaxies. MNRAS **353**, 879 (2004)
Croston, J.H., Hardcastle, M.J., Harris, D.E., Belsole, E., Birkinshaw, M., Worrall, D.M.: An X-ray study of magnetic field strengths and particle content in the lobes of FRII radio sources. ApJ **626**, 733 (2005)
Croston, J.H., Hardcastle, M.J., Kharb, P., Kraft, R.P., Hota, A.: Chandra evidence for AGN feedback in the spiral galaxy NGC 6764. ApJ **688**, 190 (2008)
Croston, J.H., Kraft, R.P., Hardcastle, M.J., Birkinshaw, M., Worrall, D.M., Nulsen, P.E.J., Penna, R.F., Sivakoff, G.R., Jordán, A., Brassington, N.J., Evans, D.A., Forman, W.R., Gilfanov, M., Goodger, J.L., Harris, W.E., Jones, C., Juett, A.M., Murray, S.S., Raychaudhury, S., Sarazin, C.L., Voss, R., Woodley, K.A.: High-energy particle acceleration at the radio-lobe shock of Centaurus A. MNRAS **395**, 1999–2012 (2009). doi:10.1111/j.1365-2966.2009.14715.x
Croston, J.H., Hardcastle, M.J., Mingo, B., Evans, D.A., Dicken, D., Morganti, R., Tadhunter, C.N.: A large-scale shock surrounding a powerful radio galaxy? ApJL **734**, L28 (2011). doi:10.1088/2041-8205/734/2/L28
Dey, A., Spinrad, H.: The radio galaxy 3C 265 contains a hidden quasar nucleus. ApJ **459**, 133 (1996). doi:10.1086/176874
Dreher, J.W., Carilli, C.L., Perley, R.A.: The faraday rotation of cygnus A: magnetic fields in cluster gas. ApJ **316**, 611 (1987)
Dressler, A.: Galaxy morphology in rich clusters - implications for the formation and evolution of galaxies. ApJ **236**, 351–365 (1980). doi:10.1086/157753
Dunn, R.J.H., Fabian, A.C., Taylor, G.B.: Radio bubbles in clusters of galaxies. MNRAS **364**, 1343 (2005)
Edge, D.O., Shakeshaft, J.R., McAdam, W.B., Baldwin, J.E., Archer, S.: A survey of radio sources at a frequency of 159 Mc/s. Mem. R. Astron. Soc. **68**, 37–60 (1959)
Evans, D.A., Hardcastle, M.J., Croston, J.H., Worrall, D.M., Birkinshaw, M.: Chandra and XMM-newton observations of NGC 6251. MNRAS **359**, 363 (2005)
Fanaroff, B.L., Riley, J.M.: The morphology of extragalactic radio sources of high and low luminosity. MNRAS **167**, 31P (1974)
Feigelson, E.D., Laurent-Muehleisen, S.A., Kollgaard, R.I., Fomalont, E.: Discovery of inverse-compton X-rays in radio lobes. ApJ **449**, L149 (1995)
Felten, J.E.: The radiation and physical properties of the M87 jet. ApJ **151**, 861 (1968)
Feretti, L., Perola, G.C., Fanti, R.: Tailed radio sources as probes of the intergalactic medium pressure. A&A **265**, 9 (1992)
Fernini, I., Burns, J.O., Bridle, A.H., Perley, R.A.: Very large array imaging of 5 FRII 3CR radio galaxies. AJ **105**, 1690 (1993)
Fernini, I., Burns, J.O., Perley, R.A.: VLA imaging of Fanaroff-Riley II 3CR radio galaxies. II. Eight new images and comparisons with 3CR quasars. AJ **114**, 2292 (1997)
Gaibler, V., Krause, M., Camenzind, M.: Very light magnetized jets on large scales - I. Evolution and magnetic fields. MNRAS **400**, 1785–1802 (2009). doi:10.1111/j.1365-2966.2009.15625.x
Gaibler, V., Khochfar, S., Krause, M.: Asymmetries in extragalactic double radio sources: clues from 3D simulations of jet-disc interaction. MNRAS **411**, 155–161 (2011). doi:10.1111/j.1365-2966.2010.17674.x

Gaibler, V., Khochfar, S., Krause, M., Silk, J.: Jet-induced star formation in gas-rich galaxies. MNRAS **425**, 438–449 (2012). doi:10.1111/j.1365-2966.2012.21479.x

Garrington, S., Leahy, J.P., Conway, R.G., Laing, R.A.: A systematic asymmetry in the polarization properties of double radio sources with one jet. Nature **331**, 147 (1988)

Gilbert, G., Riley, J.M., Hardcastle, M.J., Croston, J.H., Pooley, G.G., Alexander, P.: High-resolution observations of a complete sample of 27 FRII radio galaxies and quasars with $0.3 < z < 0.6$. MNRAS **351**, 845 (2004)

Goodger, J.L., Hardcastle, M.J., Croston, J.H., Kassim, N., Perley, R.A.: Inverse-compton emission from the lobes of 3C 353. MNRAS **386**, 337 (2008)

Goodger, J.L., Hardcastle, M.J., Croston, J.H., Kraft, R.P., Birkinshaw, M., Evans, D.A., Jordán, A., Nulsen, P.E.J., Sivakoff, G.R., Worrall, D.M., Brassington, N.J., Forman, W.R., Gilfanov, M., Jones, C., Murray, S.S., Raychaudhury, S., Sarazin, C.L., Voss, R., Woodley, K.A.: Long-term monitoring of the dynamics and particle acceleration of knots in the jet of centaurus A. ApJ **708**, 675–697 (2010). doi:10.1088/0004-637X/708/1/675

Gull, S.F., Northover, K.J.E.: Bubble model of extragalactic radio sources. Nature **244**, 80–83 (1973). doi:10.1038/244080a0

Hardcastle, M.J.: Synchrotron and inverse-Compton emission from radio galaxies with non-uniform magnetic field and electron distributions. MNRAS **433**, 3364–3372 (2013). doi:10.1093/mnras/stt1024

Hardcastle, M.J., Croston, J.H.: The Chandra view of extended X-ray emission from Pictor A. MNRAS **363**, 649 (2005)

Hardcastle, M.J., Croston, J.H.: Searching for the inverse-compton emission from bright cluster-centre radio galaxies. MNRAS **404**, 2018–2027 (2010). doi:10.1111/j.1365-2966.2010.16420.x

Hardcastle, M.J., Croston, J.H.: Modelling TeV γ-ray emission from the kiloparsec-scale jets of Centaurus A and M87. MNRAS **415**, 133–142 (2011). doi:10.1111/j.1365-2966.2011.18678.x

Hardcastle, M.J., Krause, M.G.H.: Numerical modelling of the lobes of radio galaxies in cluster environments. MNRAS **430**, 174–196 (2013). doi:10.1093/mnras/sts564

Hardcastle, M.J., Sakelliou, I.: Jet termination in wide-angle tailed radio sources. MNRAS **349**, 560 (2004)

Hardcastle, M.J., Alexander, P., Pooley, G.G., Riley, J.M.: The jets in 3C66B. MNRAS **278**, 273 (1996)

Hardcastle, M.J., Alexander, P., Pooley, G.G., Riley, J.M.: High resolution observations at 3.6 cm of seventeen FRII radio galaxies with $0.15 < z < 0.3$. MNRAS **288**, 859 (1997a)

Hardcastle, M.J., Alexander, P., Pooley, G.G., Riley, J.M.: The jets in 3C296. MNRAS **288**, L1 (1997b)

Hardcastle, M.J., Alexander, P., Pooley, G.G., Riley, J.M.: FR II radio galaxies with $z < 0.3$ – I. Properties of jets, cores and hot spots. MNRAS **296**, 445 (1998a)

Hardcastle, M.J., Birkinshaw, M., Worrall, D.M.: Magnetic field strengths in the hot spots of 3C 33 and 3C 111. MNRAS **294**, 615 (1998b)

Hardcastle, M.J., Worrall, D.M., Birkinshaw, M.: The dynamics of the radio galaxy 3C449. MNRAS **296**, 1098 (1998c)

Hardcastle, M.J., Birkinshaw, M., Worrall, D.M.: A *Chandra* detection of the radio hotspot of 3C 123. MNRAS **323**, L17 (2001)

Hardcastle, M.J., Birkinshaw, M., Cameron, R., Harris, D.E., Looney, L.W., Worrall, D.M.: Magnetic field strengths in the hotspots and lobes of three powerful FRII radio sources. ApJ **581**, 948 (2002)

Hardcastle, M.J., Worrall, D.M., Kraft, R.P., Forman, W.R., Jones, C., Murray, S.S.: Radio and X-ray observations of the jet in centaurus A. ApJ **593**, 169 (2003)

Hardcastle, M.J., Harris, D.E., Worrall, D.M., Birkinshaw, M.: The origins of X-ray emission from the hotspots of FRII radio sources. ApJ **612**, 729 (2004)

Hardcastle, M.J., Croston, J.H., Kraft, R.P.: A Chandra study of particle acceleration in the multiple hotspots of nearby radio galaxies. ApJ **669**, 893 (2007a)

Hardcastle, M.J., Kraft, R.P., Sivakoff, G.R., Goodger, J.L., Croston, J.H., Jordán, A., Evans, D.A., Worrall, D.M., Birkinshaw, M., Raychaudhury, S., Brassington, N.J., Forman, W.R., Harris, W.E., Jones, C., Juett, A.M., Murray, S.S., Nulsen, P.E.J., Sarazin, C.L., Woodley, K.A.: New results on particle acceleration in the Centaurus A jet and counterjet from a deep *Chandra* observation. ApJ **670**, L81 (2007b)

Hardcastle, M.J., Massaro, F., Harris, D.E.: X-ray emission from the extended emission-line region of the powerful radio galaxy 3C171. MNRAS **401**, 2697 (2010)

Hardcastle, M.J., Massaro, F., Harris, D.E., Baum, S.A., Bianchi, S., Chiaberge, M., Morganti, R., O'Dea, C.P., Siemiginowska, A.: The nature of the jet-driven outflow in the radio galaxy 3C 305. MNRAS **424**, 1774–1789 (2012). doi:10.1111/j.1365-2966.2012.21247.x

Harris, D.E., Carilli, C.L., Perley, R.A.: X-ray emission from the hotspots of cygnus A. Nature **367**, 713 (1994)

Harris, D.E., Nulsen, P.E.J., Ponman, T.P., Bautz, M., Cameron, R.A., Donnelly, R.H., Forman, W.R., Grego, L., Hardcastle, M.J., Henry, J.P., Jones, C., Leahy, J.P., Markevitch, M., Martel, A.R., McNamara, B.R., Mazzotta, P., Tucker, W., Virani, S.N., Vrtilek, J.: Chandra X-ray detection of the radio hotspots of 3C295. ApJ **530**, L81 (2000)

Harwood, J.J., Hardcastle, M.J.: What determines the properties of the X-ray jets in Fanaroff-Riley type I radio galaxies? MNRAS **423**, 1368–1380 (2012). doi:10.1111/j.1365-2966.2012.20960.x

Harwood, J.J., Hardcastle, M.J., Croston, J.H., Goodger, J.L.: Spectral ageing in the lobes of FR-II radio galaxies: new methods of analysis for broad-band radio data. MNRAS **435**, 3353–3375 (2013). doi:10.1093/mnras/stt1526

Heckman, T.M., Smith, E.P., Baum, S.A., van Breugel, W.J.M., Miley, G.K., Illingworth, G.D., Bothun, G.D., Balick, B.: Galaxy collisions and mergers: the genesis of very powerful radio sources? ApJ **311**, 526 (1986)

Hill, G.J., Lilly, S.J.: A change in the cluster environments of radio galaxies with cosmic epoch. ApJ **367**, 1 (1991)

Hota, A., Saikia, D.J.: Radio bubbles in the composite AGN-starburst galaxy NGC6764. MNRAS **371**, 945–956 (2006). doi:10.1111/j.1365-2966.2006.10738.x

Huarte-Espinosa, M., Krause, M., Alexander, P.: 3D magnetohydrodynamic simulations of the evolution of magnetic fields in Fanaroff-Riley class II radio sources. MNRAS **417**, 382–399 (2011). doi:10.1111/j.1365-2966.2011.19271.x

Ineson, J., Croston, J.H., Hardcastle, M.J., Kraft, R.P., Evans, D.A., Jarvis, M.: Radio-loud active galactic nucleus: is there a link between luminosity and cluster environment? ApJ **770**, 136 (2013). doi:10.1088/0004-637X/770/2/136

Isobe, N., Tashiro, M., Makishima, K., Iyomoto, N., Suzuki, M., Murakami, M.M., Mori, M., Abe, K.: A Chandra detection of diffuse hard X-ray emission associated with the lobes of the radio galaxy 3C 452. ApJ **580**, L111 (2002)

Jaffe, W.J., Perola, G.C.: Dynamical models of tailed radio sources in clusters of galaxies. A&A **26**, 423 (1973)

Jennison, R.C., Das Gupta, M.K.: Fine structure of the extra-terrestrial radio source cygnus I. Nature **172**, 996–997 (1953). doi:10.1038/172996a0

Jones, T.W., Ryu, D., Engel, A.: Simulating electron transport and synchrotron emission in radio galaxies: shock acceleration and synchrotron aging in axisymmetric flows. ApJ **512**, 105–124 (1999). doi:10.1086/306772

Kaiser, C.R., Alexander, P.: A self-similar model for extragalactic radio sources. MNRAS **286**, 215 (1997)

Kardashev, N.S.: Nonstationarity of spectra of young sources of nonthermal radio emission. Soviet Astronomy **6**, 317 (1962)

Kataoka, J., Stawarz, Ł.: X-ray emission properties of large scale jets, hotspots and lobes in active galactic nuclei. ApJ **622**, 797 (2005)

Krause, M.: Very light jets. I. Axisymmetric parameter study and analytic approximation. A&A **398**, 113–125 (2003). doi:10.1051/0004-6361:20021649

Laing, R.A.: A model of the magnetic field structure in extended radio sources. MNRAS **193**, 439 (1980)

Laing, R.A.: Magnetic fields in extragalactic radio sources. ApJ **248**, 87 (1981)

Laing, R.A.: Interpretation of radio polarization data. In: Bridle, A.H., Eilek, J.A. (eds.) Physics of Energy Transport in Radio Galaxies, NRAO Workshop No. 9, p. 90. NRAO, Green Bank, West Virginia (1984)

Laing, R.A.: The sidedness of jets and depolarization in powerful extragalactic radio sources. Nature **331**, 149 (1988)

Laing, R.A.: Radio observations of hot spots. In: Meisenheimer, K., Röser, H.-J. (eds.) Hotspots in Extragalactic Radio Sources, p. 27. Springer-Verlag, Heidelberg (1989)

Laing, R.A., Bridle, A.H.: Relativistic models and the jet velocity field in the radio galaxy 3C 31. MNRAS **336**, 328 (2002)

Laing, R.A., Bridle, A.H.: Systematic properties of decelerating relativistic jets in low-luminosity radio galaxies. MNRAS **437**, 3405–3441 (2014). doi:10.1093/mnras/stt2138

Laing, R.A., Bridle, A.H., Parma, P., Feretti, L., Giovannini, G., Murgia, M., Perley, R.A.: Multifrequency VLA observations of the FR I radio galaxy 3C 31: morphology, spectrum and magnetic field. MNRAS **386**, 657–672 (2008). doi:10.1111/j.1365-2966.2008.13091.x

Leahy, J.P., Black, A.R.S., Dennett-Thorpe, J., Hardcastle, M.J., Komissarov, S., Perley, R.A., Riley, J.M., Scheuer, P.A.G.: A study of FRII radio galaxies with $z < 0.15$ – II. High-resolution maps of eleven sources at 3.6 cm. MNRAS **291**, 20 (1997)

Ledlow, M.J., Owen, F.N., Keel, W.C.: An unusual radio galaxy in Abell 428: a large, powerful FRI source in a disk-dominated host. ApJ **495**, 227 (1998)

Lilly, S.J., Longair, M.S.: Infra-red studies of a sample of 3C radio galaxies. MNRAS **199**, 1053 (1982)

Liu, R., Pooley, G.G., Riley, J.M.: Spectral ageing in a sample of 14 high-luminosity double radio sources. MNRAS **257**, 545 (1992)

Longair, M.S.: High energy astrophysics. Cambridge University Press, Cambridge (2010)

Longair, M.S., Seldner, M.: The clustering of galaxies about extragalactic radio sources. MNRAS **189**, 433 (1979)

Macdonald, G.H., Kenderdine, S., Neville, A.C.: Observations of the structure of radio sources in the 3C catalogue -I. MNRAS **138**, 259 (1968)

Mack, K.H., Kerp, J., Klein, U.: The X-ray jet and halo of NGC 6251. A&A **324**, 870 (1997)

Marshall, H.L., Hardcastle, M.J., Birkinshaw, M., Croston, J., Evans, D., Landt, H., Lenc, E., Massaro, F., Perlman, E.S., Schwartz, D.A., Siemiginowska, A., Stawarz, Ł., Urry, C.M., Worrall, D.M.: A flare in the jet of Pictor A. ApJL **714**, L213–L216 (2010). doi:10.1088/2041-8205/714/2/L213

Marshall, H.L., Gelbord, J.M., Schwartz, D.A., Murphy, D.W., Lovell, J.E.J., Worrall, D.M., Birkinshaw, M., Perlman, E.S., Godfrey, L., Jauncey, D.L.: An X-ray imaging survey of quasar jets: testing the inverse compton model. ApJS **193**, 15 (2011). doi:10.1088/0067-0049/193/1/15

Massaglia, S., Bodo, G., Ferrari, A.: Dynamical and radiative properties of astrophysical supersonic jets. I. Cocoon Morphologies. A&A **307**, 997–1008 (1996)

McCarthy, P.J., van Breugel, W., Spinrad, H., Djorgovski, S.: A correlation between the radio and optical morphologies of distant 3Cr radio galaxies. ApJ **321**, L29–L33 (1987). doi:10.1086/185000

McCarthy, P.J., van Breugel, W., Kapahi, V.K.: Correlated radio and optical asymmetries and powerful radio sources. ApJ **371**, 478 (1991)

McCarthy, P.J., Baum, S.A., Spinrad, H.: Emission-line properties of 3CR radio galaxies. II. Velocity fields in the extended emission lines. ApJS **106**, 281 (1996). doi:10.1086/192339

McKinney, J.C., Blandford, R.D.: Stability of relativistic jets from rotating, accreting black holes via fully three-dimensional magnetohydrodynamic simulations. MNRAS **394**, L126–L130 (2009). doi:10.1111/j.1745-3933.2009.00625.x

McNamara, B.R., Nulsen, P.E.J.: Mechanical feedback from active galactic nuclei in galaxies, groups and clusters. New J. Phys. **14**(5), 055023 (2012). doi:10.1088/1367-2630/14/5/055023

Meisenheimer, K., Röser, H.J., Hiltner, P.R., Yates, M.G., Longair, M.S., Chini, R., Perley, R.A.: The synchrotron spectra of radio hotspots. A&A **219**, 63 (1989)
Meisenheimer, K., Yates, M.G., Röser, H.J.: The synchrotron spectra of radio hot spots. II. Infrared imaging. A&A **325**, 57 (1997)
Mendygral, P.J., Jones, T.W., Dolag, K.: MHD simulations of active galactic nucleus jets in a dynamic galaxy cluster medium. ApJ **750**, 166 (2012). doi:10.1088/0004-637X/750/2/166
Mignone, A., Rossi, P., Bodo, G., Ferrari, A., Massaglia, S.: High-resolution 3D relativistic MHD simulations of jets. MNRAS **402**, 7–12 (2010). doi:10.1111/j.1365-2966.2009.15642.x
Morganti, R., Fanti, R., Gioia, I.M., Harris, D.E., Parma, P., de Ruiter, H.: Low luminosity radio galaxies: effects of gaseous environment. A&A **189**, 11 (1988)
Morganti, R., Tadhunter, C., Oosterloo, T.A.: Fast neutral outflows in powerful radio galaxies: a major source of feedback in massive galaxies. A&A **441**, L9 (2005)
Mullin, L.M., Hardcastle, M.J.: Bayesian inference of jet speeds in radio galaxies. MNRAS **398**, 1989 (2009)
Mullin, L.M., Hardcastle, M.J., Riley, J.M.: High-resolution observations of radio sources with $0.6 < z < 1.0$. MNRAS **372**, 510 (2006)
Myers, S.T., Spangler, S.R.: Synchrotron aging in the lobes of luminous radio galaxies. ApJ **291**, 52 (1985)
Nesvadba, N.P.H., Lehnert, M.D., De Breuck, C., Gilbert, A.M., van Breugel, W.: Evidence for powerful AGN winds at high redshift: dynamics of galactic outflows in radio galaxies during the "Quasar Era". A&A **491**, 407 (2008)
Norman, M.L., Winkler, K.H.A., Smarr, L., Smith, M.D.: Structure and dynamics of supersonic jets. A&A **113**, 285–302 (1982)
Northover, K.J.E.: The radio galaxy 3C 66. MNRAS **165**, 369 (1973)
Nulsen, P.E.J., Hambrick, D.C., McNamara, B.R., Rafferty, D., Bîrzan, L., Wise, M.W., David, L.P.: The powerful outburst in hercules A. ApJ **625**, L9 (2005)
Omma, H., Binney, J.: Structural stability of cooling flows. MNRAS **350**, L13–L16 (2004). doi:10.1111/j.1365-2966.2004.07809.x
O'Neill, S.M., Tregillis, I.L., Jones, T.W., Ryu, D.: Three-dimensional simulations of MHD jet propagation through uniform and stratified external environments. ApJ **633**, 717–732 (2005). doi:10.1086/491618
O'Sullivan, S.P., Feain, I.J., McClure-Griffiths, N.M., Ekers, R.D., Carretti, E., Robishaw, T., Mao, S.A., Gaensler, B.M., Bland-Hawthorn, J., Stawarz, Ł.: Thermal plasma in the giant lobes of the radio galaxy centaurus A. ApJ **764**, 162 (2013). doi:10.1088/0004-637X/764/2/162
Pacholczyk, A.G.: Radio Astrophysics. Freeman, San Francisco (1970)
Parma, P., Fanti, C., Fanti, R., Morganti, R., De Ruiter, H.R.: VLA observations of low-luminosity radio galaxies. VI: discussion of radio jets. A&A **181**, 244 (1987)
Perley, R.A., Dreher, J.W., Cowan, J.J.: The jet and filaments in Cygnus A. ApJ **285**, L35 (1984)
Perley, R.A., Röser, H..J., Meisenheimer, K.: The radio galaxy pictor A – a study with the VLA. A&A **328**, 12 (1997)
Perlman, E.S., Adams, S.C., Cara, M., Bourque, M., Harris, D.E., Madrid, J.P., Simons, R.C., Clausen-Brown, E., Cheung, C.C., Stawarz, L., Georganopoulos, M., Sparks, W.B., Biretta, J.A.: Optical polarization and spectral variability in the M87 Jet. ApJ **743**, 119 (2011). doi:10.1088/0004-637X/743/2/119
Perucho, M., Martí, J.M.: A numerical simulation of the evolution and fate of a Fanaroff-Riley type I jet. The case of 3C 31. MNRAS **382**, 526–542 (2007). doi:10.1111/j.1365-2966.2007.12454.x
Perucho, M., Martí, J.M., Cela, J.M., Hanasz, M., de La Cruz, R., Rubio, F.: Stability of three-dimensional relativistic jets: implications for jet collimation. A&A **519**, A41 (2010). doi:10.1051/0004-6361/200913012
Prestage, R.M., Peacock, J.A.: The cluster environments of powerful radio galaxies. MNRAS **230**, 131 (1988)

Rafferty, D.A., McNamara, B.R., Nulsen, P.E.J., Wise, M.W.: The feedback-regulated growth of black holes and bulges through gas accretion and starbursts in cluster central dominant galaxies. ApJ **652**, 216–231 (2006). doi:10.1086/507672

Ramos Almeida, C., Bessiere, P.S., Tadhunter, C.N., Pérez-González, P.G., Barro, G., Inskip, K.J., Morganti, R., Holt, J., Dicken, D.: Are luminous radio-loud active galactic nuclei triggered by galaxy interactions? MNRAS **419**, 687–705 (2012). doi:10.1111/j.1365-2966.2011.19731.x

Reber, G.: Cosmic Static. ApJ **100**, 279 (1944). doi:10.1086/144668

Rees, M.J.: The radio/optical alignment of high-z radio galaxies - triggering of star formation in radio lobes. MNRAS **239**, 1P–4P (1989)

Rossi, P., Mignone, A., Bodo, G., Massaglia, S., Ferrari, A.: Formation of dynamical structures in relativistic jets: the FRI case. A&A **488**, 795–806 (2008). doi:10.1051/0004-6361:200809687

Rybicki, G.B., Lightman, A.P.: Radiative processes in astrophysics. Wiley, New York (1979)

Ryle, M., Longair, M.S.: A possible method for investigating the evolution of radio galaxies. MNRAS **136**, 123 (1967)

Ryle, M., Smith, F.G., Elsmore, B.: A preliminary survey of the radio stars in the Northern Hemisphere. MNRAS **110**, 508 (1950)

Sarazin, C.L.: X-ray emission from clusters of galaxies. Rev. Mod. Phys. **58**, 1 (1986)

Saslaw, W.C., Tyson, J.A., Crane, P.: Optical emission in the radio lobes of radio galaxies. ApJ **222**, 435–439 (1978). doi:10.1086/156157

Scheuer, P.A.G.: Models of extragalactic radio sources with a continuous energy supply from a central object. MNRAS **166**, 513 (1974)

Scheuer, P.A.G.: Morphology and power of radio sources. Extragalactic radio sources. In: Heeschen, D.S., Wade, C.M. (eds.) IAU Symposium 97, Kyoto, p. 163. Reidel, Dordrecht (1982)

Scheuer, P.A.G.: Lobe asymmetry and the expansion speeds of radio sources. MNRAS **277**, 331 (1995)

Scheuer, P.A.G., Readhead, A.C.S.: Superluminally expanding radio sources and the radio-quiet QSOs. Nature **277**, 182 (1979)

Schoenmakers, A.P., de Bruyn, A.G., Röttgering, H.J.A., van der Laan, H., Kaiser, C.R.: Radio galaxies with a ‘double-double morphology’ - I. Analysis of the radio properties and evidence for interrupted activity in active galactic nuclei. MNRAS **315**, 371–380 (2000). doi:10.1046/j.1365-8711.2000.03430.x

Schreier, E.J., Feigelson, E., Delvaille, J., Giacconi, R., Grindlay, J., Schwartz, D.A., Fabian, A.C.: Einstein observations of the X-ray structure of centaurus A - evidence for the radio-lobe energy source. ApJ **234**, L39 (1979)

Shelton, D.L., Hardcastle, M.J., Croston, J.H.: The dynamics and environmental impact of 3C 452. MNRAS **418**, 811–819 (2011). doi:10.1111/j.1365-2966.2011.19533.x

Smith, D.A., Wilson, A.D., Arnaud, K.A., Terashima, Y., Young, A.J.: A Chandra X-ray study of cygnus A. III. The cluster of galaxies. ApJ **565**, 195 (2002)

Sutherland, R.S., Bicknell, G.V.: Interactions of a light hypersonic jet with a nonuniform interstellar medium. ApJS **173**, 37–69 (2007). doi:10.1086/520640

Tavecchio, F., Maraschi, L., Sambruna, R.M., Urry, C.M.: The X-ray jet of PKS 0637-752: inverse compton radiation from the cosmic microwave background? ApJ **544**, L23 (2000)

Taylor, G.B., Perley, R.A.: Magnetic fields in the hydra A cluster. ApJ **416**, 554 (1993). doi:10.1086/173257

Tregillis, I.L., Jones, T.W., Ryu, D.: Simulating electron transport and synchrotron emission in radio galaxies: shock acceleration and synchrotron aging in three-dimensional flows. ApJ **557**, 475 (2001)

Tregillis, I.L., Jones, T.W., Ryu, D.: Synthetic observations of simulated radio galaxies. I. Radio and X-ray analysis. ApJ **601**, 778–797 (2004). doi:10.1086/380756

Tribble, P.C.: Radio spectral ageing in a random magnetic field. MNRAS **261**, 57 (1993)

van Breugel, W., Filippenko, A.V., Heckman, T., Miley, G.: Minkowski's object - a starburst triggered by a radio jet. ApJ **293**, 83–93 (1985). doi:10.1086/163216

Virdee, J.S., Hardcastle, M.J., Rawlings, S., Rigopoulou, D., Mauch, T., Jarvis, M.J., Verma, A., Smith, D.J.B., Heywood, I., White, S.V., Baes, M., Cooray, A., Zotti, G.D., Eales, S., Michałowski, M.J., Bourne, N., Dariush, A., Dunne, L., Hopwood, R., Ibar, E., Maddox, S., Smith, M.W.L., Valiante, E.: Herschel-ATLAS/GAMA: what determines the far-infrared properties of radio galaxies? MNRAS **432**, 609–625 (2013). doi:10.1093/mnras/stt488

Wardle, J.F.C., Aaron, S.E.: How fast are the large-scale jets in quasars? constraints on both doppler beaming and intrinsic asymmetries. MNRAS **286**, 425 (1997)

Williams, A.G., Gull, S.F.: Multiple hotspots in extragalactic radio sources. Nature **313**, 34 (1985)

Wilson, A.S., Young, A.J., Shopbell, P.L.: Chandra X-ray observations of pictor A: high energy cosmic rays in a radio galaxy. ApJ **547**, 740 (2001)

Worrall, D.M., Birkinshaw, M., Cameron, R.A.: The X-ray environment of the dumbbell radio galaxy NGC 326. ApJ **449**, 93 (1995)

Wykes, S., Croston, J.H., Hardcastle, M.J., Eilek, J.A., Biermann, P.L., Achterberg, A., Bray, J.D., Lazarian, A., Haverkorn, M., Protheroe, R.J., Bromberg, O.: Mass entrainment and turbulence-driven acceleration of ultra-high energy cosmic rays in centaurus A. A&A **558**, A19 (2013). doi:10.1051/0004-6361/201321622

Yates, M.G., Miller, L., Peacock, J.A.: The cluster environments of powerful, high-redshift radio galaxies. MNRAS **240**, 129 (1989)

Zanni, C., Bodo, G., Rossi, P., Massaglia, S., Durbala, A., Ferrari, A.: X-ray emission from expanding cocoons. A&A **402**, 949–962 (2003). doi:10.1051/0004-6361:20030302

Chapter 5
Parsec-Scale Jets in Active Galactic Nuclei

Denise C. Gabuzda

Abstract Considerable progress has been made over the past decade or so in understanding the jets of Active Galactic Nuclei on the parsec scales probed by Very Long Baseline Interferometry. The availability of multi-wavelength polarization VLBI observations for relatively large samples of objects for the first time has provided fundamental new information about the spectral indices and Faraday rotation measures in the core and jet components. Reliable estimates of the core magnetic fields and degrees of circular polarization have also become available for the first time. The jets exhibit complex behaviour, such as accelerations and non-radial motions of individual moving features, and even swings of the jet as a whole. New approaches have been developed to estimate the intrinsic speeds and viewing angles of the jets. A variety of new evidence seems to be suggesting that many of the properties we are observing are a consequence of helical magnetic fields carried outward by the jets, which should come about naturally due to the rotation of the central black hole and its accretion disk together with the jet outflow.

5.1 Introduction

The central regions of powerful Active Galactic Nuclei (AGN) radiate huge amounts of energy; the strongest AGN are about 10^5 times more luminous than an entire normal galaxy such as the Milky Way. The source of this phenomenal energy is believed to be the gravitational energy released by matter accreting onto a supermassive ($\sim 10^9$ times the mass of the Sun) black hole at the galactic centre.

About 10–15 % or so of all AGN are "radio-loud," while the remaining are "radio-quiet." The generally accepted definition of a radio-loud AGN is that the ratio of its radio (5 GHz) to its optical (B-band) flux be ≥ 10 (Kellermann et al. 1989a). The radio emission is predominantly associated with jets ejected from the central region of the AGN and the lobes that they inflate, suggesting that jets are either

D.C. Gabuzda (✉)
Department of Physics, University College Cork, Cork, Ireland
e-mail: d.gabuzda@ucc.ie

I. Contopoulos et al. (eds.), *The Formation and Disruption of Black Hole Jets*,
Astrophysics and Space Science Library 414,
DOI 10.1007/978-3-319-10356-3_5

absent or much weaker in the radio-quiet AGN. This radio emission is synchrotron radiation emitted by relativistic electrons moving through regions with magnetic (**B**) field; the jets presumably create conditions where both electrons can be accelerated to high energies and magnetic fields can be generated and amplified.

It is natural to strive to image AGN with high resolution, to obtain information about regions close to the central energy source. An extremely powerful tool for studies of radio-loud AGN is Very Long Baseline Interferometry (VLBI), in which radio telescopes around the world are used in synchrony to obtain images with angular resolutions of the order of a milliarcsecond on the sky, or of order a parsec at the cosmological distances characteristic of AGN. Various arrays of radio telescopes around the world are outfitted to carry out VLBI observations, such as the American Very Long Baseline Array (VLBA) and the European VLBI Network (EVN).

Most or all of the radio emission of AGN is associated with jets of plasma that emerge from their central regions, presumably along the rotational axis of the central supermassive black hole. These jets are present on the smallest scales that can be probed with VLBI, and sometimes extend out to scales of many kiloparsecs (see Chap. 3). The synchrotron radiation given off by each individual electron is highly concentrated in the direction of motion of the electron, and the radiation observed for an ensemble of relativistic electrons will, in general, be linear polarized if the synchrotron **B** field is at least partially ordered. Linear-polarization observations can thus provide direct information about the degree of order and orientation of the **B** field giving rise to the observed synchrotron radiation, and thus play a key role in studies of the conditions in and around AGN jets.

Section 5.2 summarizes the main relativistic effects that directly affect VLBI observations of AGN jets. Section 5.3 provides a brief overview of some basic properties of incoherent synchrotron radiation, and Sect. 5.4 a basic overview of various theoretical models and processes, that feed into interpretations of VLBI observations of AGN jets. Finally, Sect. 5.5 provides a review of the current observations of AGN jets, focusing on recent results and progress.

5.2 Relativistic Effects Influencing the Observations

5.2.1 Aberration and Relativistic Beaming

The Lorentz transformations relating the coordinates of an object in one frame (unprimed) and in another frame (primed) moving relative to the first frame with a relative velocity **v** can be used to find the corresponding transformations between the components of the object's velocity parallel and perpendicular to the direction of the relative motion of the two frames.

For light (a photon), the angles θ and θ' along which the light is observed to propagate in the two frames are related by the equations

$$\tan\theta = \frac{\sin\theta'}{\Gamma(\cos\theta' + \beta)} \qquad \cos\theta = \frac{\cos\theta' + \beta}{1 + \beta\cos\theta'} \tag{5.1}$$

where β is the jet velocity divided by the speed of light and Γ is the corresponding bulk Lorentz factor. Thus, a photon emitted at right angles to $\mathbf{v}$ in its rest frame ($\theta' = \pi/2$) is observed at the angle

$$\tan\theta = \frac{1}{\Gamma\beta} \qquad \cos\theta = \beta \qquad \sin\theta = \frac{1}{\Gamma} \tag{5.2}$$

This is the well-known *beaming* of the source radiation in the forward direction of its motion. When Γ is large, $\sin\theta$ will be small, and $\theta \simeq \frac{1}{\Gamma}$. Thus, half the radiation emitted by an isotropically radiating, relativistically moving source is observed within a narrow cone of half-angle $1/\Gamma$.

This *Doppler beaming* or *Doppler boosting* is believed to explain the one-sided appearance of the relativistic jets of AGN on VLBI scales (Sect. 5.5.1). It is believed that, in reality, there are two jets – one approaching (Doppler boosted) and one receding (Doppler dimmed) – if the boosting factor is high enough, the receding jet (whose radiation is beamed in the direction away from the Earth) becomes undetectable.

The difference in the observed direction of propagation of a photon in two frames moving relative to one another is referred to as aberration. Due to aberration, photons we detect arriving from a jet viewed at a small angle to our line of sight were actually emitted by the jet roughly perpendicular to the direction of its motion; thus, we may receive an approximately "side-on" view of the jet, even though we are viewing the jet at a small angle to its axis. This important effect is often neglected when interpreting images of VLBI jets, as will be discussed further below. In addition, when the viewing angle $\theta < 1/\Gamma$, then $\theta' < 90°$, and we have a "head-on" view of the object; whereas $\theta' > 90°$ when $\theta > 1/\Gamma$, and we have a "tail-on" view of the object. In other words, in the latter case, we are actually detecting photons emitted by the source roughly away from the direction toward the Earth. Taking into account the fact that jets viewed at smaller angles will appear brighter, and therefore be more prominently represented in observational samples, the most probable viewing angle for an AGN jet is about $0.6/\Gamma$ (Cohen et al. 2007), but some of the relativistic jets observed in AGN will also be viewed at angles $\theta > 1/\Gamma$, so that we are receiving a "tail-on" view; this will certainly affect observations of these objects – although unfortunately, we cannot be sure which ones they are.

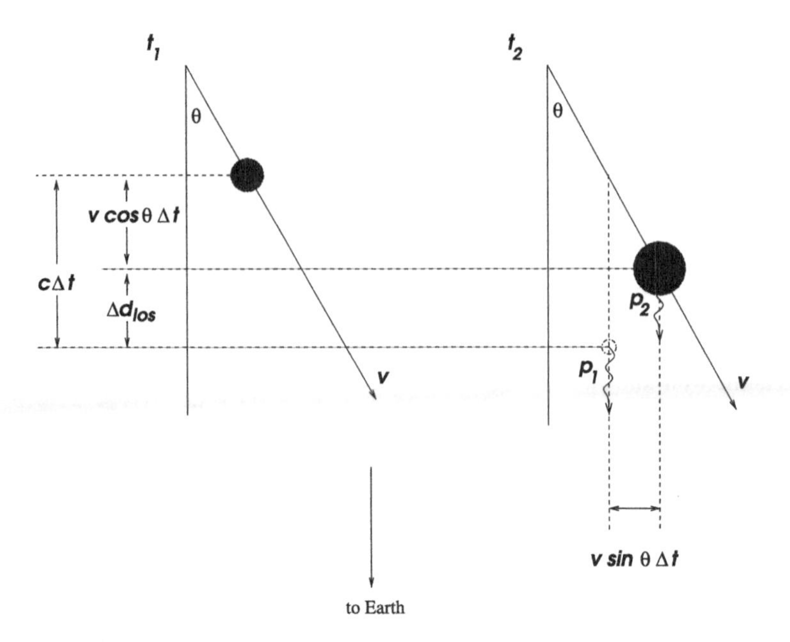

Fig. 5.1 Geometry for superluminal motion

5.2.2 Superluminal Motions

Consider the motion of a clump of plasma in a jet roughly toward the Earth at two times t_1 and t_2. The angle of the clump's motion relative to the direction toward the Earth is θ, and the clump's velocity is v (Fig. 5.1).

The distance travelled by the clump toward the Earth between the two times is $v\cos\theta\Delta t$, where $\Delta t = t_2 - t_1$. The distance travelled by a photon emitted at t_1 in the time Δt is just $c\Delta t$. The distance between the photons p_1 and p_2 will be $c\Delta t - v\cos\theta\Delta t$, and so the measured time between the arrival of photons p_1 and p_2 emitted at time t_1 and t_2 will be $\Delta t_{meas} = (1 - \beta\cos\theta)\Delta t$, where $\beta = v/c$. The observed distance travelled by the clump in the plane of the sky will be $d_{sky} = v\sin\theta\Delta t$. Thus, the apparent speed of the clump in the plane of the sky will be

$$\beta_{app} = \frac{\Delta d_{sky}}{c\Delta t_{meas}} = \frac{\beta\sin\theta}{1 - \beta\cos\theta} \tag{5.3}$$

The maximum of this function can be found by taking its derivative and setting it equal to zero. This yields a maximum when

$$\cos\theta_{max} = \beta \qquad \sin\theta_{max} = \frac{1}{\Gamma} \tag{5.4}$$

or $\theta_{max} = 1/\Gamma$ when Γ is large and θ_{max} is therefore small. The corresponding maximum apparent speed in the plane of the sky is $\beta_{max} = \beta\Gamma$.

Clearly, the observed motion will be superluminal, or apparently faster than light, for sufficiently high intrinsic speeds β and sufficiently small viewing angles $\theta \sim 1/\Gamma$. This provides a simple explanation of superluminal motions in AGN jets, which are very common; unfortunately, it is not possible to unambiguously disentangle the contributions of β and θ to the observed speeds. Nevertheless, studies of superluminal jet component speeds for samples of AGN are extremely useful in determining the nature of differences between different types of AGN, for example.

5.2.3 *Measured Variation Timescales and the Doppler Factor*

As was noted above, the measured time between the arrival of two photons emitted by a source moving toward the observer at two times separated by Δt is $\Delta t_{meas} = (1 - \beta\cos\theta)\Delta t$. Here, Δt is the time interval in the frame of the observer, and will differ from the time interval between the emission of the two photons in the rest frame of the source (which we denote with a prime) due to relativistic time dilation: $\Delta t = \Gamma\Delta t'$. Therefore,

$$\Delta t_{meas} = (1 - \beta\cos\theta)\Gamma\Delta t \equiv \frac{1}{D}t' \tag{5.5}$$

The inferred time interval between the arrival of the two photons will be too small by a factor

$$D = \frac{1}{\Gamma(1 - \beta\cos\theta)} \tag{5.6}$$

The *Doppler factor* D relates quantities in the rest frame of the source and the observer's frame, taking into account both special relativistic effects and the direction of the source's motion relative to the observer. For example, if θ is small enough that $\cos\theta \simeq 1$ and in addition $\beta \simeq 1$ (highly relativistic motion nearly directly toward the Earth), $\Delta t_{meas} \simeq \frac{1}{2\Gamma}\Delta t'$; i.e., the observer would infer a time between the emission of the two photons that is much too short, by a factor of $1/2\Gamma$.

5.3 Theory of Radio Synchrotron Emission: Potential Observational Manifestations

A review of some basic properties of synchrotron radiation is given in Sect. 4.2.1. The following sections consider some other properties of synchrotron radiation that are particularly relevant for VLBI polarization observations. More detailed information about synchrotron radiation can be found, for example, in the textbooks by Longair (2010) and Rybicki and Lightman (1979).

5.3.1 Optical Depth

Roughly speaking, if the mean-free-path of a photon in the radiating region is larger than the size of the region, the photon is likely to be able to pass through the region without being absorbed, and we say the region is *optically thin*. If the mean-free-path of a photon is appreciably less than the size of the region, the photon is likely to be absorbed before it leaves the source region, and we say the region is *optically thick*. It turns out that, after integrating over an ensemble of electrons with a power-law electron energy distribution, $N(E)dE \sim E^{-p}dE$, the power radiated at frequency ν in the optically thin part of the spectrum is also a power law proportional to ν^{α}, where $\alpha = -(p-1)/2$ is called the *spectral index*.

At low frequencies, there can be appreciable absorption of the synchrotron radiation by the radiating electrons themselves, referred to a *synchrotron self absorption*. This makes the radiating region optically thick at these frequencies; it turns out that this gives rise to a slope of +5/2 in this part of the spectrum. A sketch of the total spectrum for a single homogeneous synchrotron-radiating region is shown in Fig. 5.2. The theoretical optically thick spectral index is $+5/2$; the optically thin spectral index is $\alpha = -(p-1)/2$. The theoretical optically thick spectral index of $+5/2$ is observed very rarely; observed regions are probably usually only partially optically thick (opaque).

Generally speaking, we expect the emission in the core region to be at least partially optically thick, and the emission in the jet to be optically thin.

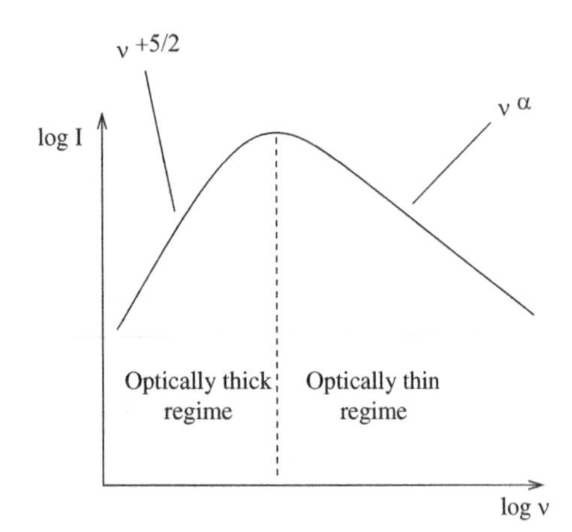

Fig. 5.2 Schematic of the spectrum of a homogeneous source of synchrotron radiation

5.3.2 Linear Polarization

Synchrotron radiation is intrinsically linearly polarized; to obtain 100 % polarization, one would have to have a perfectly ordered **B** field and an ensemble of electrons which all have the same pitch angle relative to this **B** field, corresponding to motion purely in a plane perpendicular to the field, in a region that was optically thin. Although various physical processes can give rise to highly ordered **B** fields, it is generally believed that the distribution of pitch angles for the ensemble of radiating electrons is most likely to be uniform and random, i.e. that we are dealing with "incoherent" synchrotron radiation. The degree of polarization m_{thin} expected for such an ensemble of electrons moving in a completely uniform **B** field in an optically thin region is given by

$$m_{thin} = \frac{p+1}{p+7/3} \tag{5.7}$$

where p is the index of the power law distribution of the electron energies (Pacholczyk 1970). This yields $m_{thin} \simeq 70 - 75\,\%$ for p values of $2 - 3$, which correspond to optically thin spectral indices $\alpha = -0.5$ to -1, which are fairly typical of the observed spectral indices in the VLBI jets of AGN. The maximum theoretical degree of polarization for the synchrotron radiation associated with AGN jets is usually taken to be about 75 %.

The corresponding maximum degree of polarization m_{thick} expected for an optically thin region is given by

$$m_{thick} = \frac{3}{6p+13}, \tag{5.8}$$

yielding values $m_{thick} \simeq 10 - 12\,\%$ for p values of $2 - 3$ (Pacholczyk 1970).

We can think of the radiation as having two components to its linear polarization: one with its polarization **E** vector orthogonal to the magnetic field, and the other with its polarization **E** vector aligned with the magnetic field. The former of these two dominates when the radiation is optically thin – qualitatively, this corresponds to **E** being in the plane of gyration of the electrons. Recall that the radiation will be optically thick when there is a high probability of a synchrotron photon being absorbed before it can leave the region in which it was radiated. In this regime, the component of the linear polarization that has the higher probability of being absorbed is precisely the one that had the higher probability of being emitted in the optically thin regime – the one with **E** perpendicular to the synchrotron **B** field. The other component of the polarization – with **E** parallel to the **B** field – is less likely to be absorbed, and so more likely to be able to leave the emission region.

Thus, in the optically thin regime, the observed plane of polarization is *perpendicular* to the **B** field and the degree of linear polarization can reach $\simeq$75 %, while, in the optically thick regime, the observed plane of polarization is *parallel* to the **B** field and the degree of linear polarization is of order 10–12 %. In both cases, the degree of polarization will decrease if the local **B** field is not completely ordered. The reason for the lower degree of polarization for the optically thick regime is that the component of the polarization with **E** parallel to **B** is less likely to be emitted, so the total fraction of synchrotron photons emitting this component is modest. Cawthorne and Hughes (2013) have recently pointed out that the optical depth τ at which the 90° rotation in polarization angle in the transition from optically thick to optically thin occurs is actually appreciably greater than $\tau \approx 1$.

5.3.3 *Circular Polarization*

The degree of circular polarization of incoherent synchrotron radiation is given by

$$m_c = \epsilon_\alpha^\nu \left(\frac{\nu_{B\perp}}{\nu}\right)^{0.5} \frac{B_{u,los}}{B_\perp^{rms}} \propto \nu^{-0.5} \tag{5.9}$$

where ϵ_α^ν is a constant that has values near unity for spectral indices near 0 (as is the case, for example, for most observed VLBI cores at centimeter wavelengths); $B_{u,los}$ is the component of the uniform magnetic field that is responsible for generating the circular polarization; $B_\perp^{rms}$ is the mean field component in the plane of the sky, which includes both the transverse part of the total uniform magnetic field B_u and any disordered field (which contributes to total intensity but not circular polarization); and $\nu_{B\perp} = 2.8 B_\perp^{rms}$ is the gyrofrequency for $B_\perp^{rms}$ in MHz (Legg and Westfold 1968). Thus, the degree of circular polarization associated with the synchrotron mechanism is expected to grow with decreasing frequency. Since the jets of AGN are typically linearly polarized about 10 % on parsec scales, a reasonable estimate is $B_{u,los}/B_\perp^{rms} \approx 0.10$; estimating as well $B_\perp^{rms} \approx 0.10$ G, consistent with estimates of the VLBI core magnetic fields (e.g. O'Sullivan and Gabuzda 2009; Pushkarev et al. 2012) suggests that one might obtain m_c values up to about 0.10–0.20 % at 5–15 GHz.

5.4 Standard Jet Theory: An Observer's View

5.4.1 *Blandford–Königl Jets: The Nature of the VLBI Core*

The generally accepted overall framework for the interpretation of observations of the VLBI jets of AGN is provided by the model proposed by Blandford and Königl (1979). In this model, the jets are taken to be conical, and the feature observed as the VLBI "core" is interpreted as the "photosphere" of the jet, within which the emission regions are optically thick and outside which they are optically thin. This is usually refered to as the $\tau = 1$ surface, although there is no strict reason that the optical depth at this location must be precisely equal to unity, and this term is essentially conveying the idea that the probability of an emitted photon escaping the emission region is beginning to deviate appreciably from the optically thin probability near zero in this region.

The location of this surface, i.e., the observed location of the VLBI core, depends on the observing frequency, and moves further down the jet at increasingly lower frequencies, as is shown in the schematic in Fig. 5.3. In contrast, the positions of optically thin features in the jet should coincide at different frequencies.

However, absolute position information is lost during the mapping process, and the phase center (coordinate origin) of a VLBI image will tend to coincide with the dominant partially optically thick VLBI core. Therefore, VLBI images obtained at different frequencies will, in general, not be correctly aligned in a physical sense, and cannot be directly superposed, for example, to derive spectral-index maps. To correctly align images obtained at different frequencies, it is necessary to shift them so that the positions of optically thin features coincide at the different frequencies. A variety of techniques for achieving this alignment have been exploited, most importantly model-fitting in the visibility domain to determine the positions of optically thin jet features, which are then made to coincide at different frequencies, and cross-correlation of the optically thin parts of total intensity images (e.g. Walker et al. 2000; Croke and Gabuzda 2008). In practice, different approaches may work best for different types of source structures, and both of these general approaches can yield reliable alignments in particular cases.

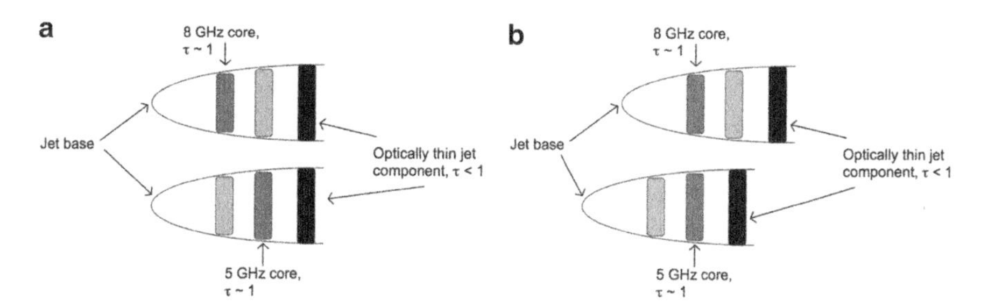

Fig. 5.3 Schematic of a Blandford–Königl jet, where the observed VLBI core corresponds approximately to the $\tau = 1$ surface in the jet at a given frequency. The same jet as observed at 8 GHz (upper jet in each panel) and 5 GHz (lower jet in each panel) is shown; in each graphic, the *black feature* represents an optically thin jet component, the *dark gray feature* the core position at the given frequency, and the *light gray feature* the core position at the other frequency. In panel (**a**), the jets observed at the two frequencies are correctly aligned, with the optically thin jet component coincident; in panel (**b**), the jets observed at the two frequencies are incorrectly aligned due to the mapping processes, which tends to place the observed core very close to the map phase center

5.4.2 Shocks

The presence of shocks in VLBI jets has often been invoked as an explanation for the sometimes rapid variability exhibited by AGN. The formation of both forward and reverse shocks has also been demonstrated in numerical simulations of the propagation of a perturbation down a relativistic jet (e.g. Aloy et al. 1999, 2000, 2003; Mimica and Aloy 2010, 2012).

In a region subject to shock compression, the **B** field is amplified in the plane of compression, giving rise to appreciable degrees of polarization and a net magnetic field lying in the plane of compression, even in the case of an initially completely tangled magnetic field (Laing 1980; Hughes et al. 1985). In a transverse shock, the plane of compression is perpendicular to the jet direction, the net magnetic field in the shocked region is also perpendicular to the jet direction, and the observed polarization is aligned with the jet direction, assuming the shock region is optically thin.

The polarization expected from oblique and conical shocks has been investigated by Cawthorne and Cobb (1990). Models in which the VLBI core is taken to be a standing shock are considered, for example, by Cawthorne et al. (2013) and Marscher (2014) [see also references therein].

5.4.3 Faraday Rotation

Faraday rotation is a rotation of the plane of polarization of an electromagnetic (EM) wave that occurs when it passes through a region with free charges and magnetic field. Any EM wave can be described as the sum of any two orthogonal components, usually considered in radio astronomy to be right circularly polarized (RCP) and left circularly polarized (LCP). Due to asymmetry in the interactions between the local free charges and the RCP and LCP components of the polarized wave, these two components have different indices of refraction, and therefore different speeds of propagation through the magnetized medium.

When the wave propagates through a vacuum, the **E** vectors for these two components rotate in opposite directions at the same rate, preserving the orientation of the plane of linear polarization, χ. When the wave propagates through a magnetized plasma, the difference in the speeds of the RCP and LCP waves induces a delay between these components, manifest as a rotation in the plane of polarization. The amount of rotation depends on the strength of the ambient magnetic field **B**, the number density of charges in the plasma n_e, the charge e and mass m of these charges, and the wavelength of the radiation λ:

$$\chi = \chi_o + RM\lambda^2 \qquad RM = \frac{e^3}{8\pi^2\epsilon_o m_e^2 c^3}\int n_e \mathbf{B}\cdot d\mathbf{l} \tag{5.10}$$

where χ is the observed polarization angle, χ_o is the intrinsic emitted polarization angle, and the integral is carried out over the line of sight from the source to the observer. Due to the inverse dependence on the square of the mass of the Faraday-rotating particles, it is usually assumed that observed Faraday rotation is due to the action of electrons; because the effective masses of relativistic electrons are higher than those of thermal electrons, these electrons are usually assumed to be thermal. The magnitude of the rotation measure RM depends on both n_e and the line-of-sight component of the magnetic field, while the sign of the Faraday rotation is determined by the direction of the line-of-sight magnetic field (either toward or away from the observer). The action of Faraday rotation can be identified through the λ^2 dependence of the observed polarization angle χ.

Note that it is the line-of-sight component of the ambient magnetic field that determines the magnitude and sign of the Faraday rotation. Essentially all observations of extragalactic sources are affected by Faraday rotation to some extent, because their radiation must always pass through our own Galaxy on its way to the Earth. Usually, these rotations are not very large at wavelengths of about 6 cm and shorter.

5.4.4 Faraday Depolarization

Faraday rotation can also give rise to *depolarization* of the radiation (e.g. Burn 1966). For example, if the source emission region is optically thin (little absorption of the radiation within the source), the radiation emitted at different depths in the source will pass through different amounts of the source volume on its way toward the observer. Therefore, if there are free electrons in the source volume, radiation emitted at different depths in the source will experience different amounts of Faraday rotation on its way through the source volume toward the observer. This causes a reduction in polarization, since it induces various offsets between emitted polarization electric vectors that are intrinsically aligned, essentially introducing a random component to the polarization. This is referred to as *front–back depolarization*.

Observations have some finite resolution, and so represent the sum of many electromagnetic waves propagating along many different lines of sight through the plasma. If appreciable inhomogeneities are present in the plasma electron density and/or ambient magnetic field on scales smaller than the resolution of the observations, different lines of sight will experience different rotation measures; this likewise leads to offsets between emitted polarization electric vectors that are intrinsically aligned, giving rise to *beam depolarization*.

Finally, if relatively large bandwidths are used at relatively long wavelengths, there can be appreciably different amounts of Faraday rotation occurring at the different frequencies in the observed band. If the signals detected within the band are averaged together, the observed polarization will be reduced, since intrinsically aligned polarization electric vectors at different frequencies within the band will become offset due to the different Faraday rotations they experience; this is referred to as *bandwidth depolarization*.

In all cases, the depolarization occurs because the total polarization is the sum of the polarizations for multiple regions, lines of sight, or nearby frequencies that experience different Faraday rotations. Because the rotation of the polarization vectors is greater at longer wavelengths, such depolarization will also increase at longer wavelengths. Depolarization is indicated by a decrease in the observed degree of polarization with increasing wavelength.

5.4.5 Faraday Conversion

As was noted in Sect. 5.3.3, incoherent synchrotron radiation produces a only very small amount of circular polarization at frequencies of several to tens of gigahertz. Another mechanism that is capable of producing higher levels of circular polarization is the Faraday conversion of linear-to-circular polarization during propagation through a magnetized plasma (Jones and O'Dell 1977; Jones 1988).

Similar to Faraday rotation, Faraday conversion can occur when a polarized electromagnetic wave passes through a magnetized plasma. In order for Faraday conversion to operate, the observed linear polarization electric (**E**) vector must have non-zero components both parallel to ($\mathbf{E}_{\parallel}$) and perpendicular to ($\mathbf{E}_{\perp}$) the magnetic field in the conversion region projected onto the sky, $\mathbf{B}_{conv}$. The electric-field component $\mathbf{E}_{\parallel}$ excites oscillations of free charges in the plasma, while $\mathbf{E}_{\perp}$ cannot, since the charges are not free to move perpendicular to $\mathbf{B}_{conv}$. This leads to a delay between $\mathbf{E}_{\parallel}$ and $\mathbf{E}_{\perp}$, manifest as the introduction of a small amount of CP; the sign of the CP depends on the relative phase of $\mathbf{E}_{\parallel}$ and $\mathbf{E}_{\perp}$.

Faraday conversion is a direct analog of Faraday rotation, but unlike Faraday rotation, Faraday conversion does not depend on the sign of the free charges involved. If the polarization **E** vector is entirely parallel to the ambient magnetic field in the conversion region, it will be entirely absorbed and re-emitted, so that no delay, and hence no circular polarization, will be generated. Similarly, if **E** is entirely orthogonal to the ambient magnetic field, it cannot be absorbed at all, so that again, no circular polarization will be generated.

5.5 What is Actually Observed?

5.5.1 Intensity Structures and Kinematics

Core-dominated AGN almost universally display one-sided core–jet structures on VLBI scales, consistent with a picture in which we are detecting only the Doppler-boosted approaching jet. Apparent superluminal motions are also extremely common, again consistent with relativistic jets oriented at relatively small angles to the line of sight.

Although there is considerable overlap between the two distributions, it has long been known that different types of AGN display somewhat different typical superluminal speeds. On average, lower apparent speeds are observed in BL Lac objects, characterized by weak optical line emission, than in quasars, characterized by stronger optical line emission (Gabuzda et al. 1994, 2000; Kellermann et al. 2004). This systematic difference could in principle be associated with a difference in either the intrinsic flow speeds in the jet or the angles made by the jets to the line of sight. If due primarily to a difference in the characteristic viewing angles of the jets, the lower speeds observed in BL Lac objects would indicate that their jets are typically viewed at angles θ that are further from the angle $1/\Gamma$ that maximises the observed apparent speed than is the case for quasars, where Γ is the bulk Lorentz factor of the jet. BL Lac objects would be observed to have relatively low superluminal speeds if θ were either appreciably *larger* or appreciably *smaller* than $1/\Gamma$. The strong variability of BL Lac objects argues against the idea that their jets are typically viewed at relatively large angles to the line of sight. Further, Gabuzda et al. (2000) pointed out that the lack of evidence that BL Lac objects are typically more beamed than quasars argues against a scenario in which the jets of BL Lac objects are viewed at angles appreciably smaller than the viewing angles of quasar jets. Accordingly, Gabuzda et al. (2000) conclude that the lower apparent superluminal speeds of BL Lac objects most likely reflect lower average bulk Lorentz factors for their jets.

VLBI jets are frequently bent, although the origin of the observed bending is usually not clear (e.g. Kellermann et al. 1998b; Zensus et al. 2002). It must also be borne in mind that relatively small intrinsic bends can appear like much larger bends when projected onto the plane of the sky. Apparent bending need not imply non-ballistic motion of individual parcels of the jet plasma. For example, changes in the direction of ejection of the "nozzle" at the base of the jet could give rise to apparent bends in the jet, as in the well known stellar system SS433, whose jets are precessing (e.g. Blundell and Bowler 2004). Periodic outbursts or evidence for periodicity in changes in the ejection position angles have been reported for a number of AGN, which has been interpreted as evidence for precession (Stirling et al. 2003; Caproni and Abraham 2004a,b; Caproni et al. 2013; Klare et al. 2005; Qian et al. 2009).

Accelerations and non-radial motions of individual components have also been observed (e.g. Homan et al. 2003; Jorstad et al. 2004; Bloom et al. 2013). These can sometimes be difficult to interpret unambiguously, since we are observing the motions projected onto the plane of the sky, but they seem to indicate changes in both the speed and direction of motion of individual features.

Another behaviour that has recently been identified is "jet swinging" – a difficult to interpret phenomenon in which correlated non-radial component motions are observed (Agudo et al. 2007). This corresponds to a superluminal change in the overall position angle of the jet, as if the jet were acting like a rigid structure, possibly indicating an important role of the magnetic field in the jet dynamics.

Stationary or nearly stationary features that have been interpreted as standing shocks have also been observed in a number of sources. These are sometimes close to the core (e.g. BL Lac: Cohen et al. 2014) and sometimes at an appreciable distance from the core (e.g. 1803+784: Britzen et al. 2005). Cohen et al. (2014) have suggested that a quasi-stationary feature observed at 15 GHz only 0.26 mas from the core is a recollimation shock, and that components emanating from this feature represent compressions established by slow and fast mode magneto-acoustic MHD waves propagating in a region of toroidal **B** field.

A problem that is continually encountered in the interpretation of multi-epoch VLBI images is disentangling the intrinsic speed of the flow from the viewing angle, as in Eq. (5.3) for the observed superluminal motions. One approach is to estimate the Doppler factor (Eq. (5.6)) based on observations of source variability and use this together with observed superluminal speeds to derive estimates of the jet speeds and jet viewing angles (Hovatta et al. 2009). Fitting of transverse polarization profiles using a helical-field model, as has been carried out by Murphy et al. (2013), can be used to estimate the viewing angle of the jet in the jet rest frame, which can then be used with measured superluminal speeds and the aberration formulas (5.1) to derive estimates of the intrinsic jet speeds and jet viewing angles in the observer's frame. Such an analysis for Mrk501 yielded the intrinsic jet speed $\beta = 0.96$ and viewing angle in the observer's frame $\theta = 15°$, consistent with but more constraining than the earlier results of Giroletti et al. (2004) [$\beta \geq 0.88$, $\theta \leq 27°$]. Transverse profile fitting for 3C 273 (Gabuzda et al., in prep) yielded $\Gamma = 13.6$ and $\theta = 3.8°$, in excellent agreement with the estimates of Hovatta et al. (2009) based on their variability analysis, $\Gamma = 14$ and $\theta = 4°$. Cross-checking estimates obtained using these two approaches for additional AGNs can potentially provide important verification of the reliability of such estimates.

5.5.2 *Magnetic-Field Structures*

The observed degrees of polarization in the jets of AGN range from a few percent up to 50–60 %, fully consistent with the jet emission being incoherent synchrotron radiation in a partially ordered **B** field. In most cases, the observed polarization vectors – and therefore the associated jet **B** fields – in the VLBI jets of AGN tend to be either parallel or perpendicular to the local jet direction. This should come about naturally if the jets locally exhibit approximately cylindrical symmetry: the **B** field can always be separated into longitudinal and transverse components projected onto the sky (e.g. Lyutikov et al. 2005).

Various polarization patterns are fairly commonly, illustrated schematically in Fig. 5.4, which shows an intensity image taken from the website of the MOJAVE project (see the Acknowledgements) with artificial polarization sticks superimposed (recall that the observed polarization is orthogonal to the underlying **B** field in optically thin regions, such as the jet). Possible origins for each of these patterns are considered below.

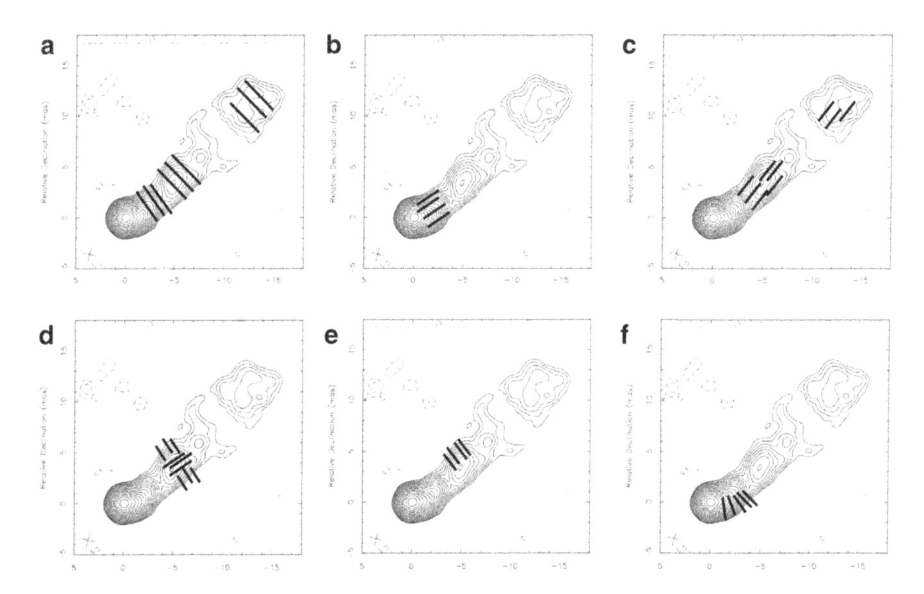

Fig. 5.4 Schematics of various polarization patterns commonly observed in AGN jets, corresponding to: (**a**) extended regions of longitudinal field; (**b**) regions of orthogonal field associated with bright, compact regions; (**c**) extended regions of orthogonal field; (**d**) spine–sheath transverse **B**-field structure; (**e**) longitudinal field offset toward one side of the jet; (**f**) longitudinal field around a bend in the jet. Recall that the **B** field is orthogonal to the observed polarization in optically thin regions. The contour map was taken from the MOJAVE website (see the Acknowledgements), and the sticks added to illustrate the various polarization configurations

Extended regions with longitudinal B field, as in Fig. 5.4a, could come about due to shear with the surrounding medium stretching out the magnetic field along the flow and thereby enhancing the longitudinal component. Alternatively, such regions could be associated with a helical **B** field with a comparatively low pitch angle (the angle between the **B** field lines and the jet axis); i.e., a "loosely wound" helical field.

Compact features with orthogonal B field, as in Fig. 5.4b, are candidates for regions of transverse shocks (Laing 1980; Hughes et al. 1985). In a region subject to shock compression, the **B** field is amplified in the plane of compression, giving rise to appreciable degrees of polarization and a net magnetic field lying in the plane of compression, even in the case of an initially completely tangled magnetic field.

Extended regions of orthogonal B field, as in Fig. 5.4c, could in principle be due to a series of transverse shocks, but this picture seems somewhat contrived. It is probably more likely that such regions are revealing the presence of a toroidal field or high-pitch-angle ("tightly wound") helical field.

Spine–sheath polarization structure implying an orthogonal **B** field near the jet ridgeline and longitudinal **B** field near one or both edges of the jet (Fig. 5.4d) could reflect the joint action of shock compression and a shear interaction with the surrounding medium (as proposed, e.g. by Attridge et al. 1999). However, this type of **B**-field structure could also be associated with a helical field in which the azimuthal component dominates near the center of the jet and the longitudinal component dominates near the jet edges, projected onto the sky (e.g. Pushkarev et al. 2005; Lyutikov et al. 2005). In the latter case, this should also give rise to an increase in the degree of polarization toward the jet edges.

Longitudinal B field offset from the jet ridgeline (Fig. 5.4e) could be associated with either a shear interaction occurring on one side of the jet or a helical **B** field. Some combinations of pitch angle and viewing angle should give rise to projected helical **B**-field configurations with longitudinal field on one side of the jet and transverse field on the other, which may be observed simply as an offset longitudinal field if the transverse field has much weaker polarization (e.g. Murphy et al. 2013).

Longitudinal B field along a bend in the jet (Fig. 5.4f) could be due to a shear interaction with the surrounding medium, or alternately to an enhancement ("stretching out") of the longitudinal component of the field due to the bending of the jet, sometimes referred to a "curvature-induced polarization".

More rarely, the observed polarization does not show an obvious relationship to the jet direction, even when the polarization angles have been corrected for Faraday rotation or the observations are at a high enough frequency that Faraday rotation is unlikely to be appreciable (examples can be found among the sources monitored by the MOJAVE project, e.g., Lister and Homan 2005). The origin of such "offset" polarization angles (**B** fields) is not clear, although one possibility may be oblique shocks or inaccurate knowledge of the local jet direction due to limited resolution. Intriguing fan-like structures have also been observed, which may represent conical shocks (e.g. Papageorgiou et al. 2006).

As is clear from the list above, the origin of the implied **B**-field structure is usually not clear from the observed polarization-angle distribution alone, and additional information is usually required as the basis for interpretation, such as the distribution of the degree of polarization, information about Faraday rotation occurring in the vicinity of the AGN etc.

5.5.3 Core Polarization

The core components are typically polarized from less than 1 % to a few percent, consistent with the emission being incoherent synchrotron radiation in a partially ordered **B** field; the spectral indices at centimeter wavelengths are typically fairly flat, indicating that this region is at least partially optically thick. In most cases, the observed polarization vectors at relatively short centimeter wavelengths tend to be either parallel or perpendicular to the direction of the inner jet, although offsets due to appreciable Faraday rotation can sometimes affect the observed core polarization directions at longer centimeter wavelengths. Whether these imply **B** fields that are aligned with or orthogonal to the jet direction depends on whether the net core polarization is predominantly from regions that are optically thin (χ perpendicular to **B**) or optically thick (χ parallel to **B**).

There is good evidence that the observed core polarizations are associated with predominantly optically thin regions in many cases. The observed core polarizations are typically a few percent, and can reach as high as 10 % (e.g. Lister and Homan 2005); this would require fairly highly ordered **B** fields in the core region if this was associated with optically thick regions capable of generating a maximum degree of polarization of 10–12 %. The orientation of the core and inner jet polarization angles is often the same, consistent with a picture in which they are both associated with predominantly optically thin regions with the same **B** field configuration on different scales (e.g. Lister and Homan 2005). In addition, the Faraday-corrected core polarization angles of a number of BL Lac objects were found to be well aligned with nearly simultaneously measured optical polarization angles, suggesting roughly co-spatial optically thin emission regions in both wavebands (Gabuzda et al. 2006).

It is an interesting question how to relate the observed VLBI cores to the theoretical concept of the core as the $\tau = 1$ surface in a Blandford–Königl jet (Blandford and Königl 1979). As is argued in the previous paragraph, there is considerable evidence that the core polarization is associated predominantly with optically thin emission. The core spectral indices α ($S_\nu \propto \nu^{+\alpha}$) are usually fairly flat, between -0.5 and $+0.5$, (e.g. Hovatta et al. 2014), with the vast majority of the most positive spectral indices extending up to $+1$, consistent with the idea that the observed cores include a mixture of optically thick and optically thin regions. Gabuzda and Gómez (2001) have reported a rotation of the observed core polarization angle in OJ287 by roughly 90° between 5 and 15 GHz, which they interpreted as an optically thick–thin transition; i.e., the observed core polarization was emitted by a region that was predominantly optically thin at 15 GHz but predominantly optically thick at 5 GHz. All this suggests that what we observe as the VLBI core represents a region that is only partially optically thick, corresponding to a mixture of the optically thick base of the jet (as in the Blandford–Königl picture) and optically thin regions in the innermost jet, with the latter often dominating the observed core polarization.

An alternate view of the VLBI core is that it represents a standing or recollimation shock. Gabuzda et al. (2013) are able to describe the fan-like polarization observed in the VLBI core of 1803+784 as a conical recollimation shock. Marscher (2014) (see also references therein) has recently developed a model for the variability of the flux and polarization of compact AGN in which turbulent plasma flowing at a relativistic speed down a jet crosses a standing conical shock, which we observe as the VLBI core.

5.5.4 Core Shifts and Core Magnetic Fields

The first study aimed at using the frequency dependence of AGN VLBI-core positions to estimate the core B-field strengths were carried out by Lobanov (1998). Based on images at only three frequencies, this analysis showed that the position of the VLBI core seemed to show the expected sort of frequency dependence, but only limited conclusions could be drawn, due to the limited number of frequencies and sources considered. Nevertheless, this work was important, as it laid out an approach to estimating the core B fields based on measurement of the core position as a function of frequency.

After an interval of more than a decade, there has recently been an explosion of new core-shift studies, most notably those of Kovalev et al. (2008), O'Sullivan and Gabuzda (2009), Sokolovsky et al. (2011) and Pushkarev et al. (2012). O'Sullivan and Gabuzda (2009) were the first to carry out a core-shift analysis for VLBI observations obtained simultaneously at more than only two to three frequencies. Their data based on eight frequencies from 4.6 to 43 GHz provided detailed and redundant measurements, which clearly demonstrated the expected power-law dependence of the core position as a function of frequency. These measurements also provided evidence that the parameter $k_r \simeq 1$ in most of the six AGN considered, where the position of the VLBI core r as a function of frequency ν is given by $r \propto \nu^{-1/k_r}$; a reasonable interpretation of the result that $k_r \simeq 1$ is that these core regions are close to equipartition, with the magnetic field $B \sim r^{-1}$ and the electron density $N \sim r^{-2}$. This is potentially a very important result, since equipartition has often assumed for various calculations, but without any observational evidence that this assumption is reasonable. Kovalev et al. (2008) were the first to derive core-shift measurements for a much larger number of sources (29), but at only two frequencies, while Sokolovsky et al. (2011) obtained results for 20 sources at nine frequencies. Most recently, Pushkarev et al. (2012) obtained core-shift data for more than a hundred well-studied sources from the MOJAVE project (web page) at four frequencies from 8.1 to 15.4 GHz. The core-region B fields inferred from these studies are generally tenths of a Gauss; Pushkarev et al. (2012) found evidence that the core B fields of quasars are somewhat higher than those in BL Lac objects, which is presumably related to the higher luminosities of quasars in some way.

5.5.5 Circular Polarization

Techniques for deriving circular-polarization (CP) information on parsec scales were pioneered by Homan and his collaborators in the late 1990s (Homan and Wardle 1999; Homan et al. 2001) using data taken on the VLBA. CP measurements for the first epoch of the MOJAVE project (monitoring of 133 AGN at 15 GHz with the VLBA) were published by Homan and Lister (2006). Circular polarization was detected in 34 of these objects at the 2σ level or higher. These results confirmed previously noted trends: the circular polarization is nearly always coincident with the VLBI core, with typical degrees of polarization m_c being a few tenths of a percent. Homan and Lister (2006) found no evidence for any correlation between m_c and any of 20 different optical, radio and intrinsic parameters of the AGN. Interestingly, five of the 34 AGN displayed CP in their *jets*, well outside the VLBI-core region, suggesting that the mechanism generating the circular polarization is capable of operating effectively in optically thin regions.

The two main mechanisms that are usually considered to be the most likely generators of the observed CP are the synchrotron mechanism and the Faraday conversion of linear to circular polarization, as described in Sect. 5.3.3. Although the intrinsic CP generated by synchrotron radiation may be able to reach a few tenths of a percent at 15 GHz for the **B**-field strengths characteristic of the observed VLBI cores of AGN (Sect. 5.5.3), the highest observed m_c values seem too high to plausibly be attributed to this mechanism. This suggests that Faraday conversion plays a significant, and possibly dominant role, since it is expected to be more efficient at generating CP than the synchrotron mechanism for the conditions in radio cores (Jones and O'Dell 1977).

Few multi-frequency CP measurements have been published, and relatively little is currently known about the frequency dependence of the observed CP. The 15+22+43-GHz results of Vitrishchak et al. (2008) showed excellent agreement with the earlier 15-GHz measurements of Homan and Lister (2006). They did not show any universal frequency dependence for m_c, with both rising and falling spectra with frequency being observed; similarly, Homan and Wardle (2004) found very different spectral indices for the degree of CP, α_c ($|m_c| \propto \nu^{\alpha_c}$), for three CP components in the inner jet of the nearby ($z = 0.017$) radio galaxy 3C84: $\alpha_c = -0.9$ for a component with positive CP in a predominantly optically thin region and $\alpha_c = +1.4$ and $+1.7$ for two components with negative CP in partially optically thick regions. Vitrishchak et al. (2008) found m_c to be higher at 43 GHz than at their lower two frequencies. These results seem contrary to expectations, since the degree of circular polarization from both synchrotron radiation and Faraday conversion should decrease towards higher frequencies if the source is homogeneous. Vitrishchak et al. (2008) also found evidence in several objects that the detected circular polarization was near, but not coincident with the core, as well as several cases of changes in sign with frequency.

The collected VLBI-scale circular-polarization measurements indicate that the sign of the circular polarization at a given observing frequency is generally consistent across epochs separated by several years or more, suggesting stability of the magnetic field orientation in the innermost jets. Overall, it seems likely that the observed CP is predominantly generated by Faraday conversion, and that the observed values of α_c are determined by effects associated with the intrinsic inhomogeneity of the jets, as well as the possible presence of several regions of CP of either one or both signs contributing to the observed "core" CP (see the discussion in Vitrishchak et al. 2008 and Wardle and Homan 2003).

5.5.6 *Parsec-Scale Faraday Rotation Measurements*

Core vs. jet rotation measures A number of early Faraday-rotation studies of AGN on parsec scales were carried out by Taylor (1998, 2000) and Zavala and Taylor (2003, 2004), who were primarily interested in looking for differences between the core and jet RM values. These Faraday-rotation maps for about 40 AGN showed a clear tendency for the core RMs to be greater in magnitude than the jet RMs, apparently due to higher electron densities and magnetic-field strengths on smaller scales in the jets. These studies also presented evidence that the core RMs of quasars tend to be higher than those of BL Lac objects; it is an obvious possibility that this is related to the stronger optical line emission of quasars, compared to BL Lac objects.

Faraday rotation as a probe of the azimuthal B field As was pointed out by Blandford (1993), if a jet has a helical **B** field, we should observe a Faraday-rotation gradient across the jet, due to the systematically changing line-of-sight component of the **B** field across the jet. Thus, the detection of transverse Faraday-rotation gradients across AGN jets could potentially provide a powerful diagnostic for the presence of an azimuthal field component associated with a helical or toroidal jet **B** field. The theoretical simulations of Broderick and McKinney (2010) directly demonstrate the development of helical jet **B** fields and the resulting presence of Faraday rotation gradients across the simulated jets.

A number of transverse Faraday-rotation gradients across AGN jets have been reported in the literature, most recently by Asada et al. (2010), Croke et al. (2010), Gómez et al. (2011), Hovatta et al. (2012), Mahmud et al. (2013) and Gabuzda et al. (2013, 2014); an example is shown in Fig. 5.5. Gómez et al. (2011) superposed

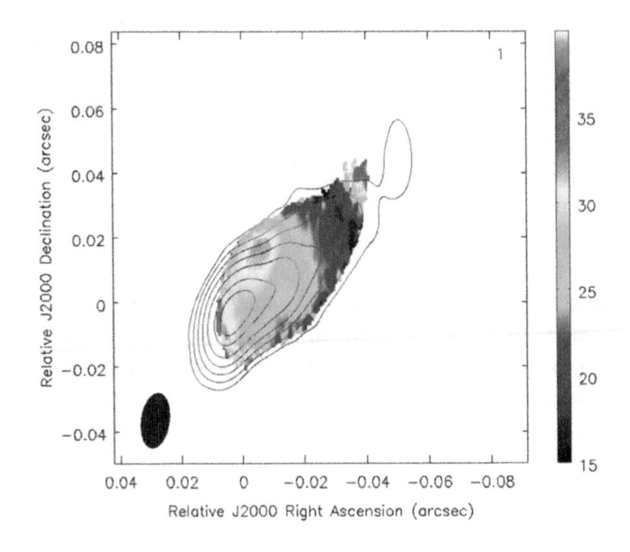

Fig. 5.5 Transverse Faraday RM gradient across the VLBI jet of 3C380 (Gabuzda et al. 2013)

degree of polarization and RM images for 3C120 obtained at multiple epochs to effectively map out these quantities along a much longer portion of the jet than was possible at any single epoch, making the transverse structure in the polarization and Faraday rotation appreciably clearer.

Of course, it is also possible to imagine situations when a gradient in the Faraday rotation roughly across a VLBI jet occurs, not due to a systematic variation in the line-of-sight **B**-field component, but due to a gradient in the density of thermal electrons in the region surrounding the jet. This could be the case, for example, if the jet were propagating through a non-uniform medium that was denser on one side of the jet than on the other. In this regard, it is important to point out a crucial discriminator: while transverse RM gradients with RM values of a single sign that increase toward one edge of the jet could potentially be caused by either a helical jet **B** field or a density gradient in the surrounding medium, transverse RM gradients that display one sign at one edge of the jet, pass through zero and then display the other sign at the other edge of the jet are a clear and unambiguous sign of a helical jet **B** field. It is therefore highly significant that transverse RM gradients displaying *both* signs have been observed for a number of AGN, after subtracting the effect of the RM arising in our Galaxy; in such cases, the only plausible explanation for the observed transverse RM gradients is that the corresponding jets have helical **B** fields.

As Sikora et al. (2005) have pointed out, the lack of deviations from a λ^2 wavelength dependence for the observed polarization angles in some case indicates that the Faraday rotation must be external, suggesting that the associated helical **B** field may surround the jet, but not necessarily fill the jet volume; alternatively, it may be that the helical field fills the jet volume, but the thermal electrons required for the Faraday rotation do not.

Reliability of observed Faraday-rotation gradients Thus, a number of studies carried out in different groups around the world have reported detections of transverse Faraday-rotation gradients across AGN jets on parsec scales, interpreted as evidence for helical jet **B** fields. Further, the theoretical simulations of the propagation of jets from a rotating black hole/accretion disk system carried out by Broderick and McKinney (2010) clearly showed the presence of transverse RM gradients, including across the VLBI core regions, and also showed that these could sometimes remain visible even when convolved with a beam that was much larger than the intrinsic jet width (e.g. a 0.9-mas beam and an intrinsic jet width at its base of about 0.05 mas; Fig. 8 in Broderick and McKinney 2010).

At the same time, it seems somewhat counter-intuitive that it could be possible to detect transverse polarization and Faraday rotation structure across poorly resolved jets, whose intrinsic widths may be much narrower than the beam. This led Taylor and Zavala (2010) to suggest that an observed Faraday-rotation gradients should span at least three "resolution elements" (usually taken to mean three beamwidths) across the jet in order to be considered reliable, although they did not present any demonstration that this was the case. In addition, there had been no clear agreement in early analyses of transverse Faraday-rotation gradients about the most correct way to estimate the uncertainties on the RM values at specified locations in an RM image. Thus, there was a need to clarify whether there was a minimum resolution required to reliably detect Faraday-rotation structure, and identify a suitable approach for estimating uncertainties in quantities at individual locations in VLBI images.

A first important step was taken by Hovatta et al. (2012), who carried out Monte Carlo simulations based on realistic "snapshot" baseline coverage for VLBA observations at 7.9, 8.4, 12.9 and 15.4 GHz aimed at investigating the statistical occurrence of spurious RM gradients across jets with intrinsically constant polarization profiles. Inspection of the right-hand panel of Fig. 30 of Hovatta et al. (2012) shows that the fraction of spurious 3σ gradients was no more than about 1 %, even for the smallest observed RM-gradient widths they considered, about 1.4 beamwidths. Relatively few 2σ gradients were also found, although this number reached about 7 % for observed jet widths of about 1.5 beamwidths; nevertheless, Hovatta et al. (2012) point out that 2σ gradients are potentially also of interest if confirmed over two or more epochs. The results of Hovatta et al. (2012) have also now been confirmed by similar Monte Carlo simulations carried out by Algaba (2013) for simulated data at 12, 15 and 22 GHz and by Murphy and Gabuzda (2013) for simulated data at 1.38, 1.43, 1.49 and 1.67 GHz and at the same frequencies as those considered by Hovatta et al. (2012). The 1.38–1.67 GHz frequency range considered by Murphy and Gabuzda (2013) yielded a negligible number of spurious 3σ gradients and fewer than 1 % spurious 2σ gradients, even for observed jet widths of only 1 beamwidth (i.e., for poorly resolved jets).

Two more sets of Monte Carlo simulations based on realistic snapshot VLBA baseline coverage adopted a complementary approach: instead of considering the occurrence of spurious RM gradients across jets with constant polarization, they considered simulated jets with various widths and with transverse RM gradients of various strengths, convolved with various size beams. Mahmud et al. (2013) carried out such simulations for 4.6, 5.0, 7.9, 8.4, 12.9 and 15.4 GHz VLBA data, and Murphy and Gabuzda (2013) for 1.38, 1.43, 1.49 and 1.67 GHz VLBA data. These simulations clearly showed that, with realistic noise and baseline coverage, the simulated RM gradients could remain clearly visible, even when the jet width was as small as 1/20 of a beam width.

All these simulations clearly demonstrate that the width spanned by an RM gradient is not a crucial criterion for its reliability.

Another key outcome of the Monte Carlo simulations of Hovatta et al. (2012) is an empirical formula that can be used to estimate the uncertainties in intensity (Stokes I, Q or U) images, including the uncertainty due to residual instrumental polarization ("D-terms") that has been incompletely removed from the visibility data. In regions of source emission where the contribution of the residual instrumental polarization is negligible, the typical uncertainties in individual pixels are approximately 1.8 times the rms deviations of the flux about its mean value far from regions of source emission. This is roughly a factor of two greater than the uncertainties that have usually been assigned to the intensities measured in individual pixels in past studies, indicating that the past uncertainties have been somewhat underestimated. These results have recently been confirmed by the Monte Carlo simulations of Coughlan (2014), which likewise indicate that the typical intensity uncertainties in individual pixels are of order twice the off-source rms, with significant pixel-to-pixel variations in the uncertainties appearing in the case of well resolved sources.

The reliability of transverse RM gradients observed across the partially optically thick core regions remains an open question, and further simulations are required to investigate this in more detail. However, since (as is argued in Sect. 5.5.3) what we observe as the "VLBI core" with beamwidths of the order of a milliarcsecond is actually a blend of the optically thick base of the jet and unresolved optically thin regions in the innermost jet, it is likely that the polarization of the observed core components is dominated by optically thin emission in many cases. In addition, the theoretical simulations of Broderick and McKinney (2010) show smooth, monotonic RM gradients across the VLBI core regions when they are convolved with a 0.9-mas beam. This all suggests that transverse RM gradients across core regions may often be reliable, but further simulations are required to test this more directly.

Evidence for the return B-field Mahmud et al. (2009, 2013) and Gabuzda et al. (2014) have reported reversals in the directions of observed transverse Faraday rotation gradients, either with distance along the jet (four AGNs) or with time (two AGNs).

These results at first seem puzzling, since the direction of a transverse RM gradient associated with a helical **B** field is essentially determined by the direction of rotation of the central black hole and accretion disk and the initial direction of the poloidal field component that is "wound up" by the rotation. It is difficult in this simplest picture to imagine how the direction of the resulting azimuthal field component could change with distance along the jet or with time. However, in a picture with a nested helical field structure, similar to the "magnetic tower" model of Lynden–Bell (1996), but with the direction of the azimuthal field component being different in the inner and outer regions of helical field, such a change in the direction of the net observed RM gradient could be due to a change in dominance from the inner to the and outer region of helical field in terms of their overall contribution to the Faraday rotation (Mahmud et al. 2013).

Thus, the detection of such reversals in the directions of transverse RM gradients may provide the first observational evidence for the presence of the return **B** field.

One theoretical picture that predicts a nested helical-field structure with oppositely directed azimuthal field components in the inner and outer regions of helical field is the "Poynting–Robertson cosmic battery" model of Contopoulos et al. (2009) discussed in the following subsection, although other plausible systems of currents may also give rise to similar field configurations.

Evidence for a Cosmic Battery? Any transverse Faraday-rotation gradient can be described as being directed either clockwise (CW) or counter-clockwise (CCW) on the sky, relative to the base of the jet. Contopoulos et al. (2009) reported evidence for a significant excess of CW transverse Faraday-rotation gradients for parsec-scale AGN jets, based on transverse Faraday-rotation gradients identified in maps from the literature. This is an extremely counterintuitive result, since the direction of the helical field threading an AGN jet should essentially be determined by the direction of rotation of the central black hole and accretion disc, together with the direction of the poloidal "seed" field that is wound up, and our instincts tell us that both of these should be random. A re-analysis of the transverse Faraday-rotation gradients considered by Contopoulos et al. (2009) to verify their reliability applying the new error-estimation procedure of Hovatta et al. (2012) is currently underway (Gabuzda et al. 2014), but it appears that the reported asymmetry in the collected data will probably be confirmed.

Contopoulos et al. (2009) suggest that this non-intuitive result can be explained in a straightforward way via the action of a mechanism they call the "Poynting–Robertson cosmic battery." The essence of this mechanism is the Poynting–Robertson drag experienced by charges in the accretion disc, which absorb energy emitted by the active nucleus and re-radiate this energy isotropically in their own rest frames. Because these charges are rotating with the accretion disc, this radiation will be beamed in the forward direction of their motion, i.e., in the direction of the disc rotation. Due to conservation of momentum, the charges then feel a reaction force opposite to the direction of their motion; since the magnitude of this force exhibits an inverse dependence on the mass of the charge, this leads to a difference in the deceleration experienced by the protons and electrons in the disc. Since the electrons are decelerated more strongly, this leads to a net current in the disc, in the direction of rotation. This current, in turn, gives rise to a net poloidal magnetic field whose direction is coupled to the direction of the current in the disc, i.e., to the direction of the disc rotation. This coupling of the disc rotation and the direction of the poloidal field that is "wound up" breaks the symmetry in the direction of the toroidal field component, and predicts that the observed Faraday-rotation gradients should be predominantly CW on the sky, independent of the direction of the disc rotation as seen by the observer. In a "nested helical field" type picture such as that described in the previous subsection, the inner/outer regions of helical field should give rise to CW/CCW Faraday-rotation gradients; therefore, the observed excess of CW implies that the inner region of helical **B** field dominates on parsec scales.

If this excess of CW Faraday-rotation gradients is indeed confirmed by further studies, this will have cardinal implications for our understanding of AGN jets. If the Poynting–Robertson battery does not operate sufficiently efficiently to provide the observed excess of CW gradients, then some other mechanism generating a suitable system of **B** fields and currents must be identified.

5.5.7 *Helical Jet Magnetic Fields: Basis for a New Paradigm?*

From a theoretical point of view, it would be very natural if many, or possibly even all, AGN jets are associated with helical **B** fields, produced essentially by the "winding up" of an initial "seed" field threading the accretion disc by the combination of the rotation of the central black hole and accretion disc and the jet outflow (e.g. Nakamura et al. 2001; Meier et al. 2001; Lovelace et al. 2002; Lynden–Bell 2003; Tsinganos and Bogovalov 2002). The possibility that toroidal field components develop in association with currents flowing in the jet has also been considered (Pariev et al. 2003; Lyutikov 2003). Indeed, these two ideas are not unrelated: whether or not it is related to a helical **B** field, if there is a dominant toroidal **B**-field component, basic physics tells us that there must be currents flowing in the region enclosed by the field.

This very plausible theoretical picture provides strong motivation to consider interpretations of observed properties of AGN jets as possible consequences of helical/toroidal **B**-field structures. The role of transverse Faraday rotation gradients in revealing the presence of a toroidal or helical **B** field has already been discussed; several other pieces of observational evidence supporting the idea that many AGN jets carry helical **B** fields are considered below.

B-field configurations In general, a helical **B** field can be described as a superposition of a toroidal (azimuthal) and a longitudinal component. The larger the pitch angle (angle between the helix axis (jet axis) and **B** vector), i.e. the more tightly wound the helix, the stronger the toroidal relative to the longitudinal component. Depending on the combination of the pitch angle and viewing angle, an overall helical jet **B** field can give rise to an observed field structure that is everywhere orthogonal, everywhere longitudinal, orthogonal and longitudinal on either side of the jet, or spine–sheath-like, with a region of orthogonal field near the jet axis flanked on one or both sides with regions of longitudinal field. Thus, many of the polarization patterns observed for AGN jets can be understood as consequences of helical jet **B** fields with various combinations of helical pitch angles and viewing angles.

Transverse B-field structure The best case for polarization configurations best explained by helical jet **B** fields are asymmetric transverse structures, such as spine–sheath polarization structures or longitudinal polarization that is not centered on the jet ridgeline. When an AGN also displays a transverse Faraday-rotation gradient, consistency between the properties of this gradient and the observed transverse polarization structure can be checked. Here, we must recall that the degree of polarization is determined by the component of the jet **B** field in the plane of the sky (e.g., a **B** field directed precisely toward the observer would give no linear polarization). As is shown by the helix in Fig. 5.6, the poloidal component of a helical field projected onto the sky makes the overall field lie predominantly along

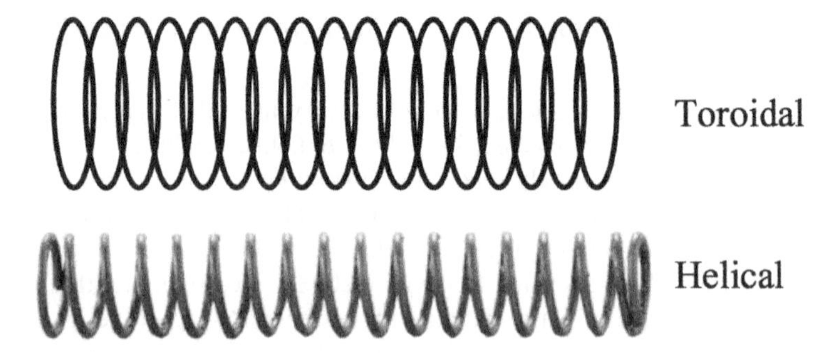

Fig. 5.6 Schematic illustrating how a toroidal jet **B**-field geometry should give rise to symmetrical polarization structure across the jet, while a helical jet **B**-field geometry should give rise to an asymmetrical transverse polarization structure

the line of sight at one edge of the helix, giving rise to higher Faraday rotation and lower degrees of polarization (top edge of the helix shown), and predominantly in the plane of the sky at the other edge of the helix, giving rise to lower Faraday rotation and higher degrees of polarization (bottom edge of the helix shown).

Murphy et al. (2013) have recently carried out fitting of the transverse total intensity and polarization profile across the jet of Mrk501 on parsec scales, which display appreciable asymmetry. The profiles were fit well using a simple helical-field model with a pitch angle of about 53° and a viewing angle in the jet rest frame of about 83° (recall that this corresponds to a much smaller viewing angle in the observer's frame due to aberration). In addition, the parsec-scale jet of Mrk501 exhibits a transverse RM gradient (Gabuzda et al. 2004; Croke et al. 2010), with the larger-magnitude RM values present on the side of the jet with a lower degree of polarization, as is expected for a helical jet B field. Although such results are available for only one or two sources, this approach could potentially be quite valuable when applied to a larger number of AGN of different types. Profile fitting can also be used to investigate change in the pitch angle and viewing angle of the helical field along the jet.

Circular polarization As is indicated above, it is generally agreed that the most plausible mechanism for the generation of the circular polarization observed on VLBI scales in AGN is the Faraday conversion of linear-to-circular polarization during the propagation of a linearly polarized electromagnetic wave through a magnetized plasma (Jones and O'Dell 1977; Jones 1988). In order for Faraday conversion to operate, the **B** field in the conversion region $\mathbf{B}_{conv}$ must have a non-zero component parallel to the plane of linear polarization. For this reason, as has been pointed out by Wardle and Homan (2001), a helical **B** field provides an interesting example of an overall ordered **B** field geometry that can potentially facilitate linear-to-circular conversion: synchrotron radiation emitted at the "far" side of the helical field relative to the observer can undergo conversion in the "near" side of the helical field. This raises the question of whether the CP detected in

AGN on VLBI scales could be generated in helical **B** fields associated with the corresponding AGN jets. Inspection of the MOJAVE sources in which CP was detected (Homan and Lister 2006) reveals that a number of them show various properties that can be associated with the presence of helical jet B fields, such as extended regions of transverse **B** field, spine+sheath jet **B**-field structures and transverse RM gradients, providing indirect evidence that at least some of the detected CP may be associated with helical jet **B** fields. It is also intriguing that extended regions of CP clearly in the VLBI jet, well away from the optically thick core region, have now been detected in several cases (Homan and Lister 2006; Vitrishchak and Gabuzda 2007) – linear to circular conversion in helical **B** fields associated with these jets could explain this result in a natural way.

"Inverse depolarization" A surprising result to come out of an analysis of multi-wavelength data for the MOJAVE AGN sample was the identification of a number of AGN with components displaying degrees of polarization that increase with increasing wavelength (Hovatta et al. 2012). This is counter to the usual expectation that depolarization should correspond to decreasing polarization with increasing wavelength, and has accordingly been referred to as "inverse depolarization." Homan (2012) has developed a model that can explain this unexpected behaviour as an effect of internal Faraday rotation occurring in a jet whose **B** field displays structural inhomogeneities, such as are naturally produced in helical jet **B** fields.

Superluminal speeds trends It has independently been suggested by Gabuzda et al. (2000) and Asada et al. (2002) that the observed difference in the typical superluminal speeds of BL Lac objects and quasars referred to in Sect. 5.5.1 could be related to another systematic difference in the VLBI properties of these two types of AGN: the VLBI jets of BL Lac objects/quasars most often display predominantly **B** fields that are orthogonal to/aligned with the jet direction. Let us suppose that both quasars and BL Lac objects characteristically have helical jet **B** fields that come about due to the combination of rotation of the central black hole/accretion disc and the jet outflow. If the intrinsic outflow speeds of quasars are, on average, higher than those of BL Lac objects, this might mean that the ratios of their outflow speeds to their central rotational speeds are also higher, giving rise to helical jet **B** fields that are less tightly wound. On the other hand, lower outflow speeds in BL Lac objects might lead to the helical **B** fields associated with their jets being more tightly wound. Thus, it may be possible to understand the observed systematic differences in the VLBI properties of BL Lac objects and quasars within a single scenario in which a tendency for lower outflow speeds in BL Lac objects compared to quasars leads to a tendency for more tightly wound helical jet **B** fields in BL Lac objects, which are manifest as a predominance for jet **B** fields orthogonal to the jet direction in these sources.

Can we distinguish between helical and toroidal fields? The detection of a transverse Faraday-rotation gradient across an AGN jet can reflect the presence of an azimuthal field component, but this component could be associated with either a helical or a toroidal B field. Generally speaking, it is only asymmetry of the intensity and polarization profiles across the jet that can distinguish observationally between a helical field (with an ordered poloidal field component) and a toroidal field (with a disordered poloidal field component). Here, we must recall that the degree of polarization is determined by the component of the jet magnetic field in the plane of the sky (e.g., a magnetic field directed precisely would the observer would give no linear polarization). Figure 5.6 shows qualitatively an example in which the poloidal field component of a helical field makes the overall field predominantly along the line of sight at one edge of the helix (giving rise to higher Faraday rotation and lower degrees of polarization), and predominantly in the plane of the sky at the other edge of the helix (giving rise to lower Faraday rotation and higher degrees of polarization). In contrast, in the absence of other factors, a toroidal field should give rise to a symmetric polarization profile across the jet, independent of the viewing.

Thus, the firm detection of asymmetric transverse polarization structure with a configuration consistent with one of those expected for helical fields provides a good diagnostic for the presence of a helical, as opposed to toroidal, **B** field.

Acknowledgements This review has made use of data from the MOJAVE database that is maintained by the MOJAVE team (Lister et al., 2009, AJ, 137, 3718). The image used to compose Fig. 5.4 was taken from the MOJAVE website, http://www.physics.purdue.edu/astro/MOJAVE/allsources.html.

References

Agudo, I., Bach, U., Krichbaum, T.P., Marscher, A.P., Gonidakis, I., Diamond, P.J., Perucho, M., Alef, W., Graham, D.A., Witzel, A., Zensus, J.A., Bremer, M., Acosta-Pulido, J.A., Barrena, R.: Superluminal non-ballistic jet swing in the quasar NRAO 150 revealed by mm-VLBI. A&A **476**, 17 (2007)

Algaba, J.C.: High-frequency very long baseline interferometry rotation measure of eight active galactic nuclei. MNRAS **429**, 3551 (2013)

Aloy, M.A., Ibánez, J. M., Martí, J. M., Gómez, J.-L., Müller, E.: High-resolution three-dimensional simulations of relativistic jets. ApJ **523**, 125 (1999)

Aloy, M.A., Gómez, J.-L., Ibánez, J.M., Martí, J.M., Müller, E.: Radio emission from three-dimensional relativistic hydrodynamic jets: Observational evidence of jet stratification. ApJ **528**, 95 (2000)

Aloy, M.A., Martí, J.M., Gómez, J.-L., Agudo, I., Müller, E., Ibánez, J.M.: Three-dimensional simulations of relativistic precessing jets probing the structure of superluminal sources. ApJ **585**, 109 (2003)

Asada, K., Inoue, M., Uchida, Y., Kameno, S., Fujiawa, K., Iguchi, S., Mutoh, M.: A helical magnetic field in the jet of 3C 273. PASJ **54**, L39 (2002)

Asada, K., Nakamura, M., Inoue, M., Kameno, S., Nagai, H.: Multi-frequency polarimetry toward S5 0836+710: A possible spine-sheath structure for the jet. ApJ **720**, 41 (2010)

Attridge, J.M., Roberts, D.H., Wardle, J.F.C.: Radio jet-ambient medium interactions on parsec scales in the blazar 1055+018. ApJ **518**, 87 (1999)

Blandford, R.D.: Astrophysical Jets, p. 26. Cambridge University Press, Cambridge (1993)

Blandford, R.D., Königl, A.: Relativistic jets as compact radio sources. ApJ **232**, 34 (1979)

Bloom, S.D., Fromm, C.M., Ros, E.: The accelerating jet of 3C 279. AJ **145**, 12 (2013)

Blundell, K.M., Bowler, M.G.: Symmetry in the changing jets of SS 433 and its true distance from us. ApJ **616**, 159 (2004)
Britzen, S., Witzel, A., Krichbaum, T.P., Beckert, T., Campbell, R.M., Schalinski, C., Campbell, J.: The radio structure of S5 1803+784. MNRAS **362**, 966 (2005)
Broderick, A.E., McKinney, J.C.: Parsec-scale faraday rotation measures from general relativistic magnetohydrodynamic simulations of active galactic nucleus jets. ApJ **725**, 750 (2010)
Burn, B.J.: On the depolarization of discrete radio sources by Faraday dispersion. MNRAS **133**, 67 (1966)
Caproni, A., Abraham, Z.: Precession in the inner jet of 3C 345. ApJ **602**, 625 (2004a)
Caproni, A., Abraham, Z.: Can long-term periodic variability and jet helicity in 3C 120 be explained by jet precession? MNRAS **349**, 1218 (2004b)
Caproni, A., Abraham, Z., Monteiro, H.: Monteiro, parsec-scale jet precession in BL lacertae (2200+420). MNRAS **428**, 280 (2013)
Cawthorne, T.V., Cobb, W.K.: Linear polarization of radiation from oblique and conical shocks. ApJ **350**, 536 (1990)
Cawthorne, T.V., Hughes, P.A.: The radiative transfer of synchrotron radiation through a compressed random magnetic field. ApJ **77**, 60 (2013)
Cawthorne, T.V., Jorstad, S.G., Marscher, A.P.: Polarization structure in the core of 1803+784: A signature of recollimation shocks? ApJ **772**, 14 (2013)
Cohen, M. H., Lister, M. L., Homan, D. C., Kadler, M., Kellermann, K.I., Kovalev Y. Y, Vermeulen, R. C.: Relativistic beaming and the intrinsic properties of extragalactic radio jets. ApJ, **658**, 232 (2007)
Cohen, M.H., Meier, D.L., Arshakian, T.G., Homan, D.C., Hovatta, T., Kovalev, Y.Y., Lister, M.L., Pushkarev, A.B., Richards, J.L., Savolainen, T.: Studies of the jet in bl lacertae. I. recollimation shock and moving emission features. ApJ **787**, 151 (2014)
Coughlan, C.P.: The development of new methods for high resolution radio astronomy imaging, PhD Thesis, University College Cork, Ireland (2014)
Contopoulos, I., Christodoulou, D.M., Kazanas, D., Gabuzda, D.C.: The invariant twist of magnetic fields in the relativistic jets of active galactic nuclei. ApJL **702**, L148 (2009)
Croke, S.M., Gabuzda, D.C.: Aligning VLBI images of active galactic nuclei at different frequencies. MNRAS **386**, 619 (2008)
Croke, S.M., O'Sullivan, S.P., Gabuzda, D.C.: The parsec-scale distributions of intensity, linear polarization and Faraday rotation in the core and jet of Mrk501 at 8.4-1.6 GHz. MNRAS **402**, 259 (2010)
Gabuzda, D.C., Gómez, J.-L.: iVSOP polarization observations of the BL Lacertae object OJ 287. MNRAS **320**, 49 (2001)
Gabuzda, D.C., Mullan, C.M., Cawthorne, T.V., Wardle, J.F.C., Roberts, D.H.: Evolution of the milliarcsecond total intensity and polarization structures of BL Lacertae objects. ApJ **435**, 140 (1994)
Gabuzda, D.C., Pushkarev, A.B., Cawthorne, T.V.: Analysis of $\lambda = 6$ cm VLBI polarization observations of a complete sample of northern BL lacertae objects. MNRAS **319**, 1109 (2000)
Gabuzda, D.C., Murray, E., Cronin, P.J.: Helical magnetic fields associated with the relativistic jets of four BL Lac objects. MNRAS **351**, L89 (2004)
Gabuzda, D.C., Rastorgueva, E.A., Smith, P.S., O'Sullivan, S.P.: O'Sullivan, Evidence for cospatial optical and radio polarized emission in active galactic nuclei. MNRAS **369**, 1596 (2006)
Gabuzda, D.C., Cantwell, T.M., Cawthorne, T.V.: Cawthorne, magnetic field structure of the extended 3C 380 jet. MNRAS **438**, 1 (2013)
Gabuzda, D.C., Reichstein, A.R., O'Neill, E.L.: Are spine-sheath polarization structures in the jets of active galactic nuclei associated with helical magnetic fields? MNRAS (2014, in press)
Giroletti, M., Giovannini, G., Feretti, L., Cotton, W.D., Edwards, P.G., Lara, L., Marscher, A.P., Mattox, J.R., Piner, B.G., Venturi, T.: Parsec-scale properties of markarian 501. ApJ **600**, 127 (2004)
Gómez, J.-L., Roca–Sogorb, M., Agudo, I., Marscher, A.P., Jorstad, S.G.: On the source of faraday rotation in the jet of the radio galaxy 3c 120. ApJ **733**, 11 (2011)

Homan, D.C.: Inverse Depolarization: A potential probe of internal faraday rotation and helical magnetic fields in extragalactic radio jets. ApJ **747**, 24 (2012)
Homan, D.C., Lister, M.L.: MOJAVE: Monitoring of jets in active galactic nuclei with vlba experiments. II. First-Epoch 15 GHz circular polarization results. AJ **131**, 1262 (2006)
Homan, D.C., Wardle, J.F.C.: Detection and measurement of parsec-scale circular polarization in four AGNS. AJ **118**, 1942 (1999)
Homan, D.C., Wardle, J.F.C.: High levels of circularly polarized emission from the radio jet in NGC 1275 (3C 84). ApJ **602**, 13 (2004)
Homan, D.C., Lister, M.L., Kellermann, K.I., Cohen, M.H., Ros, E., Zensus, J.A., Kadler, M., Vermeulen, R.C.: Jet collimation in action: Realignment on kiloparsec scales in 3C 279. ApJ **589**, 9 (2003)
Homan, D.C., Attridge J.M., Wardle, J.F.C.: Parsec-scale circular polarization observations of 40 blazars. ApJ **556**, 113 (2001)
Hovatta, T., Valtaoja, E., Tornikoski, M., Lḧteenmäki, A.: Doppler factors, Lorentz factors and viewing angles for quasars, BL Lacertae objects and radio galaxies. A&A **494**, 527 (2009)
Hovatta, T., Lister, M.L., Aller, M.F., Aller, H.D., Homan, D.C., Kovalev, Y.Y., Pushkarev, A.B., Savolainen, T.: MOJAVE: Monitoring of jets in active galactic nuclei with vlba experiments. VIII. Faraday rotation in Parsec-scale AGN jets. AJ **144**, 105 (2012)
Hovatta, T., Aller, M.F., Aller, H.D., Clausen–Brown, E., Homan, D.C., Kovalev, Y.Y., Lister, M.L., Pushkarev, A.B., Savolainen, T.: MOJAVE: Monitoring of jets in active galactic nuclei with VLBA experiments. XI. Spectral distributions. AJ **147**, 143 (2014)
Hughes, P.A., Aller, H.D., Aller, M.F.: Polarized radio outbursts in bl-lacertae - part two - the flux and polarization of a piston-driven shock. ApJ **298**, 301 (1985)
Jones, T.W.: Polarization as a probe of magnetic field and plasma properties of compact radio sources - simulation of relativistic jets. ApJ **332**, 678 (1988)
Jones, T.W., O'Dell, S.L.: Transfer of polarized radiation in self-absorbed synchrotron sources. I. Results for a homogeneous source. ApJ **214**, 522 (1977)
Jorstad, S.G., Marscher, A.P., Lister, M.L., Stirling, A.M., Cawthorne, T.V., Gómez, J.-L., Gear, W.K.: Change in speed and direction of the jet near the core in the quasar 3C 279. AJ **127**, 3115 (2004)
Kellermann, K.I., Sramek, R., Schmidt, M., Shaffer, D.B., Green, R.: VLA observations of objects in the palomar bright quasar survey. AJ **98**, 1195 (1989a)
Kellermann, K.I., Vermeulen, R.C., Zensus, J.A., Cohen, M.H.: Sub-milliarcsecond imaging of quasars and active galactic nuclei. AJ **115**, 1295 (1998b)
Kellermann, K.I., Lister, M.L., Homan, D.C., Vermeulen, R.C., Cohen, M.H., Ros, E., Kadler, M., Zensus, J.A., Kovalev, Y.Y.: Sub-milliarcsecond imaging of quasars and active galactic nuclei. III. Kinematics of parsec-scale radio jets. ApJ **609**, 539 (2004)
Klare, J., Zensus, J.A., Lobanov, A.P., Ros, E., Krichbaum, T.P., Witzel, A.: in *Future Directions in High Resolution Astronomy: The 10th Anniversary of the VLBA*, vol. 340, p. 40. Astronomical Society of the Pacific Conference Proceedings, San Francisco (2005)
Kovalev, Y.Y., Lobanov, A.P., Pushkarev, A.B., Zensus, J.A.: Zensus, opacity in compact extragalactic radio sources and its effect on astrophysical and astrometric studies. A&A **483**, 759 (2008)
Laing, R.A.: A model for the magnetic-field structure in extended radio sources. MNRAS **193**, 439 (1980)
Legg, M.P.C., Westfold, K.C.: Elliptic polarization of synchrotron radiation. ApJ **154**, 499 (1968)
Lister, M.L., Homan, D.C.: MOJAVE: Monitoring of jets in active galactic nuclei with VLBA Experiments. I. First-Epoch 15 GHz linear polarization images. AJ **130**, 1389 (2005)
Lister, M.L., Aller, H.D., Aller, M.F., cohen, M.H., Homan, D.C., Kadler, M., Kellermann, K.I., Kovalev, Y.Y., Ros, E. , Savolainen, T., Zensus J.A., Vermeulen, R.C.: MOJAVE: Monitoring of jets in active galactic nuclei with VLBA experiments. V. Multi-Epoch VLBA images. AJ **137**, 3718 (2009)
Lobanov, A.P.: Ultracompact jets in active galactic nuclei. A&A **330**, 79 (1998)
Longair, M.S.: High Energy Astrophysics. Cambridge University Press, Cambridge (2010)

Lovelace, R.V.E., Li, H., Koldoba, A.V., Ustyugova, G.V., Romanova, M.M.: Poynting jets from accretion disks. ApJ **572**, 445 (2002)
Lynden–Bell, D.: Magnetic collimation by accretion discs of quasars and stars. MNRAS **279**, L389 (1996)
Lynden-Bell, D.: On why discs generate magnetic towers and collimate jets. MNRAS **341**, 1360 (2003)
Lyutikov, M.: Role of reconnection in AGN jets. New Astron. Rev. **47**, 513 (2003)
Lyutikov, M., Pariev, V.I., Gabuzda, D.C.: Polarization and structure of relativistic parsec-scale AGN jets. MNRAS **360**, 869 (2005)
Mahmud, M., Gabuzda, D.C., Bezrukovs, V.: Surprising evolution of the parsec-scale Faraday Rotation gradients in the jet of the BL Lac object B1803+784. MNRAS **400**, 2 (2009)
Mahmud, M., Coughlan, C.P., Murphy, E., Gabuzda, D.C., Hallahan, D.R.: Connecting magnetic towers with Faraday rotation gradients in active galactic nuclei jets. MNRAS **431**, 695 (2013)
Marscher, A.P.: Turbulent, extreme multi-zone model for simulating flux and polarization variability in blazars. ApJ **780**, 87 (2014)
Meier, D.L., Koide, S., Uchida, Y.: Magnetohydrodynamic production of relativistic jets. Science **291**, 84 (2001)
Mimica, P., Aloy, M.A.: On the dynamic efficiency of internal shocks in magnetized relativistic outflows. MNRAS **401**, 525 (2010)
Mimica, P., Aloy, M.A.: Radiative signature of magnetic fields in internal shocks. MNRAS **421**, 2635 (2012)
Murphy, E., Cawthorne, T.V., Gabuzda, D.C.: Analysing the transverse structure of the relativistic jets of active galactic nuclei. MNRAS **430**, 1504 (2013)
Murphy, E., Gabuzda, D.C.: The innermost regions of relativistic jets and their magnetic fields. EPJ Web Conf. **61**, id.07005 (2013)
Nakamura, M., Uchida, Y., Hirose, S.: Production of wiggled structure of AGN radio jets in the sweeping magnetic twist mechanism. New Astron. **6**, 61 (2001)
O'Sullivan, S.P., Gabuzda, D.C.: Magnetic field strength and spectral distribution of six parsec-scale active galactic nuclei jets. MNRAS **400**, 260 (2009)
Pacholczyk, A.G.: Radio Astrophysics. Freeman, San Francisco (1970)
Papageorgiou, A., Cawthorne, T.V., Stirling, A., Gabuzda, D., Polatidis, A.G.: Space very long baseline interferometry observations of polarization in the jet of 3C380. MNRAS **373**, 449 (2006)
Pariev, V.I., Istomin, Ya.N., Beresnyak, A.R.: Relativistic parsec-scale jets: II. Synchrotron emission. A&A **403**, 805 (2003)
Pushkarev, A.B., Gabuzda, D.C., Vetukhnovskaya, Yu.N., Yakimov, V.E.: Spine-sheath polarization structures in four active galactic nuclei jets. MNRAS **356**, 859 (2005)
Pushkarev, A.B., Hovatta, T., Kovalev, Y.Y., Lister, M.L., Lobanov, A.P., Savolainen, T., Zensus, J.A.: MOJAVE: Monitoring of jets in active galactic nuclei with VLBA experiments. IX. Nuclear opacity. A&A **545**, 113 (2012)
Qian, S.-J., Witzel, A., Zensus, J.A., Krichbaum, T.P., Britzen, S., Zhang, X.-Zh.: Periodicity of the ejection of superluminal components in 3C345. Res. Astron. Astrophys. **9**, 137 (2009)
Rybicki, G.B., Lightman, A.P.: Radiative processes in astrophysics. Wiley, New York (1979)
Sikora, M., Begelman, M.C., Madejski, G.M., Lasota, J.-P.: Lasota, are quasar jets dominated by poynting flux? ApJ **625**, 62 (2005)
Sokolovsky, K.V., Kovalev, Y.Y., Pushkarev, A.B., Lobanov, A.P.: A VLBA survey of the core shift effect in AGN jets. I. Evidence of dominating synchrotron opacity. A&A **532**, 38 (2011)
Stirling, A.M., Cawthorne, T.V., Stevens, J.A., Jorstad, S.G., Marscher, A.P., Lister, M.L., Gómez, J.-L., Smith, P.S., Agudo, I., Gabuzda, D.C., Robson, E.I., Gear, W.K.: Discovery of a precessing jet nozzle in BL Lacertae. MNRAS **341**, 405 (2003)
Taylor, G.B.: Magnetic fields in quasar cores. ApJ **506**, 637 (1998)
Taylor, G.B.: Magnetic fields in quasar cores. II. ApJ **533**, 95 (2000)
Taylor, G.B., Zavala, R.: Are there rotation measure gradients across active galactic nuclei jets? ApJ **722**, 183 (2010)

Tsinganosn, K., Bogovalov, S.: Magnetic collimation of relativistic outflows in jets with a high mass flux. MNRAS **337**, 553 (2002)

Vitrishchak, V.M., Gabuzda, D.C.: New measurements of the circular polarization of the radio emission of active galactic nuclei on parsec scales, Astron. Rep. **51**, 695 (2007)

Vitrishchak, V.M., Gabuzda, D.C., Algaba, J.C., Rastorgueva, E.A., O'Sullivan, S.P., On'Dowd, A.: The 15-43 GHz parsec-scale circular polarization of 41 active galactic nuclei. MNRAS **391**, 124 (2008)

Walker, R.C., Dhawan, V., Romney, J.D., Kellermann, K.I., Vermeulen, R.C.: VLBA absorption imaging of ionized gas associated with the accretion disk in NGC 1275. ApJ **530**, 233 (2000)

Wardle, J.F.C., Homan, D.C.: Laing, R.A., Blundelln, K. (eds.) Particles and Fields in Radio Galaxies. Astronomical Society of the Pacific Conference Series vol. 250, p. 15 (2001)

Wardle, J.F.C., Homan, D.Cn.: Circular polarisation from relativistic jet sources. Astrophys. Space Sci. **288**, 143 (2003)

Zavala, R.T., Taylor, G.B.: A view through faraday's fog: Parsec-scale rotation measures in active galactic nuclei. ApJ **589**, 126 (2003)

Zavala, R.T., Taylor, G.B.: A View through faraday's fog. II. Parsec-scale rotation measures in 40 active galactic nuclei. ApJ **612**, 749 (2004)

Zensus, J.A., Rosn, E., Kellermann, K.I., Cohen, M.H., Vermeulen, R.C., Kadler, M.: Sub-milliarcsecond imaging of quasars and active galactic nuclei. II. Additional sources. AJ **124**, 662 (2002)

Chapter 6
Black Hole Magnetospheres

Brian Punsly

Abstract This chapter compares and contrasts winds and jets driven by the two distinct components of the black magnetosphere: the event horizon magnetosphere (the large scale magnetic field lines that thread the event horizon) and the ergospheric disk magnetosphere associated with poloidal magnetic flux threading plasma near the equatorial plane of the ergosphere. The power of jets from the two components as predicted from single-fluid, perfect MHD numerical simulations are compared. The decomposition of the magnetosphere into these two components depends on the distribution of large scale poloidal magnetic flux in the ergosphere. However, the final distribution of magnetic flux in a black hole magnetosphere depends on physics beyond these simple single-fluid treatments, non-ideal MHD (eg, the dynamics of magnetic field reconnection and radiation effects) and two-fluid effects (eg, ion coupled waves and instabilities in the inner accretion flow). In this chapter, it is emphasized that magnetic field line reconnection is the most important of these physical elements. Unfortunately, in single-fluid perfect MHD simulations, reconnection is a mathematical artifact of numerical diffusion and is not determined by physical processes. Consequently, considerable calculational progress is required before we can reliably assess the role of each of these components of black hole magnetospheres in astrophysical systems.

6.1 Introduction

Black hole magnetospheres are potential sites for launching relativistic jets and hence their interest in astrophysics (Blandford and Znajek 1977). Multiple processes for launching jets are possible depending on initial conditions and assumptions regarding the mathematical model used to approximate the microscopic physics of the plasma (Meier 1999; Punsly 2008). It was noted in Punsly and Coroniti (1990a,b) that the ultimate topology of a black hole magnetosphere and the physical

---

B. Punsly (✉)
1415 Granvia Altamira, Palos Verdes Estates, CA, USA 90274
ICRANet, Piazza della Repubblica 10, Pescara 65100, Italy
e-mail: brian.punsly1@verizon.net; brian.punsly@comdev-usa.com

I. Contopoulos et al. (eds.), *The Formation and Disruption of Black Hole Jets*,
Astrophysics and Space Science Library 414,
DOI 10.1007/978-3-319-10356-3_6

mechanism and magnitude of power extraction depend upon two factors, the nature of the plasma injection mechanism and to a greater extent, the fate of magnetic field lines that accrete toward the black hole. The field lines that thread the event horizon are drained of plasma by gravitational attraction toward the black hole at small radii and by the outflow driven along rotating field lines at large distances. Thus, plasma needs to be continually created on the field lines, or the outflow of energy will stop. The method of injection can profoundly influence the field line angular velocity and therefore the total available power in a magnetic flux tube in the event horizon magnetosphere (EHM, hereafter). Secondly, as large scale poloidal magnetic flux accretes, it is not clear where it ends up physically and how long it resides in a particular location. In a time evolving black hole accretion system, this will determine the energy output of the system. In particular, one needs to know how much flux is trapped in the EHM and how much threads the equatorial plane outside of the horizon, but within the ergosphere. The latter location creates an ergospheric disk magnetosphere (EDM, hereafter) that can be a source of powerful jets. Hence, the motivation here is to delineate the physical processes that determine the poloidal magnetic flux distribution in black hole magnetospheres. The chapter addresses the successes and shortcomings of single-fluid, perfect magnetohydrodynamic (MHD) simulations as a model of the relevant physics for determining the topology of the black hole magnetospheres.

In the past 10 years, single-fluid, perfect MHD numerical simulations have been developed in the scientific community to help understand the two issues noted above. In order to establish and maintain the event horizon magnetosphere, the simulations must rely on the numerical artifice of a mass floor, local mass injection that is designed to establish a minimum density. In spite of some discussion of anecdotal mass floor examples in the literature, perfect MHD simulations are not likely to resolve the first issue (the physical implications of mass injection into the EHM), since a mass floor explicitly violates rest-mass and energy – momentum conservation and necessarily contradicts the perfect MHD assumption upon which the simulations are predicated (McKinney 2006).

In terms of the second issue, it was noted in Punsly and Coroniti (1990b) that if vertical, large scale magnetic flux accretes, then reconnection of vertical flux would be determinant to the final magnetic field configuration since the black hole is effectively a sink with infinite capacity for mass, but with a very limited capacity to accept magnetic flux. However, there is much uncertainty in the reconnection microphysics and the expected reconnection rate in an accretion flow near a rapidly rotating black hole, since our experimental experience is based on very different environments, the solar corona, the solar wind, the Earth magnetosphere and magneto-tail, and magnetic confinement devices for thermonuclear fusion.

The development of the subject of the dynamics of magnetic flux accretion was impeded for many years by a series of persuasive arguments that the existence of coherent vertical flux within the dense accreting gas is prohibited by the turbulent magnetic diffusivity of the plasma (van Ballegooijen 1989; Lubow et al. 1994). Thus, it was not even clear if the notion of a large scale field associated with an accretion flow was viable. In spite of these concerns, the issue of flux transport seems to have been circumvented in 3-D perfect MHD simulations of black holes. In Beckwith et al. (2009) a "coronal mechanism" was discovered in the family of 3-D simulations. The first step is the transport towards the black hole of an inwardly stretched, disk-anchored, poloidal loop through the low turbulence, coronal layer just above the disk, the "hairpin" field in Beckwith et al. (2009). This coronal transport is similar to the mechanism proposed in Rothstein and Lovelace (2008), but see Beckwith et al. (2009) for some dynamical differences. Then loops of magnetic field in the inner accretion flow reconnect with the half of the hairpin at mid-latitudes, allowing it to contract into the black hole, leading to one sign of field threading the black hole. The other half of the reconnected hairpin field line is buried in the accretion flow. Reconnection is apparently "necessary" for vertical flux accretion in all the numerical simulations of this family (Beckwith et al. 2009; Hawley and Krolik 2006; McKinney and Blandford 2009). Reconnection near the black hole in a time dependent environment was proposed as the determining factor in the fate of the accreted flux in Punsly and Coroniti (1990b). However the geometry of the reconnection site in Beckwith et al. (2009) is very different than what was expected in Punsly and Coroniti (1990b) in which an ergospheric disk (a plasma disk in the ergosphere that is threaded by large scale magnetic flux) was predicted (see Sect. 6.4). In spite of this geometric difference, an ergospheric disk does form in some 3-D perfect MHD simulations (Punsly 2008).

Alternatively, adjusting the initial conditions in the simulations as in McKinney et al. (2012), Tchekhovskoy et al. (2011), and Tchekhovskoy and McKinney (2012) apparently overcomes the turbulent diffusion issue of magnetic transport. By accreting large amounts of flux, they obtain a direct advection of flux in what they call magnetically choked flows (MCAFs) and magnetically arrested accretion disks (MADs) that they liken to what was seen years earlier in the numerical work of Igumenshchev et al. (2003) and Igumenshchev (2008). In Sect. 6.4, it is discussed how the MCAF stability depends on reconnection rates. The accretion flow is un-physically hot near the black hole and is pinched by radial fans of magnetic field of opposite polarity above and below the equatorial mass sheet. The resulting equilibrium is one in which the enormous outward force associated with gas pressure in the mass sheet is balanced by inward directed magnetic pressure gradients above and below the accreting gas. This configuration is indicative of a pinch sheet instability that would naturally create magnetic reconnection sites (Syrovatskii 1981).

The implication is that the global topology of the black hole magnetosphere is highly dependent on the time evolution that is driven by reconnection. This is not a trivial circumstance because in single-fluid, perfect MHD simulations, reconnection is created by numerical diffusion. The situation is rendered even more ambiguous by the complicated twisted magnetic field line geometries in 3-D simulations around rapidly rotating black holes. The complications of far less intricate 3-D field line topology have been recognized in solar and planetary physics (Pontin 2011; Wilmot-Smith et al. 2010).

This chapter explores the implications of particle injection and reconnection around rotating black holes and their relationship to the topology of astrophysical black hole magnetospheres. The discussion begins with an exposition of the differences in energetics and topology of the EHM and the EDM. The jet power is described in terms of the field line angular velocity and the total magnetic flux in the jet. There is a particular emphasis on the EDM and its elevated jet efficiency. This chapter is not a review of Blandford and Znajek (1977) in the EHM. For more details on this topic please consult the chapter of A. Tchekhovskoy and the references therein. Section 6.3 highlights the details of the dynamics of the EDM in the context of 3-D accretion simulations. The connection between magnetically arrested accretion and the creation of "negative energy" plasma in isolated islands is shown in detail. It is demonstrated that these islands are isolated pockets of a "Penrose process" that exist nested within an enveloping accretion environment. Section 6.4 discusses how reconnection can drastically alter the global topology of the black hole magnetosphere. It is emphasized that reconnection near a black hole is yet to be understood in a convincing way. The last discussion investigates some topics that are related to plasma injection in black hole magnetospheres.

6.2 The Field Line Angular Velocity

The physical interest in the field line angular velocity, Ω_F, is that it is enters into the poloidal Poynting flux, S^P, quadratically. For any MHD Poynting flux dominated jet, regardless of the source, the total integrated electromagnetic poloidal energy flux is

$$\int S^P \mathrm{d}A_{\perp} = k\frac{\Omega_F^2 \Phi^2}{2\pi^2 c} , \tag{6.1}$$

where Φ is the total magnetic flux enclosed within the jet, $\mathrm{d}A_{\perp}$ is the cross-sectional area element and k is a geometrical factor that equals 1 for a uniform highly collimated jet (Punsly 2008). In this section, Ω_F, the field line angular velocity in Boyer-Lindquist coordinates, will be studied near a rapidly rotating black hole in the Kerr space-time. This is also known as the field line angular velocity as viewed by an observer that is stationary with respect to asymptotic infinity.

6.2.1 Ω_F of Ergospheric Disk Jets in 3-D Numerical Simulations

The following nomenclature is implemented in the discussion of accretion flows near black holes. The metric of the Kerr spacetime (that of a rotating uncharged black hole) in Boyer-Lindquist coordinates, $g_{\mu\nu}$, is given by the line element

$$ds^2 \equiv g_{\mu\nu}\, dx^\mu dx^\nu = -\left(1 - \frac{2Mr}{\rho^2}\right) dt^2 + \rho^2 d\theta^2$$

$$+ \left(\frac{\rho^2}{\Delta}\right) dr^2 - \frac{4Mra}{\rho^2} \sin^2\theta$$

$$d\phi\, dt + \left[(r^2 + a^2) + \frac{2Mra^2}{\rho^2} \sin^2\theta\right] \sin^2\theta\, d\phi^2 \,, \tag{6.2}$$

$$\rho^2 = r^2 + a^2 \cos^2\theta, \tag{6.3}$$

$$\Delta = r^2 - 2Mr + a^2 \equiv (r - r_+)(r - r_-) \,. \tag{6.4}$$

In the expressions above, M is the mass and a is the angular momentum per unit mass of the black hole in geometrized units. There are two event horizons given by the roots of the equation $\Delta = 0$. The outer horizon at r_+ is of physical interest

$$r_+ = M + \sqrt{M^2 - a^2} \,. \tag{6.5}$$

The stationary limit surface, at $r = r_s$ occurs when $\partial/\partial t$ changes from a time-like vector field to a space-like vector field.

$$r_s = M + \sqrt{M^2 - a^2 \cos^2\theta} \,. \tag{6.6}$$

This change happens when g_{tt} switches sign. The region in which $g_{tt} > 0$ is the ergosphere. The outer boundary of the ergosphere is the stationary limit surface.

The nonlinear dynamics and evolution of vertical flux distributions near black holes are best studied with long term 3-D MHD numerical simulations in the Kerr space-time as originally developed in De Villiers et al. (2003), Hirose et al. (2004), Krolik et al. (2005), Hawley and Krolik (2006), and Beckwith et al. (2008). These simulations evolve from gaseous tori that are destabilized by loops of poloidal magnetic flux. We caution the reader that the results of these types of simulations are highly sensitive to the initial conditions that are imposed. Of all the loop configurations that are tried, only poloidal loops of the same orientation in a dipolar configuration produce strong Poynting jets. In Beckwith et al. (2008), it was shown that if the loops are in a quadrupolar orientation in the initial state, the Poynting jet power coming from near the black hole is reduced to about 1 % of the value obtained from a common orientation of the dipolar loops. If the loops are toroidal in the initial state, the Poynting jet power coming from near the black hole is reduced to about 0.1 % of the value obtained from a common orientation of the dipolar loops. Finally, Beckwith et al. (2008) showed that if the initial state of the torus is threaded by dipolar loops with more than one orientation, a strong Poynting jet forms only if a coherent packet of loops of like orientation accretes. The nature of the Poynting jet in this family of simulations is clearly strongly dependent on initial conditions. In spite of this, these simulations are an interesting virtual laboratory for studying the physics of distant, unreachable astrophysical black holes. The self consistent magnetic structures that form near the black hole in this "numerical laboratory" are potentially related to phenomena that can occur close to an astrophysical black hole.

One of the highest spin simulations in this family is from Hawley and Krolik (2006) and is known as KDJ, $a/M = 0.99$. The last three time steps of KDJ with data collection were generously shared by J. Krolik and J. Hawley. Expressing the mass in geometrized units, the data dumps were performed at t=9,840 M, t=9,920 M and t=10,000 M. The ergospheric disk in these simulations was studied in Punsly (2007, 2008) and Punsly et al. (2009).

The conservation of global, redshifted, or equivalently the Boyer-lindquist coordinate evaluated energy flux, defined in terms of the stress-energy tensor, is shown in Punsly (2007) to simply reduce to

$$\frac{\partial(\sqrt{-g}\, T_t^{\ \nu})}{\partial(x^{\nu})} = 0 \,. \tag{6.7}$$

The four-momentum $-T_t^{\ \nu}$ has two components: one from the fluid, $-(T_t^{\ \nu})_{\text{fluid}}$, and one from the electromagnetic field, $-(T_t^{\ \nu})_{\text{EM}}$. The quantity $g = -(r^2 + a^2 \cos^2\theta)^2 \sin^2\theta$ is the determinant of the metric. The integral form of the conservation law arises from the trivial integration over Boyer – Lindquist coordinates of the partial differential in Eq. (6.7) (Punsly 2007). It follows that the redshifted Poynting flux vector can defined in Boyer-Lindquist coordinates as

$$\mathbf{S} = (S^r,\ S^\theta,\ S^\phi) = (-\sqrt{-g}\,(T_t^{\ r})_{\text{EM}},\ -\sqrt{-g}\,(T_t^{\ \theta})_{\text{EM}},\ -\sqrt{-g}\,(T_t^{\ \phi})_{\text{EM}}) \tag{6.8}$$

After averaging over azimuth, we use the poloidal projection of the Poynting vector, $\mathbf{S}^P$, to display the propagation of electromagnetic energy flux out of the ergosphere in KDJ. Figure 6.1 is a false color contour plot of the azimuthal average of S^r. Notice that the flow of S^r across the white boundary (the stationary limit) is dominated by a narrow channel that seems to initiate suspended outside the event horizon ($r_+ = 1.175$ M) and the inner calculation boundary (shaded black) at r=1.203 M. In order to explain the origin of this energy beam, the direction of the poloidal Poynting vector is overlaid on the false color contour map. The white arrows represent the poloidal direction of the azimuthally averaged poloidal component of the Poynting flux, $\mathbf{S}^P$. The length of the arrows is proportional to the grid spacing and is not related to the magnitude of $\mathbf{S}^P$. The black contours on the plots are of the Boyer-Lindquist evaluated density, scaled from the peak value within the frame at relative levels of 0.5 and 0.1. The integral curves that are approximated by the white arrows show that the primary source of S^r is Poynting flux emerging from the accretion flow near the equatorial plane of the ergosphere (the EDM). The direction of the white arrows indicate that $\mathbf{S}$ has a large S^θ component at its point of origin near the equatorial plane. In Punsly (2007), it was shown that there is a strong vertical magnetic field component at these locations as well, which is consistent with the arrow directions. The white arrows and the false color contours show the ergospheric disk jet (EDJ, hereafter) emerging from the environs of the black hole.

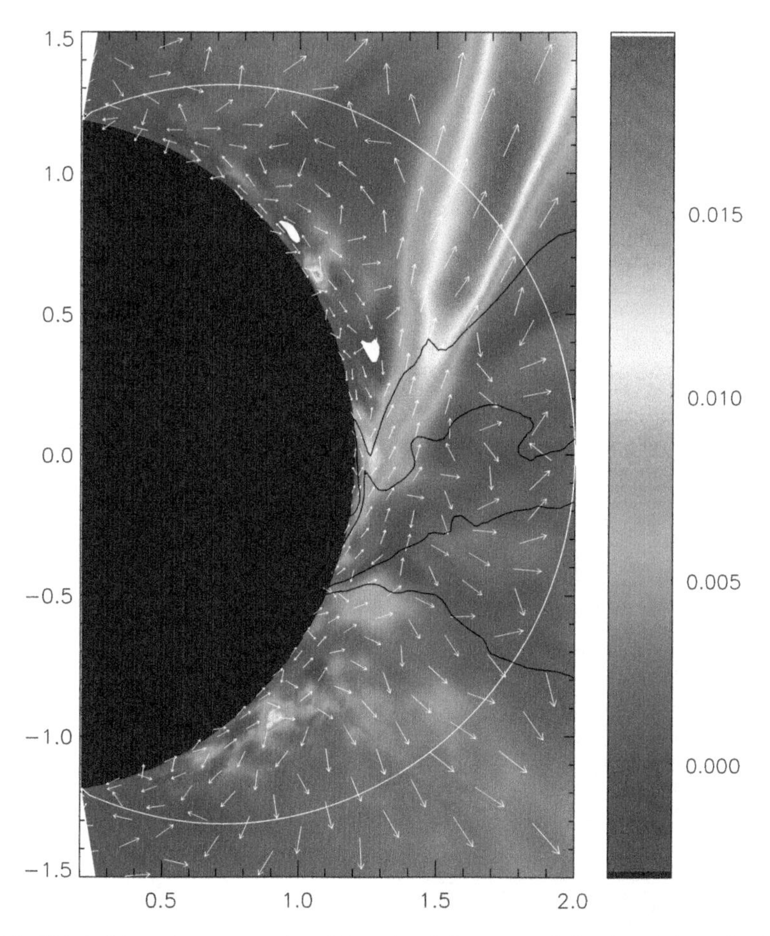

Fig. 6.1 This is false color contour plot of the azimuthal average of S^r. The relative units (based on code variables) are in a color bar to right. Superimposed are *white arrows* that represent the direction of the azimuthally averaged poloidal component of the Poynting flux, $\mathbf{S}^P$. The length does not represent the strength of the poloidal Poynting flux, but is proportional to the grid spacing. The white contour is the stationary limit surface. The black contours that are superimposed on the false color contours are the Boyer-Lindquist evaluated mass density, scaled from the peak value within the frame at relative levels 0.5 and 0.1. A low density of 0.1 times the peak density might be consider to be a crude indicator of the transition from the accretion flow to the corona. The inside of the inner calculational boundary (r = 1.203 M) is *black*. This designation will often be used interchangeably with "the event horizon" in the following. The calculational boundary near the poles is at 8.1° and 171.9°. There is no data clipping, so saturated regions appear *white*

The simulation KDJ can be used to study Ω_F in the associated EDM from which strong Poynting jets emerge. In a non-axisymmetric, non-time stationary flow, there is still a well defined notion of Ω_F: the rate at which a frame of reference at fixed r and θ would have to rotate so that the poloidal component of the electric field, $E^{\perp}$, that is orthogonal to the poloidal magnetic field, B^P, vanishes. This was first derived in Punsly (1991) (see the extended discussion in Punsly (2008) for the various physical interpretations), and has recently been written out in B-L coordinates in Hawley and Krolik (2006) in terms of the plasma three-velocity, v^i and the Faraday tensor as

$$\Omega_F = v^\phi - F_{\theta r}\frac{g_{rr}v^r F_{\phi\theta} + g_{\theta\theta}v^\theta F_{r\phi}}{(F_{\phi\theta})^2 g_{rr} + (F_{r\phi})^2 g_{\theta\theta}} \,. \tag{6.9}$$

Figure 6.2 is a false color contour plot of the field line angular velocity as seen from asymptotic infinity in units of the event horizon angular velocity, Ω_H averaged over azimuthal angle at t = 9,840 M. Comparing Figs. 6.1 and 6.2, it is clear that Ω_F is

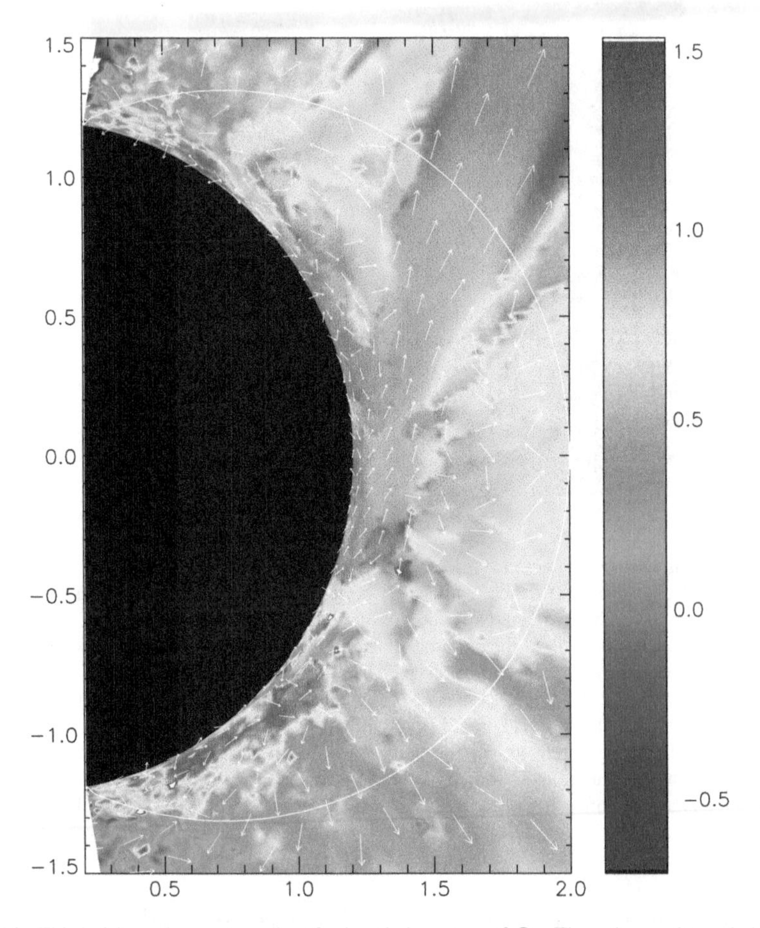

Fig. 6.2 This is false color contour plot of azimuthal average of Ω_F. The units are the scaled to the horizon angular velocity, Ω_H. Superimposed are *white arrows* that represent the direction of the azimuthally averaged poloidal component of the Poynting flux, $\mathbf{S}^P$. The length does not represent the strength of the poloidal Poynting flux, but is proportional to the grid spacing. The white contour is the stationary limit surface

highly elevated along the conduits that carry the enhanced energy flux from the EDM. The magnitude of Ω_F is consistent with theory of the EDJ in Punsly and Coroniti (1990b), $\Omega_F \lesssim \Omega_H$.

6.2.2 Comparing Ω_F in Event Horizon and Ergospheric Disk Jets

It was mentioned earlier that by adjusting initial conditions and time dependent boundary conditions that a maximal magnetic flux can be imparted to the event horizon, the MCAF solutions. These solutions almost completely thread the full 4π steradians of the horizon with large scale magnetic flux. The large amounts of flux and EHM jet power emanating from near equatorial latitudes is what distinguishes these types of solutions from other EHM solutions. The EDJ also has large jet power emanating form near the equator and close to the event horizon as evidenced by Fig. 6.1. Thus, it is of interest to compare the efficiency of the MCAF jet and the EDJ. Equation (6.1) shows the poloidal Poynting flux, S^P, in any relativistic MHD jet for a given magnetic flux contained within the jet, Φ, is proportional to Ω_F^2. Figure 6.3 compares Ω_F^2 in the MCAF simulation in McKinney et al. (2012) and

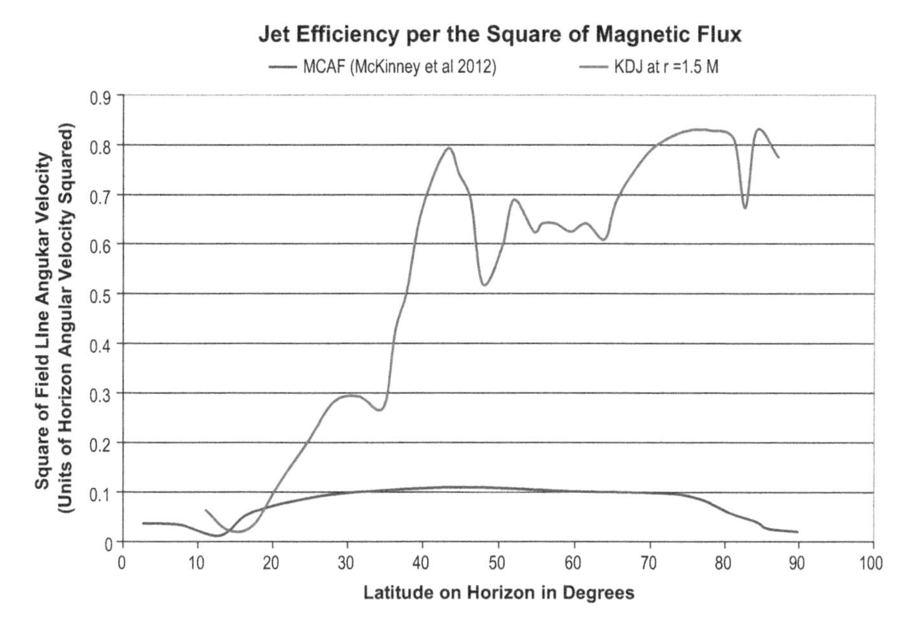

Fig. 6.3 A comparison of the time and azimuth average of Ω_F^2 in the MCAF solution of McKinney et al. (2012) and the azimuth average of Ω_F^2 of the EDJ in the KDJ simulation at t = 9,840 M as plotted in Fig. 6.2. Notice that the magnetic field rotates much faster in the EDJ than in an MCAF and by Eq. (6.1) provides much more power per unit magnetic flux in the jet

the EDJ simulation data taken from Fig. 6.2. Notice that the EDJ can create much larger Poynting fluxes per unit magnetic flux squared. We continue to compare and contrast these solutions in Sect. 6.4. It is worth noting that $\Omega_F < 0$ in the MCAF solution to the left of the minimum in Ω_F^2.

6.3 The Manifestation of the Ergospheric Disk in an Accretion Environment

This section describes the interaction of the accretion flow with the large scale magnetic field in the ergopshere near the equatorial plane in the simulation KDJ. This interaction resembles magnetically arrested accretion in many respects. As the gas is arrested, it is driven onto "negative energy" orbits by magnetic torques and this simultaneously converts the plasma energy into Poynting flux in the form of a relativistic jet. This type of interaction is referred to as a "gravitohydromagnetic" or GHM interaction in Punsly (2008).

6.3.1 Magnetically Arrested Streamlines

Figure 6.4 depicts a strong interaction exerted by the magnetic pressure from the vertical flux forming the base of the jet and the accreting gas in the simulation, KDJ, at t = 9,840 M. Two strong, twisted magnetic flux tubes are plotted, one intersects the false color density contour plot (restricted to the plane of the figure) on the left and one intersects the density contours on the right. The blue magnetic field lines terminate on the event horizon and were designated as type II magnetic field lines in Punsly et al. (2009). They traverse the mid-plane of the accretion flow, connecting the jet in one hemisphere to their point of origin on the event horizon in the opposite hemisphere. The red magnetic field lines, type I field lines, are distinguished by connecting to the Poynting jet in one hemisphere only, with the other end spiraling around within the accreting gas in the opposite hemisphere. Another distinguishing feature of type I field lines is that the azimuthal direction of the magnetic field changes its direction as the field line crosses the midplane of the accretion flow.

The Boyer-Lindquist evaluated gas density is seen in cross-section by means of a false color contour map in Fig. 6.4. The color bar on the right hand side of the figure is in code units and the scale was optimized to give the best contrast between the field lines, the accretion flow gas density and the background. The figure highlights two "magnetic walls," one on the left and one on the right. The "wall-like" nature of these flux bundles will be developed below. The figure attempts to capture all the details of magnetically arrested accretion at these magnetic walls. The choice of plotting preferences has the tremendous advantage that one can see the density of the gas in which the magnetic wall is embedded. In order to elucidate the structure of

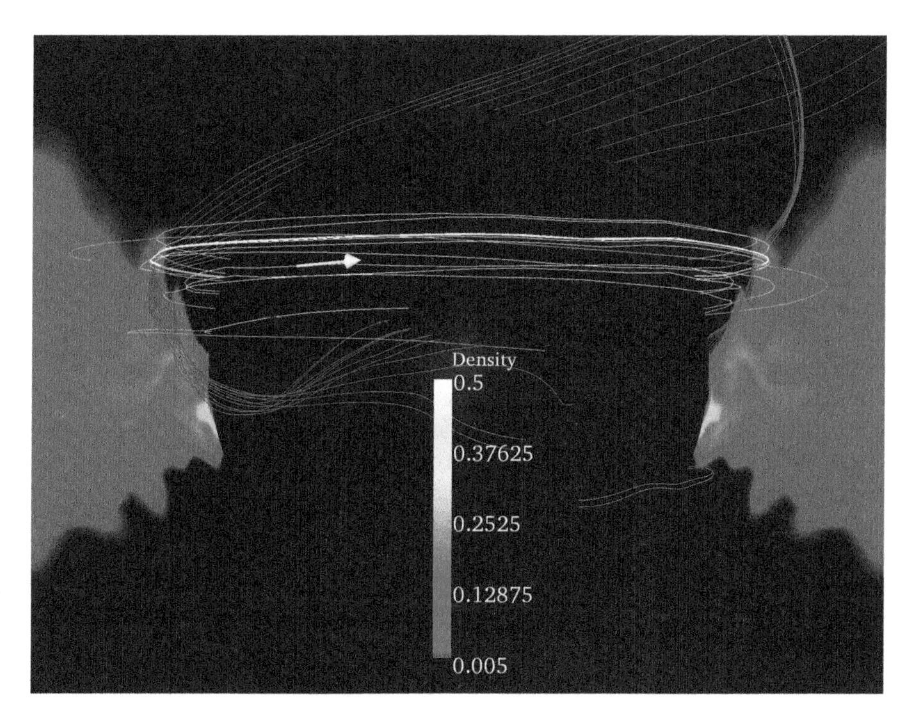

Fig. 6.4 The figure indicates a region of magnetically arrested accretion in the ergosphere (the GHM dynamo for the toroidal magnetic field and the Poynting jet). The *white* and *green curves* are integral curves of the Boyer-Lindquist evaluated velocity at t = 9,840 M, i.e., gas streamlines. The false color contour map of a cross-section of the density is in code units. The *blue* (type II) and *red* (type I) curves are magnetic field lines with the color coding described in the text. The *white curve* is a magnetically arrested streamline

the magnetic wall, the opacity of the gas was chosen to be relatively high (85 %), so that it was obvious when the field lines penetrated the gas within the cross-sectional slice. In particular, the field lines comprising the magnetic wall on the left hand side of Fig. 6.4, suddenly almost completely disappear when they intersect the red false color gas density contour. On the left hand side, we see a bundle of type II field lines. On the right hand side, the magnetic flux bundle appears to be a roughly equal mix of type I and type II field lines. They overlay almost on top of each other, so the red and blue meld into a maroon color.

On the backdrop of the magnetic flux bundles and the accretion flow, two "streamlines" of the accreting gas are plotted. The steamlines are integral curves of the local velocity vector field at the instant of time, t = 9,840 M. It is important to clarify for the reader that these are not true streamlines in the time dependent sense. The green and white lines do not represent global trajectories into the black hole, since the vertical magnetic field is dynamic and therefore so are the magnetic forces in KDJ. Thus, the magnetic forces could be significantly altered after a few revolutions of the black hole. For this reason, the trajectories are only representative of the gas propagation in a local sense. Two starting points for the integration of the gas streamlines were chosen. The first integration resulted in the white integral curve that was selected to interact strongly with the magnetic flux bundle on the right hand side and at the inner edge of the magnetic flux bundle at the left hand side. Only one arrow (to avoid cluttering the figure) is shown to indicate the direction of motion in both streamlines. The white curve shows us two things. First of all, the interaction of the gas with the field is significant as the integral curve is closely aligned with the kinks seen in the magnetic flux bundle. These kinks seem to represent local enhanced magnetic stresses that arise to counter-balance the gravitational forces and pressure gradients within the gas. Such a force balance resembles a magnetic Rayleigh-Taylor interface. These equilibria are not stable in general, so it is likely prone to significant magnetic deformations. Secondly, the accretion of the plasma that is represented by this streamline is clearly magnetically arrested. It takes many spirals around the black hole before it crosses the magnetic barrier by winding through small gaps and finally reaches the inner calculational boundary. Hence, the use of the term "magnetic wall." The magnetic wall wraps around the black hole azimuthally. The two flux bundles indicated in Fig. 6.4 represent two disjoint sections of the magnetic wall, so as not to leave the figure cluttered with field lines. The magnetic wall is very non-uniform with gaps and magnetic flux enhancements. Fortunately two of the strongest flux bundles intersect the plane of the figure (the plane of the false color density contour) separated by $\approx 180°$ in azimuth, providing maximum clarity.

To show that this streamline actually represents magnetically arrested gas, we plotted a "control" streamline in green. This streamline is chosen at a similar disk height, but farther out in the disk. It never strongly interacts with the magnetic wall. It reaches the event horizon by flowing always outside the magnetic wall. It penetrates the inner calculational boundary at an azimuthal coordinate just beyond the "footpoints" of the section of magnetic wall on the left. Thus, it flows toward the inner calculational boundary virtually unimpeded by magnetic forces. The green streamline in Fig. 6.4 is typical of most streamlines in the body of the accretion disk and indicates that a direct inflow takes only ≈ 1.5 spirals to reach the inner calculational boundary, a much shorter journey than the white streamline.

In summary, the slow inward spiraling white streamline is clearly magnetically arrested by the magnetic wall. Even though the white streamline does not represent a global time dependent path for the gas (because of the time variations noted above), it demonstrates that a strong interaction exists along the length of a slow inward spiral. The magnetic kinks and the low pitch angle spiral of the streamline indicates that the gas is being held up by magnetic forces virtually over the entire interface with the magnetic wall. The magnetically arrested property is depicted by Fig. 6.4 (even though the number of spirals around the black hole that it takes for a given fluid element to reach the black hole is not determined by this integration). The strongest interaction is on the right hand side. The details are highlighted in the Fig. 6.5. The false color gas density contours have been removed so that one can clearly see the magnetic wall bending in the poloidal plane around the pinching

Fig. 6.5 A closeup of the strong interaction of the plasma and field at the magnetic wall on the right hand side of Fig. 6.4 without the false color gas density contour in order to increase the clarity

coils of the white streamline. The curvature of the poloidal magnetic field creates an outward directed stress in the poloidal plane that opposes the gravitational force.

Even though some gas is magnetically arrested, the majority of the accreting gas finds its way to the black hole by maneuvering around the strong patches of vertical flux, passing through the azimuthal gaps in the vertical magnetic flux distribution and/or the vertical gaps in the azimuthal twists in the type I field lines near the equatorial plane. Dynamically, it is plasma on streamlines, like the white one, that create the (GHM) interaction that drains energy from the plasma and transfers it to the EDJ. This dynamic is developed in the next subsection.

6.3.2 Review of the Idealized Ergopsheric Disk and GHM

It is worth reviewing some of the connections that have been made between the Poynting jets in KDJ and the ergospheric disk model that were noted in

Punsly (2007) and in Chapter 11 of Punsly (2008). The model describes an ideal mechanism for initiating Poynting jets from an equatorial disk of plasma in the ergosphere if that plasma is threaded by vertical magnetic flux. What differentiates this mechanism from an accretion disk Poyntng jet, is that the power delivered by the jet is independent of the amount of energy that is stored within the disk plasma. For example, in the idealized model of Punsly and Coroniti (1990b), the equatorial plasma was positronic, with an energy density comparable to that of the magnetic field. The accretion rate in the equatorial plane was slow and balanced by annihilation. There was no turbulence and it did not relate to a quasar environment. The advantage of the idealized model is that the physics of the GHM Poynting jet creation was clearly evident. In this extreme instance the jet extracted more energy than existed in the plasma. We review the details of this phenomenon below.

To understand this, note that in analogy to the redshifted Poynting flux, there is a mechanical energy flux associated with the accreting plasma and the two are coupled by the energy conservation equation. In Boyer-Lindquist coordinates one can use the four-momentum, P^{μ} of the plasma to define the mechanical energy, ω, and the angular momentum, m;

$$\omega = -P \cdot \partial/\partial t \ , \tag{6.10}$$

$$m = P \cdot \partial/\partial \phi \ . \tag{6.11}$$

The quantities, ω and m, are sometimes called the redshifted energy and angular momentum density of the gas. These are the energy and angular momentum densities that are relevant to the global energy conservation law. Within the ergosphere, there are physical states (i.e., the energy is positive in a local inertial frame) in which $\omega < 0$. The Penrose process of energy extraction from a black hole operates by splitting a hypothetical particle into two pieces, one with $\omega < 0$ and one with $\omega > 0$ (Penrose 1969). By sending $\omega < 0$ particles or plasma into the black hole and sending $\omega > 0$ particles or plasma outward, the black hole is effectively powering an outflow of mechanical energy (Penrose 1969). As such, this process can extract energy from the black hole if it exists in isolation. All $\omega < 0$ states are also negative angular momentum states, $m < 0$. Thus, the Penrose process extracts rotational inertia from the black hole. This process only occurs near rotating black holes where there is an ergosphere and a frame dragging force associated with gradients in the metric coefficient $g_{\phi t}$.

The ergospheric disk driven jet operates like a Penrose process. The tenuous plasma within the atmosphere of positronic gas threads the large scale magnetic field lines to distances that extend far from the black hole. The frame dragging forces in the ergospheric disk cause the plasma to corotate with the event horizon, even $m < 0$ states are forward rotating with respect to infinity. This is actually a gradual condition that occurs because of the dragging of inertial frames throughout the ergosphere (the region in which $g_{tt} > 0$). In particular, the plasma angular velocity, $d\phi/dt \equiv \Omega_p$, is bounded by

$$\Omega_{\min} \leq \Omega_p \leq \Omega_{\max} , \tag{6.12}$$

$$\Omega_{\min} = \frac{-g_{\phi t}}{g_{\phi\phi}} - \frac{\sqrt{\Delta}\sin\theta}{g_{\phi\phi}} , \tag{6.13}$$

$$\Omega_{\max} = \frac{-g_{\phi t}}{g_{\phi\phi}} + \frac{\sqrt{\Delta}\sin\theta}{g_{\phi\phi}} , \tag{6.14}$$

$$\lim_{r\to r_+} \Omega_{\min} = \Omega_H , \qquad \lim_{r\to r_+} \Omega_{\max} = \Omega_H , \tag{6.15}$$

$$\Omega_{\min} > 0 \quad \Rightarrow \quad g_{tt} > 0 . \tag{6.16}$$

The condition that is described by Eq. (6.12) is equivalent to bounding the magnitude of the azimuthal velocity less than the speed of light in any local physical frame. In fact, according to Eq. (6.15), within the inner ergosphere, all the disk plasma essentially corotates with the horizon ($d\phi/dt \approx \Omega_H$), even if $m \ll 0$. The gist of this is that the ergospheric disk plasma is rotating much faster than the distant positronic plasma that threads the large scale magnetic field. Thus, the large scale magnetic field (the vertical flux through the equatorial plane of the ergosphere) transmits large magnetic stresses between the ergospheric plasma and the distant atmosphere. The azimuthal drag imposed by the positronic magnetosphere twists up the magnetic field and torques the ergospheric disk plasma. Eventually, the angular momentum becomes very negative and $\omega < 0$ in the ergospheric disk. The field is approximately frozen into the plasma and because of the aforementioned dragging of inertial frames, the field lines are rotating as well, $\Omega_F \approx d\phi/dt \approx \Omega_H$. Note that this occurs in Figs. 6.2 and 6.3 for the KDJ simulation. A nested set of rotating coiled magnetic field lines is essentially a Poynting jet. This process is the GHM version of the Penrose process. In the analogy to the Penrose process, the ingoing particle with $\omega < 0$ is the ergospheric disk plasma and the outgoing $\omega > 0$ particle is the Poytning jet.

6.3.3 Comparison of Ergospheric Disk to the Accretion in KDJ

Negative energy plasma occurs in the magnetically arrested accretion flow in KDJ. From Eq. (6.10), one can write the condition for negative redshifted specific mechanical energy condition as

$$\omega = \mu(nu^t)(-u_t) < 0 , \tag{6.17}$$

where μ is the enthalpy per particle, n is the proper density, (nu^t) is the Boyer-Lindquist evaluated density and u_t is the covariant component of the Boyer-Lindquist evaluated plasma bulk flow velocity. For example, in a cold gas released at rest from asymptotic infinity, $-u_t = 1$. The effect of kinetic energy is to make $-u_t > 1$. In fact, the original Penrose process was defined in terms of an ingoing $-u_t < 0$ particle and an outgoing $-u_t > 0$ particle. The plot, in Fig. 6.6, is provided

Fig. 6.6 This figure shows the creation of negative energy on magnetically arrested streamlines. The *white curve* is the magnetically arrested streamline from Fig. 6.4. The false color plot is the contour of the specific energy as observed form asymptotic infinity, $-u_t$, with a value $-u_t = -0.1$. The *blue curves* are the magnetic field (type II) lines from Fig. 6.4

to show that the magnetically arrested plasma in Fig. 6.4 attains negative energy inside of the magnetic wall. Figure 6.6 shows a closeup of the same magnetically arrested streamline that appears in Fig. 6.4 as it approaches the event horizon. The false color plot is the contour of the specific energy with a value of –0.1 as observed from asymptotic infinity (i.e., $-u_t = -0.1$). The opacity of the contour is chosen to be 85 %. Notice that on final approach to the event horizon, the magnetically arrested streamline penetrates the $-u_t = -0.1$ contour. By Eq. (6.17), this gas is of negative energy just as it is in GHM and in the Penrose process. In Chapter 11 of Punsly (2008), it was shown that the plasma loses energy continuously, but rather abruptly as in flows through the putative GHM dynamo for the toroidal magnetic field (equivalently, the source of the Poynting flux). To appreciate this fact, note that just beyond the outer edge of the GHM dynamo, at $r \approx 1.6M$, $-u_t \approx 1$. In the dynamo region, $-u_t$ decreases. In less dense regions of the accretion flow $-u_t < 0$ as the plasma traverses the dynamo at $r < 1.6M$ (as in Fig. 6.6). In the densest regions of the GHM dynamo, $-u_t \approx 0.25$ by the time that it reaches the inner calculational boundary. In summary, as the plasma accretes radially inward in the dynamo region, it crosses the Poynting flux generation region (see Fig. 6.1) and simultaneously experiences a decrease in specific mechanical energy. This decrease continues as the power is extracted electromagnetically even until the mechanical energy becomes negative for some streamlines in KDJ. Thus, the extracted energy is more than the energy stored in this plasma in localized regions. Therefore, the process creating the Poynting jet is independent of the stored energy within the plasma and is driven by the only other dynamic element available, frame dragging. It should be noted that in KDJ, there is a flood of accreting gas from the MRI (magneto-rotational instability) driven accretion flow, so there is more positive energy than negative energy gas that reaches the inner calculational boundary. It should also be noted that the accretion flows in the existing 3-D MHD simulations around rotating black holes is single-fluid and protonic. Consequently, the $-u_t = -0.1$ contour in Fig. 6.6 corresponds to plasma with an energy per particle of $< -100\,\mathrm{MeV}$!

The GHM process is more general than the simplified ergospheric disk. The basic idea is that any plasma that is held up by "magnetic ropes" deep in the ergosphere (such as the vertical field) will experience a torque and if the gas is of low enough density and if the magnetically arrested state lasts long enough then an $\omega < 0$ state will be attained. The longer that the ergospheric plasma resides in a magnetically arrested state, the more energy that is extracted by the GHM mechanism. The energy generated by this process is not limited by the inertia of the plasma. Consequently, magnetically arrested states in the ergosphere are directly related to GHM.

An effective ergospheric disk exists within the inner accretion flow of the KDJ simulation as demonstrated by the following points:

1. Just as in the ergospheric disk, the Poynting flux emerges from the ergospheric equatorial accretion flow.
2. The GHM dynamo is triggered by the ergospheric plasma accretion towards the black hole being impeded by a large scale poloidal magnetic flux barrier. Within KDJ there are strong patches of vertical flux coincident with the base of the Poynting jets.
3. For a putative Blandford–Znajek process within a magnetosphere shaped by the accretion vortex, the field line angular velocity is, $\Omega_F \approx \Omega_H/2$ near the pole and decreases with latitude to $\approx \Omega_H/5$ near the equatorial plane of the inner ergosphere (Phinney 1983; McKinney et al. 2012). In a GHM ergospheric disk, since the magnetic flux is anchored by the inertia of the accretion flow in the inner ergosphere, frame dragging enforces $d\phi/dt \approx \Omega_H$. One therefore has the condition, $\Omega_F \approx \Omega_H$ in the inner regions of the ergosphere. This GHM condition of $\Omega_F \approx \Omega_H$ was shown to hold in KDJ in regions that are spatially and temporally coincident with the base of the Poynting jet in Sect. 6.2, Figs. 6.2 and 6.3.
4. The torsional magnetic stress between the plasma-filled vertical flux tubes and the equatorial plasma creates ergospheric disk plasma with negative mechanical energy (the Penrose process). It is shown that this occurs as the plasma accretes

through the dynamo for toroidal magnetic field at the base of the Poynting jet, even in the presence of an intense bath of accreting positive energy protonic plasma.

Figures 6.4 and 6.5 describe the relationship between the white magnetically arrested streamline and GHM. The magnetically arrested gas interacts with the magnetic field for a prolonged period and the results are accumulative. The whole time the interaction is occurring, the GHM interaction is removing energy and angular momentum from the gas and transferring it to the magnetic field. By the time that the streamline passes through the magnetic barrier into the low density gap between the magnetic wall and the black hole its energy, as observed from asymptotic infinity, has become negative, $\omega < 0$. The inner region of the accretion flow and corona in KDJ is the high accretion rate analog of the ergospheric disk.

An important aspect that has been typically missed in the discussion of the EDJ in the context of intense accretion is the following. A strong GHM interaction is concentrated predominantly in discrete magnetic islands. Only within these magnetic islands is negative energy plasma created. The entire accretion flow is not negative energy. Even though the magnetic islands in which the GHM interaction occurs might be small individually, the cumulative effect of many islands within the ergosphere results in a prodigious jet as indicated in the plot from KDJ in Fig. 6.7.

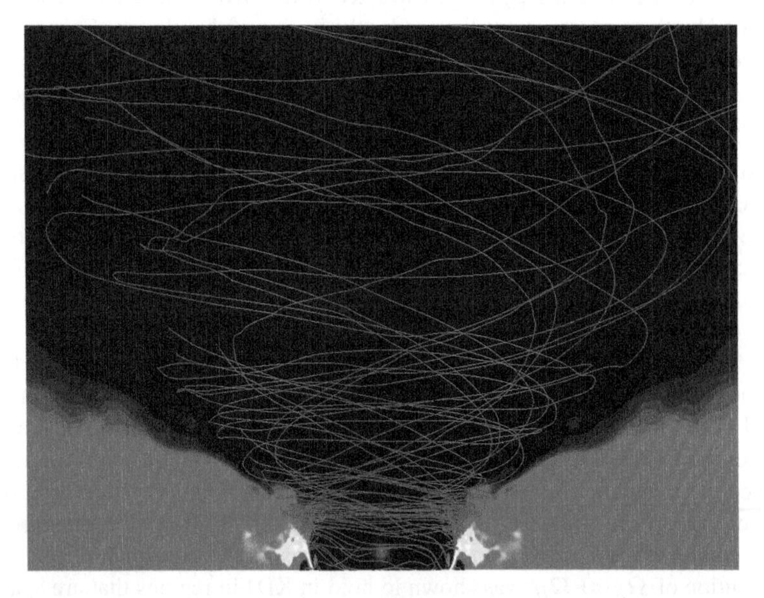

Fig. 6.7 The net effect of isolated islands of GHM interactions can be quite prodigious as illustrated by the EDJ at t = 10,000 M in KDJ above. This figure shows that the magnetic flux that supports the powerful jet emerges from the coronal and equatorial gas in the ergosphere

6.4 Fate of Accreted Flux: Reconnection

In this section, the critical role of reconnection in the determination of the global magnetospheric topology is examined. It is striking to realize that the entire topology can be determined by the microphysics of reconnection in a very small region near the black hole. This is explicitly illustrated in the example of Fig. 6.8. This issue has been known as critical for a quarter of a century when it was asked in Punsly and Coroniti (1990b), where does the magnetic flux go when it is transported inward through the radial extent of the accretion disk? A related issue is the fate of the flux generated at the inner edge of an accretion disk by the Poynting-Robertson effect (Contopoulos and Papadopoulos 2012).

The status of modeling the EHM has evolved greatly in the first decade of long duration numerical simulations. Initial claims of the discovery of robust physics were based on the independence of the final state from a set of "natural" initial conditions. It was later demonstrated that the jets from such an EHM were necessarily weak by astrophysical standards and were not likely to relate to the set of problems for which they were intended. Thus, the "naturalness" condition was abandoned in favor of adjusting initial conditions until a maximum jet strength was achieved. The current sentiment is that the end result of this search for initial conditions and boundary conditions has been found and this is the set of simulations that describe relativistic, astrophysical jets – the MCAFs (McKinney et al. 2012; Sikora and Begelman 2013). The MCAF poloidal field line geometry near the equatorial plane of the ergospheric accretion flow is plotted in the top frame of Fig. 6.8. Below this is the inner accretion flow envisioned in Punsly and Coroniti (1990b) that describes the formation of a time dependent magnetosphere of vertical magnetic flux through the equatorial plane of the ergosphere – an EDM. The comparison of the two scenarios in Fig. 6.8 highlights the similarity of an MCAF and an EDM poloidal field configuration near the equatorial plane in the ergosphere. The two frames also contrasts the different physics that occurs.

First, consider the MCAF. The initial conditions and time dependent boundary conditions are chosen to maximize the magnetic flux in the EHM. As such, the field fans out to low latitudes thereby compressing the accretion flow to a geometrically thin cusp. The compressed thin accretion flow is shown schematically as the gray shading. The density is very high. The gas is also very hot as there is no radiation in the simulations. Since the simulations are performed in the single-fluid approximation, the gas temperature $\sim$1 GeV! Maximum field strength and maximum jet power are achieved in the EHM just above the equatorial plane. Equilibrium is obtained vertically near the equatorially plane by the balance of high gas pressure with high magnetic pressure that occurs both above and below the thin equatorial accretion cusp. This configuration appears to be both pinch and Rayleigh-Taylor unstable (Syrovatskii 1981). Rayleigh-Taylor instability does occur in the numerical simulations, allowing gas to accrete more efficiently. However, the pinch instability rate and the associated reconnection rate is so slow that there is never significant poloidal flux through the equatorial plane of the ergosphere as was

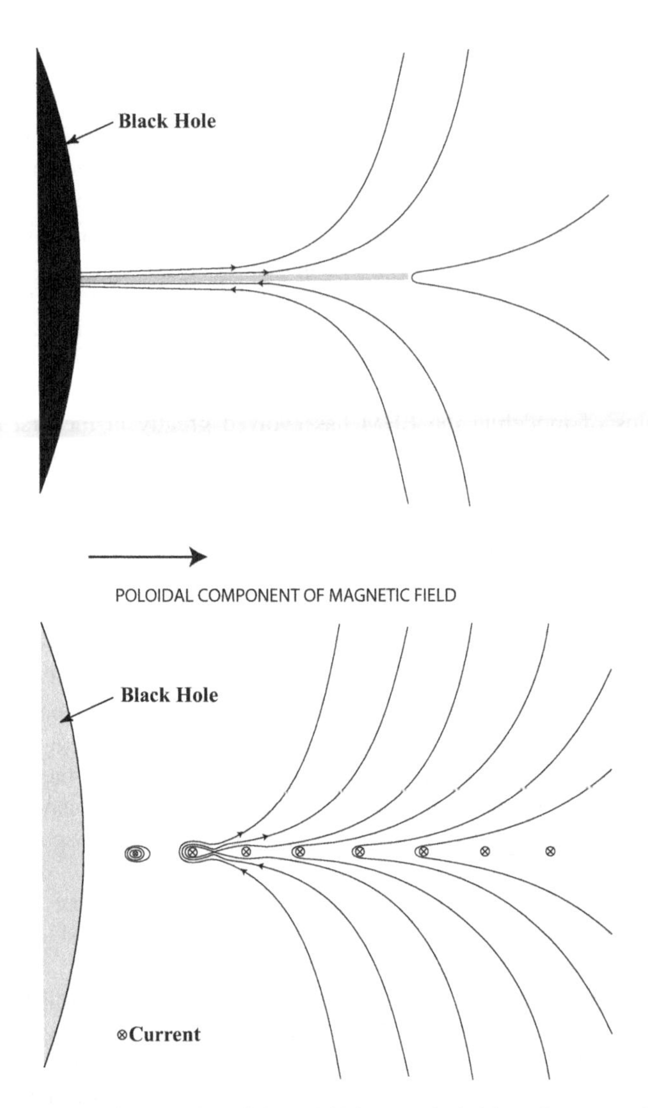

Fig. 6.8 The *top frame* is indicative of the poloidal magnetic field topology near the equatorial plane in the ergosphere in the MCAF solutions as illustrated in the magnetic field plots of McKinney et al. (2012). The *black curves* are schematic representations of the poloidal magnetic field near the equatorial plane of the ergosphere and the *gray shading* represents the pinched accreting gas in an equatorial mass sheet. The *bottom frame* is a figure from Punsly and Coroniti (1990b) that describes a different dynamic in this same region. The *bottom frame* leads to an ergospheric disk. The MCAFs arise in single-fluid perfect MHD simulations. A more complete physical treatment should render the MCAF configuration as being spontaneously unstable to reconnection in the equatorial plane of the ergosphere. If so, this would reconcile the physical picture in the *top frame* with that on the *bottom*

anticipated in the scenario depicted in the bottom frame of Fig. 6.8 (McKinney et al. 2012).

Now consider the proposed dynamics from Punsly and Coroniti (1990b). It is assumed that the pinch sheet is very unstable deep in the ergosphere (Syrovatskii 1981). The instability geometry is conducive to the occurrence of reconnection along a 1 dimensional curve of X-points (i.e., in an axisymmetric configuration this would be a circle). The pinched off poloidal field forms a 1 dimensional set of O-points. The field contracts into this "O-line" as the "O-line" shrinks into the event horizon. The overall dynamic is sometimes called a tearing mode instability (Syrovatskii 1981). After the excision of the dense equatorial plasma, the large scale flux tubes become buoyant within the accretion flow. As the flux tubes become buoyant within the accretion flow. they work their way outward by means of interchange instabilities and are re-threaded by plasma on a time scale that depends on the numerical recipe for resistivity (if there is one) and numerical diffusion. Depending on the diffusion time scale, the added inertia imparted to the flux tubes allows gravity the ability to drag this field line into the ergosphere again, repeating the cycle. This means that a time dependent EDM forms with a significant time averaged poloidal flux. Unfortunately, the diffusion time scale in perfect MHD simulations is an artifact of numerical diffusion and is not representative of a physical time scale. Thus, the current suite of single-fluid perfect MHD simulations are incapable of determining the relevant time scales correctly and therefore the time evolution of buoyant flux tubes cannot be modeled accurately. Thus, existing numerical work is incapable of determining the two relevant dynamical elements that form an ergospheric disk: the rate at which buoyant flux tubes are created by reconnection, nor determine the time evolution of buoyant flux tubes in the ergospheric plasma.

Considering the difference between MCAF and EDJ jet efficiency illustrated in Fig. 6.3, it is worth understanding which physical scenario is more physical. In the MCAF, MHD simulations magnetic field reconnection proceeds by numerical diffusion. It is not even a physical process, yet it is determining the global field topology. It is proposed here that radiation losses make the equilibrium depicted in the top frame of Fig. 6.8 very unstable. The un-physical gas temperatures should make the gas prone to radiative losses. Given the non-stationary highly variable field line distribution and field line motion seen in the 3-D simulations of the EHM (see for example the movies linked to McKinney et al. 2012; Tchekhovskoy et al. 2011; Tchekhovskoy and McKinney 2012), one finds large and rapid variation of the magnetic pressure that is responsible for compressing the ergospheric mass sheet. Thus, the compressive heating is variable as well. Therefore, one expects corresponding sporadic releases of radiation (with realistic physics) that collapse the pinched accretion flow and thereby drive the X-type of reconnection seen in the bottom frame of Fig. 6.8. After reconnecting, the footpoints of the open field lines will discontinuously jump from the event horizon to the ergospheric disk. This should be happening continually. The radial inflow induced by gravity keeps replenishing the field and plasma that creates this very unstable situation.

In the spirit of trying to find the actual physical configuration, we point out the various elements that are missing from the analysis.

1. The most obvious missing ingredient in the numerical recipe is resistivity that is required for reconnection (Baumann et al. 2013; Yamada 2007). However, this is likely the most complicated aspect to implement. A realistic Ohm's law is not at hand. A complicating factor is that the conductivity tensor is not isotropic as a consequence of the magnetic field. A proper derivation requires a kinetic treatment and possibly a particle in cell analysis perhaps ameliorated by a multi-fluid approximation (Zenitani et al. 2011, 2013; Zocco and Schekochihin 2011). An accurate model of resistivity is also required for the late stages of the interaction – the fate of magnetic flux tubes after they reconnect within the inner disk and plasma gets re-threaded on the flux tubes by diffusion.
2. Two-fluid effects are often considered in reconnection dynamics. Treating the protons and electrons equivalently in a very hot environment is not justified. In particular, the Hall term that arises in a two-fluid treatment is considered important in solar system reconnection (Malakit et al. 2009). Even in the two-fluid formalism, a high fidelity model of resistivity is crucial (Threlfall et al. 2012).
3. Radiation effects are critical. As mentioned above radiation losses will be an important driver in initiating the pinch instability that leads to a reconnection environment. Secondly, radiation effects are often considered an important dynamic in establishing in the reconnection rate (Uzdensky 2011).

It is likely that all of these elements need to be incorporated before a realistic modeling of the EHM can be considered to have been achieved. As a more general consideration of point 2, involving two-fluid MHD and reconnection, the single-fluid approximation is not likely to be robust in a system that has nonuniform physical properties. For example, in astrophysical accretion systems, the thermal conduction in the corona (two-temperature plasma) and the accretion disk (single temperature plasma) are conjectured to be of different origins (Esin et al. 1997). Thus, the coupling of the ionic and electronic components is not uniform. The ionic coupling creates a rich and varied array of plasma waves and related instabilities that simply do not occur in a single-fluid approximation (Stix 1992). An accurate account of the dynamics in the disk-corona interface is critical to understanding potential magnetic field reconnection sites and rates. This cannot be accomplished with single-fluid perfect MHD. Kinetic theory must also be considered in the very hot coronal region of a realistic plasma. This will alter the wave propagation speeds and damp certain plasma modes in the high temperature corona that is not considered to be in thermodynamic equilibrium (Stix 1992). In summary, it seems premature to claim that the plasma dynamics is accurate enough in existing single-fluid numerical codes to definitively determine the distribution of magnetic flux around a black hole. We might not be as close as advertised to knowing the magnetospheric topology around black holes.

6.5 Particle Injection and Mass Floors

The relationship of the mass floor to true particle injection is not clear. Since the process resides outside the realm of ideal single-fluid MHD, it cannot be resolved within the confines of ideal single-fluid MHD numerical simulations. We explore some interesting theoretical issues associated with this topic in this section.

6.5.1 *Simulating* $\Omega_F(\theta) \ll \Omega_H/2$ *Winds from the EHM*

It is worth commenting on physical particle injection and numerical mass floors in the EHM. For many years, the creators of the numerical simulations of jets from the EHM intimated that the solutions were representative of nature because they were independent of the net magnetic flux in the initial state. We were also being told that the use of a mass floor was justified. The first claim seems to have been abandoned by many. The earliest numerical attempts would crash after a finite time as the plasma was drained off the field lines into the black hole and slung out to infinity (Meier et al. 2001). A mass floor resolves this issue and allows the simulations to continue in perpetuity. It was also stated that the results of the simulations did not depend on the mass floor and the system naturally evolves to $\Omega_F(\theta) \approx \Omega_H/2$ as in the Blandford and Znajek (1977) analytic solution (McKinney and Gammie 2004; McKinney 2006). However, consider Fig. 6.3. The MCAF solution has $\Omega_F(\theta) \ll \Omega_H/2$ for $\theta < 25°$. The explanation in McKinney et al. (2012) is "... we cannot compare the full $\Omega_F(\theta)$ because in the simulations this is affected by ideal MHD effects and numerical floor mass injection near the polar axes." So apparently, the particle injection mechanism (even if un-physical such as a mass floor) can affect the power output and Ω_F.

Thus, the numerical simulations have rediscovered a few facts shown analytically years before. One of the main conclusions of Punsly and Coroniti (1989) was that the event horizon will passively any value of Ω_F imposed on it. Consequently, it is not surprising that various numerical simulations have found values ranging from $-0.1\Omega_H \leq \Omega_F \leq \Omega_H$ (Punsly 2007; McKinney et al. 2012).

The solution space analysis of Punsly (1998) demonstrated that there were other MHD solutions besides those proposed in Blandford and Znajek (1977) and Phinney (1983) with much smaller values of Ω_F, like what is found at $\theta < 25°$ in the MCAF. Can these be obtained numerically as a global solution? If we follow the logic of the numerical simulations presented in Tchekhovskoy et al. (2010) then the answer must be yes. They claimed that the accretion flow and corona can be replaced by a boundary surface and accurate results can be obtained for the EHM jet. In fact they used these results in an attempt to explain real physical systems, AGN jets. Thus, it follows that the MCAF solution in McKinney et al. (2012) can be recreated for field lines emanating from $\theta < 25°$ at the horizon by the following prescription.

- The same mass floor as the MCAF is imposed.
- The boundary surface conforms to the time averaged poloidal field lines (flux tube) emerging from the event horizon at $\theta = 25°$. The boundary surface would be used to specify the time averaged MCAF values of the magneto-fluid restricted to this surface.

Following Tchekhovskoy et al. (2010), this should successfully recreate the MCAF solution interior to the boundary surface. In particular, it would simulate an EHM jet with an average $\Omega_F \approx 0.1\Omega_H$ (see Fig. 6.3) in accordance with the analytic results of Punsly (1998). The general setup of this solution and the relevant boundary surface is indicated by the diagram in Fig. 6.9. The reason that this needed to be rediscovered is that the perfect MHD versions of the Blandford and Znajek (1977) solutions are the unique ideal MHD, axisymmetric solutions to the trans-field momentum equation (pulsar equation) for which there are no boundaries in the upper $r - \theta$ half-plane (Phinney 1983). There is no explicit plasma injection either. When the EHM is shaped by an external boundary and plasma injection is a free parameter more MHD solutions are allowed. An important aspect of these low Ω_F solutions is the strong collimation of the wind zone toward the polar axis (Punsly 1998).

6.5.2 Protonic Ergospheric Disk Jets

On an unrelated note, the solution KDJ as depicted in Fig. 6.1 has taught us something interesting about the EDJ. Notice that it emerges from the corona above the inner accretion flow. The plasma in the EDJ can be protonic (or positronic) since the flow division point can be inside of the corona. Thus, the corona can feed protonic material to the out-flowing relativistic jet. By contrast, a relativistic EHM jet is charged starved and must be positronic near the black hole. Entraining significant protonic material farther out would likely diminish the relativistic speed. Furthermore, highly supersonic flows have highly suppressed rates of entrainment (Bicknell 1994). Thus, if an ultra-relativistic jet is born positronic, it should remain positronic unless it sheds its relativistic nature. The clear discovery of protonic relativistic jets in AGN or Galactic black holes would be strong evidence in favor of the EDJ.

6.5.3 Dissipative EHM Winds

In Punsly (1991), it was shown that another class of wind solutions can exist in the EHM besides those of Blandford and Znajek (1977). These winds arise as a consequence of the injection of very low angular velocity plasma near the black hole. The particle angular velocity as viewed from asymptotic infinity is such that

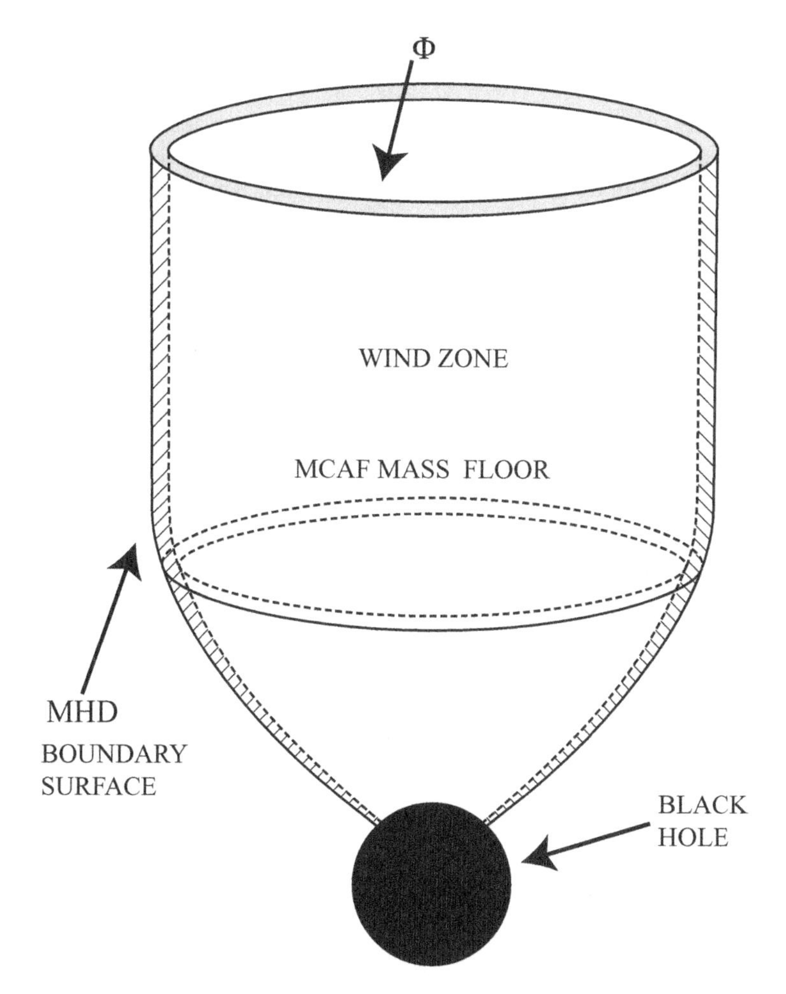

Fig. 6.9 A boundary value problem equivalent to the polar wind of the MCAF solution of McKinney et al. (2012) based on the methods implemented in Tchekhovskoy et al. (2010). The wind zone is created by am MHD boundary surface that is contoured to the time averaged poloidal flux tube that emanates from the event horizon at $\theta = 25°$. The wind contains a total magnetic flux Φ. The information on the boundary surface is realized by the time averaged MHD data from the MCAF simulation that is restricted to this flux tube. The mass floor is chosen to be identical to the prescription of McKinney et al. (2012). The resultant EHM jet has a time and spatially averaged value of $\Omega_F \approx 0.1\Omega_H$. An important aspect of these solutions is the strong collimation of the wind zone

$d\phi/dt$ is much less than the value of Ω_F in the corresponding field line geometry based on the methods of Blandford and Znajek (1977). These winds can violate MHD in the ergosphere due to radiation losses. The resultant EHM winds are weak with a final small value of Ω_F that is similar to that of the perfect MHD winds described at the beginning of this section.

6.6 Discussion

In this chapter, the potential role of the ergospheric disk jet was promoted for astrophysical systems. This was accomplished in the framework of 3-D single fluid MHD simulations around rotating black holes. These are potentially the most energetic black hole jets because of the elevated Ω_F as compared to EHM jets. Secondly, we pointed out that the difference between the MCAF and EDJ solutions appears to be related to reconnection in the equatorial regions near the event horizon. Since reconnection is a numerical artifact in the current suite of simulations, the actual magnetic field topology in the equatorial regions near the event horizon must be considered an unknown to date. It is well understood in solar astronomy that simulating reconnection accurately is extremely difficult and is not likely accomplished with simplified treatments of the particle kinetics. Many of the complications were highlighted in Sect. 6.4.

The fact that the EDJ can dominate the jet launching dynamics after $\sim 10^6$ time steps in a high resolution 3-D numerical simulation is something that cannot really be ignored. Considering that state of the art super-computing has found these solutions in single-fluid MHD and the tremendous uncertainty in the field configuration that results from inadequate modeling of reconnection in the present suite of simulations, any claims that jets only arise in the EHM seems premature. Considering the complex, unknown microphysics of the high temperature reconnection, accurate modeling of the phenomena cannot be verified in the near future. We might need to rely on empirical information from observation to determine the jet origin. An "event horizon telescope" such as a submm array will not be able to discern a difference. Figures 6.2, 6.3 and 6.5 illustrate that given the uncertainty in supermassive black hole masses, spins and orientations, the difference between a radius of 1 and 1.4 M (the event horizon size versus the ergospheric disk size) is not a realistic item to test. A possible difference between the EDJ and jets from the EHM was noted in Sect. 6.5. The EHM jet is charged starved and is necessarily positronic if it is relativistic. In an accretion scenario, the EDJ jet emanates from the corona of the innermost accretion disk. Thus, it can not only transport positronic plasma, but protonic plasma as well at ultra-relativistic speeds.

References

Baumann, G., Galsgaard, K. Norlund, A.: 3D Solar null point reconnection MHD simulations. Sol. Phys. **284**, 467 (2013)

Beckwith, K., Hawley, J., Krolik, J.: The influence of magnetic field geometry on the evolution of black hole accretion flows: similar disks, drastically different jets. ApJ **678**, 1180 (2008)

Beckwith, K., Hawley, J., Krolik, J.: Transport of large-scale poloidal flux in black hole accretion. ApJ **707**, 428 (2009)

Bicknell, G.: On the relationship between BL lacertae objects and fanaroff-riley I radio galaxies. ApJ **422**, 542 (1994)

Blandford, R., Znajek, R.: Electromagnetic extraction of energy from kerr black holes. MNRAS. **179**, 433 (1977)
Contopoulos, I., Papadopoulos, D.: The cosmic battery and the inner edge of the accretion disc. MNRAS. **425**, 147 (2012)
De Villiers, J-P., Hawley, J., Krolik, J.H.: Magnetically driven accretion flows in the kerr metric. I. models and overall structure. ApJ **599**, 1238 (2003)
Esin, A., McClintock, J.E., Narayan, R.: Advection-dominated accretion and the spectral states of black hole X-ray binaries: application to nova muscae 1991. ApJ **489**, 865 (1997)
Hawley, J., Krolik, J.: Magnetically driven jets in the kerr metric. ApJ **641**, 103 (2006)
Hirose, S., Krolik, K., DeVilliers, J., Hawley, J.: Magnetically driven accretion flows in the kerr metric. II. structure of the magnetic field. ApJ **606**, 1083 (2004)
Igumenshchev, I.V.: Magnetically arrested disks and the origin of poynting jets: a numerical study. ApJ **677**, 317 (2008)
Igumenshchev, I.V., Narayan, R., Abramowicz, M.A.: hree-dimensional magnetohydrodynamic simulations of radiatively inefficient accretion flows. ApJ **592**, 1042 (2003)
Krolik, J., Hawley, J., Hirose, S.: Magnetically driven accretion flows in the kerr metric. IV. dynamical properties of the inner disk. ApJ **622**, 1008 (2005)
Lubow, S.H., Papaloizou, J.C.B., Pringle, J.E.: Magnetic field dragging in accretion discs. MNRAS **267**, 235 (1994)
Malakit, K., Cassak, P., Shav, M., Drake, F.: The hall effect in magnetic reconnection: hybrid versus hall-less hybrid simulations. Geosphys. Res. Lett. **36**, L07107 (2009). doi:10.1029/2009GL037538
McKinney, J.: General relativistic magnetohydrodynamic simulations of the jet formation and large-scale propagation from black hole accretion systems. MNRAS **368**, 1561 (2006)
McKinney, J., Blandford, R.: Stability of relativistic jets from rotating, accreting black holes via fully three-dimensional magnetohydrodynamic simulations. MNRAS Lett. **394**, 126 (2009)
McKinney, J., Gammie, C.: A measurement of the electromagnetic luminosity of a kerr black hole. ApJ **611**, 977 (2004)
McKinney, J., Tchekhovskoy, A., Blandford, R.: General relativistic magnetohydrodynamic simulations of magnetically choked accretion flows around black holes. MNRAS **423**, 3083 (2012)
Meier, D.L.: A magnetically switched, rotating black hole model for the production of extragalactic radio jets and the fanaroff and riley class division. ApJ **522**, 753 (1999)
Meier, D.L., Koide, S., Uchida, Y.: Magnetohydrodynamic production of relativistic jets. Science **291**, 84 (2001)
Penrose, R.: Extraction of rotational energy from a black holes. Nuovo Cimento Nuovo Cimento Rivista Serie **1**, 252 (1969)
Phinney, E.S.: Ph.D. Dissertation, A theory of radio sources. University of Cambridge (1983)
Pontin, D.I.: Three-dimensional magnetic reconnection regimes: a review, Adv. Space Res. **47**, 1508 (2011)
Punsly, B.: Magnetically dominated accretion onto black holes. ApJ **372**, 424 (1991)
Punsly, B.: Minimum torque and minimum dissipation black hole-driven winds. ApJ **506**, 790 (1998)
Punsly, B.: Three-dimensional simulations of ergospheric disk-driven poynting jets. ApJL **661**, 21 (2007)
Punsly, B.: Black hole gravitohydromagnetics, 2nd edn. Springer, New York (2008)
Punsly, B., Coroniti, F.V.: Electrodynamics of the event horizon. Phys. Rev. D **40**, 3834 (1989)
Punsly, B., Coroniti, F.V.: Relativistic winds from pulsar and black hole magnetospheres. ApJ **350**, 518 (1990a)
Punsly, B., Coroniti, F.V.: Ergosphere-driven winds. ApJ **354**, 583 (1990b)
Punsly, B., Igumenshchev, I.V., Hirose, S.: Three-dimensional simulations of vertical magnetic flux in the immediate vicinity of black holes. ApJ **704**, 1065 (2009)
Rothstein, D., Lovelace, R.V.E.: Advection of magnetic fields in accretion disks: not so difficult after all. ApJ **677**, 1221 (2008)

Sikora, M., Begelman, M.: Magnetic flux paradigm for radio loudness of active galactic nuclei. ApJL **764**, 24 (2013)

Stix, T.: Waves in plasmas. American Institue of Physics, New York (1992)

Syrovatskii, S.: Pinch sheets and reconnection in astrophysics. ARA& A **19**, 163 (1981)

Tchekhovskoy, A., McKinney, J.: Prograde and retrograde black holes: whose jet is more powerful?. MNRAS Lett. **423**, 55 (2012)

Tchekhovskoy, A., Narayan, R., McKinney, J.: Black hole spin and the radio loud/quiet dichotomy of active galactic nuclei. ApJ **711**, 50 (2010)

Tchekhovskoy, A., Narayan, R., McKinney, J.: Efficient generation of jets from magnetically arrested accretion on a rapidly spinning black hole. MNRAS Lett. **418**, 79 (2011)

Threlfall, J. et al.: Nonlinear wave propagation and reconnection at magnetic X-points in the hall MHD regime. A& A **544**, 24 (2012)

Uzdensky, D.: Magnetic reconnection in extreme astrophysical environments. Space Sci Rev. (2011). doi:10.1007/s11214-011-9744-5. http://xxx.lanl.gov/abs/1101.2472v1. Published online 25 Feb 2011

van Ballegooijen, A.A.: In: Belvedere, G. (ed.) Accretion disks and magnetic fields in astrophysics. ASSL, vol. 156, p. 99. Kluwer, Dordrecht (1989)

Wilmot-Smith, A.L., Pontin, D.I., Hornig, G.: Dynamics of braided coronal loops. I. onset of magnetic reconnection, A& A **516**, A5 (2010)

Yamada, M.: Progress in understanding magnetic reconnection in laboratory and space astrophysical plasmas. Phys. Plasmas **14**, 058102 (2007)

Zenitani, S., Hesse, M., Klimas, A., Kuznetsova, M.: New measure of the dissipation region in collisionless magnetic reconnection. Phys. Rev. Lett. **106**, 195003 (2011)

Zenitani, S., Shinohara, I., Nagai, T., Wada, T.: Kinetic aspects of the ion current layer in a reconnection outflow exhaust. Phys. Plasmas **20**, 092120 (2013). doi:10.1063/1.4821963

Zocco, A., Schekochihin, A.: Reduced fluid-kinetic equations for low-frequency dynamics, magnetic reconnection, and electron heating in low-beta plasmas. Phys. Plasmas **18**, 102309 (2011)

Chapter 7
Theory of Relativistic Jets

Nektarios Vlahakis

Abstract Relativistic jets can be modeled as magnetohydrodynamic flows. We analyze the related equations and discuss the involved acceleration mechanisms, their relation to the collimation, to the jet confinement by its environment, and to possible rarefaction waves triggered by pressure imbalances.

7.1 Introduction

Relativistic jets is one of the most interesting class of astrophysical outflows, connected to some of the most energetic phenomena in the Universe. Their velocities are observed either directly through superluminal apparent motions, or indirectly through the Doppler boosted radiation that we receive from them. The associated Lorentz factors range from hundreds in GRBs to tens in AGNs or mildly relativistic in microquasars.

A question arises on how these jets are accelerated to such high bulk Lorentz factors. One possible answer is related to thermal pressure gradients. However, it is unlikely that thermal driving alone can explain the terminal speeds of relativistic jets. The required high initial temperatures point towards an initially magnetically dominated outflow. Magnetic fields can tap the rotational energy of the compact object or disk, providing the most plausible mechanism of energy extraction. The extraction mechanism is acting like a unipolar inductor, see Figs. 7.1 and 7.2, with the rotating "disk" being the accretion disk, or the central object. Qualitatively it works in the same way no matter if the central source is a protostar and/or its accretion disk, a stellar mass compact object and/or its accretion disk – as in the GRB and microquasar cases – or an active galactic nucleus. The only requirement is to have a rotating conducting source threaded by a large scale magnetic field. In the case of a rapidly rotating black hole the spacetime itself plays the role of the "disk"; this is the so-called Blandford and Znajek process (Blandford and Znajek 1977).

N. Vlahakis (✉)
Faculty of Physics, Department of Astrophysics, Astronomy and Mechanics, University of Athens, Panepistimiopolis 15784 Zografos, Athens, Greece
e-mail: vlahakis@phys.uoa.gr

I. Contopoulos et al. (eds.), *The Formation and Disruption of Black Hole Jets*, Astrophysics and Space Science Library 414,
DOI 10.1007/978-3-319-10356-3_7

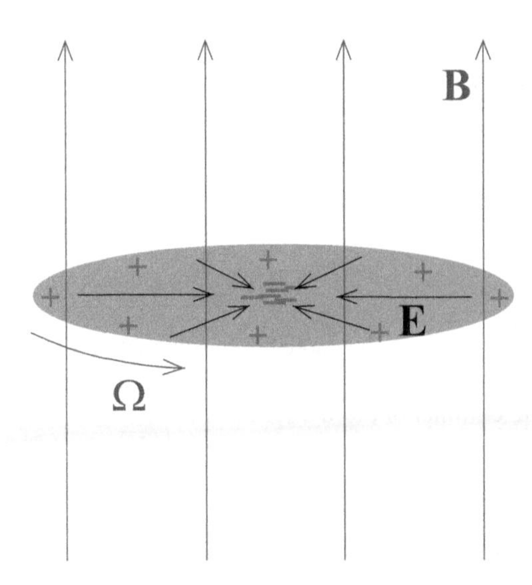

Fig. 7.1 A unipolar inductor (or Faraday disk). An electromotive force is created when a conducting disk is rotated inside a magnetic field. The Lorentz force induces negative charges at the central region of the disk close to the rotation axis, and positive charges at its perimeter. The charge density and electric field is shown in the figure

Guided by the Blandford and Payne model of nonrelativistic jets (Blandford and Payne 1982) one could expect that magnetic forces could efficiently accelerate the outflow, reaching kinetic energy flux comparable to the ejected Poynting flux at Alfvénic distances. It turns out, however, that a relativistic flow is still Poynting flux dominated at these distances and the magnetic acceleration should work at much larger scales. In fact, there was a general believe for decades, based on a model of Michel (1969) on conical flows, that the mechanism of magnetic acceleration is in general inefficient. This corresponds to an outflow with streamlines along the current lines, the so-called force-free case, in which the Lorentz force $\boldsymbol{J} \times \boldsymbol{B}_\phi / c$ has no component along the flow (see Fig. 7.2). The extracted energy remains in the electromagnetic field up to large distances from the source and a dissipation mechanism (e.g. reconnection or instabilities) is needed in order to somehow transfer the energy to the observed radiation. An alternative way to transfer the energy to radiation is through shocks (and associated particle acceleration in them), or some instability after first creating a relativistically moving jet. This requires to accelerate the plasma by efficiently transferring the Poynting flux to matter kinetic energy flux. The first works showing that this is indeed possible were based on self-similar solutions of the magnetohydrodynamic (MHD) equations in the cold limit (Li et al. 1992; Contopoulos 1994), and including a hot electron/positron/photon fluid (Vlahakis and Königl 2003a). Komissarov et al. (2009) generalized these results with the help of numerical simulations (see also Komissarov et al. 2007; Tchekhovskoy et al. 2011), giving also analytical scalings that relate the spatial distribution of the Lorentz factor and the flow shape with the pressure of the

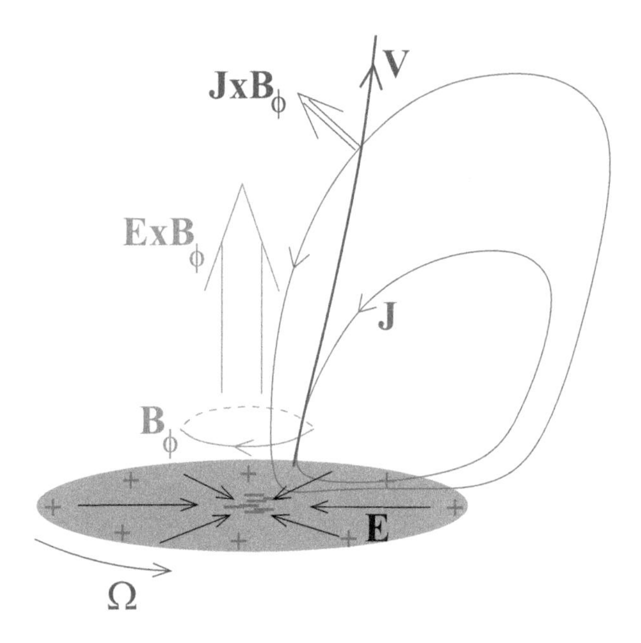

Fig. 7.2 A "wire" of plasma charges closes the circuit. The current $\boldsymbol{J}$ and its associated azimuthal magnetic field $\boldsymbol{B}_\phi$ is shown in the figure. An outflowing Poynting flux $(c/4\pi)\boldsymbol{E} \times \boldsymbol{B}_\phi$ is naturally formed, and thus energy and angular momentum is extracted from the disk. If the streamlines are more collimated than the current lines then the $\boldsymbol{J} \times \boldsymbol{B}_\phi/c$ force (shown in the figure) has a component along the bulk speed, accelerating the outflow, and a component normal to it, affecting its collimation

environment of the jet. Tyically the magnetic acceleration leads to Lorentz factors $\sim 0.5E/Mc^2$ or even larger for not extremely relativistic flows, where E is the extracted energy flux (that initially resides in the electromagnetic field) and M is the baryonic mass flux of the jet.

In the next section these results will be extensively analyzed. We start by presenting the MHD equations in Sect. 7.2 and the associated integrals of motion for steady and axisymmetric flows in Sect. 7.3. Then we describe the collimation-acceleration mechanism in Sect. 7.4, the rarefaction acceleration in Sect. 7.5, and the impulsive acceleration in Sect. 7.6.

7.2 MHD Equations

A magnetized outflow is described by the set of the well known ideal magnetohydrodynamic equations. In order to focus on the interaction of matter bulk motion with the electromagnetic field and simplify as much as possible the picture, we neglect thermal effects and consider the flow as cold. This is a reasonable assumption especially for relativistic flows, since the possible thermal content by itself is unlikely to solve the problem of reaching high bulk Lorentz factors (this requires huge temperatures as discussed in Sect. 7.1); it only affects the evolution of the flow near the source.

In the framework of special relativity the stress-energy tensor of the system consists of two parts, one for the cold matter and a second for the electromagnetic field.

The matter part is given by $\rho_0 U^\kappa U^\nu$ (with $\kappa, \nu = 0, 1, 2, 3$). Here ρ_0 is the rest-mass density, and $U^\nu = (\gamma c, \gamma \boldsymbol{V})$ is the fluid four-velocity, with $\boldsymbol{V}$ being the three-velocity of plasma measured in the frame of the central object, and $\gamma = 1/(1 - V^2/c^2)^{1/2}$ the corresponding Lorentz factor.

Including the contribution of the electric ($\boldsymbol{E}$) and magnetic ($\boldsymbol{B}$) fields (measured in the central-object frame), the components of the total stress-energy tensor in a flat spacetime with metric tensor $\eta^{\kappa\nu} = \text{diag}\,(-1\;1\;1\;1)$ and Cartesian space coordinates x_j take the form ($j, k = 1, 2, 3$)

$$T^{00} = \gamma^2 \rho_0 c^2 + \frac{E^2 + B^2}{8\pi}, \tag{7.1}$$

$$T^{0j} = T^{j0} = \left(\rho_0 c \gamma^2 \boldsymbol{V} + \frac{\boldsymbol{E} \times \boldsymbol{B}}{4\pi}\right) \cdot \hat{x}_j, \tag{7.2}$$

$$T^{jk} = \rho_0 \gamma^2 V_j V_k - \frac{E_j E_k + B_j B_k}{4\pi} + \frac{E^2 + B^2}{8\pi} \eta^{jk}, \tag{7.3}$$

with T^{00}, $cT^{0j}\hat{x}_j$, and T^{jk} representing the energy density, energy flux, and spatial stress contributions, respectively.

The electromagnetic field obeys Maxwell's equations

$$\nabla \cdot \boldsymbol{B} = 0, \quad \nabla \cdot \boldsymbol{E} = \frac{4\pi}{c} J^0,$$
$$\nabla \times \boldsymbol{B} = \frac{1}{c}\frac{\partial \boldsymbol{E}}{\partial t} + \frac{4\pi}{c}\boldsymbol{J}, \quad \nabla \times \boldsymbol{E} = -\frac{1}{c}\frac{\partial \boldsymbol{B}}{\partial t}, \tag{7.4}$$

where $J^\nu = \left(J^0, \boldsymbol{J}\right)$ is the four-current (with J^0/c representing the charge density).

Under the assumption of ideal MHD, the electric field in the frame of the moving plasma is zero, which implies

$$\boldsymbol{E} = -\frac{\boldsymbol{V}}{c} \times \boldsymbol{B}. \tag{7.5}$$

Maxwell's and Ohm's equations together with the conservation of mass

$$(\rho_0 U^\nu)_{,\nu} = 0 \Leftrightarrow \frac{\partial (\gamma \rho_0)}{\partial t} + \nabla \cdot (\gamma \rho_0 \boldsymbol{V}) = 0, \tag{7.6}$$

and energy-momentum, in the absence of a gravitational field or any other external force,

$$T^{\kappa\nu}_{,\nu} = 0\,, \tag{7.7}$$

fully describe a MHD flow.

7.3 Steady-State, Axisymmetric Case

In many astrophysical settings observations justify to assume to zeroth order that the flow is steady-state, i.e., it has no direct dependence on time ($\partial/\partial t = 0$), and axisymmetric around the axis of rotation of the central source, i.e. $\partial/\partial\phi = 0$ in cylindrical $(z\,, \varpi\,, \phi)$ or spherical $(r\,, \theta\,, \phi)$ coordinates. These symmetries imply that there are integrals of motion that greatly simplify the equations (Bekenstein and Oron 1978; Okamoto 1978; Lovelace et al. 1986; Camenzind 1986) and allow to understand the behavior of the solutions even without fully solving the problem.

The partial integrations can be done by first splitting each vector into poloidal and azimuthal components. For example for the magnetic field $\boldsymbol{B} = \boldsymbol{B}_p + \boldsymbol{B}_\phi$ (the subscripts p and ϕ denote the poloidal and azimuthal components, respectively), where $\boldsymbol{B}_p = B_\varpi(\varpi\,, z)\hat{\varpi} + B_z(\varpi\,, z)\hat{z}$, and $\boldsymbol{B}_\phi = B_\phi(\varpi\,, z)\hat{\phi}$.

The solenoidal condition for the magnetic field, $\nabla \cdot \boldsymbol{B} = 0$, or $\nabla \cdot \boldsymbol{B}_p = 0$, implies that there is a vector potential corresponding to the poloidal component of the magnetic field which can be conveniently written as

$$\boldsymbol{B}_p = \nabla \times \left(\frac{A(\varpi\,, z)}{\varpi} \hat{\phi} \right) = \frac{\nabla A \times \hat{\phi}}{\varpi}\,. \tag{7.8}$$

The scalar function $A(\varpi\,, z)$, the so-called poloidal magnetic flux function, corresponds to the magnetic flux through a surface enclosed by the curve $\varpi = \text{const}$, $z = \text{const}$ (a circle around the symmetry axis passing through the point (ϖ, z) of the poloidal plane), since Eq. (7.8) implies $\iint \boldsymbol{B} \cdot d\boldsymbol{S} = 2\pi A$. Each poloidal magnetic field line can be described by its value of the function A, which remains the same in all its points. Including the azimuthal magnetic field component, the field line is of course three dimensional, but its projection on the poloidal plane is $A(\varpi\,, z) = \text{const}$.

For steady-state electromagnetic field the electric field obeys $\nabla \times \boldsymbol{E} = 0$, and thus can be written as gradient of a scalar potential, $\boldsymbol{E} = -\nabla\Phi$. Axisymmetry enforces that $E_\phi = 0$, and thus Ohm's equation (7.5) implies $\boldsymbol{V}_p \parallel \boldsymbol{B}_p$, from which it follows that there are functions Ψ_A and Ω (whose coordinate dependence is discussed below) such that

$$V = \frac{\Psi_A}{4\pi\gamma\rho_0}B + \varpi\Omega\hat{\phi}, \quad \frac{\Psi_A}{4\pi\gamma\rho_0} = \frac{V_p}{B_p}, \quad V_\phi = \varpi\Omega + V_p\frac{B_\phi}{B_p}. \tag{7.9}$$

This is a manifestation of the flux-freezing condition in ideal MHD flows, according to which the magnetic flux passing through a surface that moves with the plasma remains constant. Note however, that the velocity and magnetic fields need to be parallel only on the poloidal plane.

The function Ψ_A is a constant of motion, as can be proven by combining equation (7.9) with the continuity equation (7.6) $\boldsymbol{B}_p \cdot \nabla\Psi_A = 0 \Leftrightarrow \Psi_A = \Psi_A(A)$. In fact the continuity equation can by itself be integrated to give an expression of the poloidal speed through a mass flux function Ψ which remains constant along streamlines, $4\pi\gamma\rho_0\boldsymbol{V}_p = \nabla\Psi \times \hat{\varpi}/\varpi$. The requirement that $\boldsymbol{V}_p \parallel \boldsymbol{B}_p$ implies that streamlines and field lines coincide on the poloidal plane, or $\Psi = \Psi(A)$. Ψ_A is just the derivative $d\Psi/dA$ and expresses the mass-to-magnetic flux ratio.

Combining the velocity expression (7.9) with Ohm's law (7.5) and Eq. (7.8) we find the expression for the electric field

$$\boldsymbol{E} = -\frac{\Omega}{c}\nabla A, \quad E = \frac{\varpi\Omega}{c}B_p. \tag{7.10}$$

Substituting the previous expression for the electric field in Faraday's law we find $\nabla\Omega \times \nabla A = 0 \Leftrightarrow \Omega = \Omega(A)$, i.e., the quantity Ω is a constant of motion.

Since the electric field is normal to $\boldsymbol{B}_p$ (and $\boldsymbol{V}_p$), poloidal magnetic field/streamlines are equipotentials. Indeed, from Eq. (7.10) we get $\nabla\Phi = \Omega\nabla A/c$ which means that $\Phi = \Phi(A)$ and $\Omega = cd\Phi/dA$.

Besides its relation with the electric potential, the integral Ω also represents some angular velocity at the base of the flow and is often called field angular velocity. Indeed Eq. (7.9) gives $V_\phi \approx \varpi\Omega$ in regions where $V_p \ll V_\phi$ or $|B_\phi| \ll B_p$. This is usually the case near the base of an outflow and implies that the flow corotates with its base. At larger distances however, this cannot remain true since for sufficient expansion (which is always realized in relativistic MHD flows) the $\varpi\Omega$ equals the speed of light at some surface, called light surface.[1] At this distance the contribution of the other term in the expression (7.9) of V_ϕ cannot be ignored; in fact the sign of B_ϕ is such that to keep V_ϕ always subluminal (negative if $\Omega > 0$ and $B_z > 0 \Leftrightarrow \Psi_A > 0$). The same sign can be found if we think that the magnetic field lines due to their inertia are "left behind" rotation (for $B_z > 0$ the base rotates faster than tip of the $\boldsymbol{B}$ arrow). This is also equivalent with having outflowing poloidal Poynting flux $(c/4\pi)\boldsymbol{E} \times \boldsymbol{B}_\phi = -(\varpi\Omega B_\phi/4\pi)\boldsymbol{B}_p = -(\varpi\Omega B_\phi/\Psi_A)\gamma\rho_0\boldsymbol{V}_p$. In the following we assume without loss of generality that we have an outflow in the

[1] If Ω is the same in all streamlines as is often the case in pulsar winds, this surface is a cylinder, called light cylinder.

"northern" hemisphere $z > 0$ with $\Omega > 0$ and $B_z > 0 \Leftrightarrow \Psi_A > 0$ (since $V_z > 0$ for outflows). A change of coordinate system covers all other cases. Note also that the MHD equations remain the same if we change the sign of all components of the electromagnetic field.

The momentum conservation equation is given by the $\kappa = 1, 2, 3$ components of $T^{\kappa\nu}_{,\nu} = 0$. In cylindrical coordinates we get

$$\left[\frac{1}{\varpi}\frac{\partial(\varpi T^{z\varpi})}{\partial\varpi} + \frac{\partial T^{zz}}{\partial z}\right]\hat{z} + \left[\frac{1}{\varpi}\frac{\partial(\varpi T^{\varpi\varpi})}{\partial\varpi} - \frac{T^{\phi\phi}}{\varpi} + \frac{\partial T^{z\varpi}}{\partial z}\right]\hat{\varpi}$$
$$+\left[\frac{1}{\varpi^2}\frac{\partial(\varpi^2 T^{\varpi\phi})}{\partial\varpi} + \frac{\partial T^{z\phi}}{\partial z}\right]\hat{\phi} = 0 \quad (7.11)$$

It can also be written in a form resembling Newton's law

$$\gamma\rho_0 \left(\boldsymbol{V}\cdot\nabla\right)\left(\gamma\boldsymbol{V}\right) = \frac{J^0\boldsymbol{E} + \boldsymbol{J}\times\boldsymbol{B}}{c}. \quad (7.12)$$

The momentum equation gives two additional integrals of motion. The first, related to the assumption of axisymmetry, can be obtained from the $\hat{\phi}$ component. By dotting equation (7.12) with $\hat{\phi}$ and after some manipulation we find that

$$\varpi\gamma V_\phi - \frac{\varpi B_\phi}{\Psi_A} = L(A). \quad (7.13)$$

This integral represents the total (matter + electromagnetic) angular momentum flux to mass flux ratio.[2] Indeed, the part of the stress-energy tensor representing flux of $\hat{\phi}$ momentum in the $\boldsymbol{V}_p$ direction is $\rho_0\gamma^2 V_\phi V_p - B_\phi B_p/4\pi$. By multiplying with ϖ to get angular momentum and dividing with the mass flux $\gamma\rho_0 V_p$ ($\gamma\rho_0$ is the density in the laboratory frame) we find the expression of L. Angular momentum can be transferred from the electromagnetic field to the plasma or vice-versa, but the sum remains always constant.

The second integral which is the result of the steady-state assumption, can be found by dotting equation (7.12) with $\boldsymbol{V}$. (It can also be obtained from the "0"th component of the equations of motion $T^{0j}_{,j} = 0$.) After some manipulation using vector identities we find

$$\gamma c^2 - \frac{\varpi\Omega B_\phi}{\Psi_A} = \mu(A)c^2. \quad (7.14)$$

μc^2 represents the total (matter + electromagnetic) energy-to-mass flux ratio.[3] Indeed, it can be easily checked that it equals the total energy flux given by the

[2] Note that $B_\phi < 0$ so that both parts of the angular momentum are positive.

[3] As in the case of the angular momentum, both parts are positive since $B_\phi < 0$.

poloidal parts of the cT^{0j} elements of the stress-energy tensor, divided by $\gamma\rho_0 V_p$. As in the case of the angular momentum, energy can be transferred between the plasma and the electromagnetic field, keeping however the sum constant. The maximum possible value for the plasma part of the integral is obviously μc^2, corresponding to a full transformation of electromagnetic energy to matter kinetic energy.

The ratio of the two parts of energy fluxes is the so-called magnetization function σ

$$\sigma = \frac{\mu - \gamma}{\gamma} = -\frac{\varpi \Omega B_\phi}{\gamma \Psi_A c^2} = -\frac{(c/4\pi) E B_\phi}{\gamma^2 \rho_0 V_p c^2} . \tag{7.15}$$

Note that the electromagnetic parts of the energy and angular momentum fluxes are proportional to each other, see Eqs. (7.14) and (7.13). Their ratio equals Ω, hence the name "field angular velocity" for this integral.

The electromagnetic part of Eq. (7.14) is proportional to ϖB_ϕ (Ω and Ψ_A are constants of motion). This quantity is directly related to the current I passing through a surface around the symmetry axis, enclosed by a circle ϖ =cosnt, z =const. Indeed Apmére's law yields $2\pi\varpi B_\phi = (4\pi/c)I$, and thus the lines of poloidal current $\boldsymbol{J}_p = (c/4\pi)\nabla \times \boldsymbol{B}_\phi$ are ϖB_ϕ =const. The plasma is accelerated if $|\varpi B_\phi|$ drops as we move downstream, meaning that the poloidal streamlines cross current lines of progressively smaller values of I.

Two equations remain to be integrated: the transfield component of the momentum equation (its projection normal to the flow and the magnetic field, i.e., along $\boldsymbol{E}$), and the Bernoulli (or "wind") equation which is obtained after substituting all quantities in the identity $\gamma^2 - (\gamma V_\phi/c)^2 = 1 + (\gamma V_p/c)^2$. There are correspondingly two unknown functions, which are usually chosen to be the magnetic flux function A and the density, or equivalently the square of the "Alfvénic" Mach number (Michel 1969)

$$M^2 = \frac{4\pi\rho_0\gamma^2 V_p^2}{B_p^2} = \frac{\Psi_A^2}{4\pi\rho_0} . \tag{7.16}$$

The expressions for all the rest physical quantities in terms of the defined variables A, M^2 and the constants of motion (Ω, Ψ_A, L, μ), as well as the explicit expressions for the Bernoulli and transfield equations can be found e.g. in Vlahakis and Königl (2003a) and Beskin (2010). Although some solutions will be presented in the following sections, here we only define some variables that will be used in the general discussion that follow. It is convenient to scale the cylindrical radius ϖ in units of the "light surface" radius c/Ω, and also define some combinations of integrals

$$x = \frac{\varpi\Omega}{c} , \quad \sigma_{\rm M} = \frac{A\Omega^2}{c^3\Psi_A} , \quad \varpi_{\rm A} = \left(\frac{L}{\mu\Omega}\right)^{1/2} , \quad x_{\rm A} = \frac{\varpi_{\rm A}\Omega}{c} . \tag{7.17}$$

Using these quantities we can write the magnetization function as

$$\sigma = \frac{x^2 - x_{\rm A}^2}{M^2 + x^2 - 1} . \tag{7.18}$$

7.4 The Collimation-Acceleration Mechanism

Suppose a Poynting-dominated outflow is ejected from the vicinity of a compact object or the surrounding accretion disk. If $\gamma_i \sim 1$ and $\sigma_i \gg 1$ are the initial Lorentz factor and magnetization, respectively, and $\dot{M}$ the ejected mass per time, then the electromagnetic energy per time is $\gamma_i \sigma_i \dot{M}$. The most important parameter of the flow is the ejected energy per mass, which equals the integral $\mu = \gamma_i(1+\sigma_i)$ (times c^2). For highly relativistic outflows this is $\gg 1$. Although μ corresponds to the maximum possible Lorentz factor of the flow, it is not certain that this value will be reached, since it is possible that the electromagnetic field retains its energy flux up to large distances without an efficient transfer of energy to the plasma.

During its first steps the plasma is dominated by the poloidal magnetic field and is accelerated magnetocentrifugally, like beads on rotating wires (Blandford and Payne 1982). It passes the light surface $\varpi = c/\Omega$ where $|B_\phi| \approx B_p$, such that the two terms in the expression (7.9) of V_ϕ are comparable in order to avoid a superluminal V_ϕ. At distances well beyond the light surface $x \gg 1$, these terms are approximately equal (since V_ϕ being $<c$ is much smaller than $\varpi\Omega$ which is $\gg c$). Thus the azimuthal magnetic field dominates, with $B_\phi/B_p \approx -\varpi\Omega/V_p$ and $E = xB_p \approx -B_\phi V_p/c$. We can greatly simplify the MHD equations in this regime and get insights on the acceleration process and the factors that affect its final outcome.

Using Eq. (7.15) we find a simplified expression for the magnetization

$$\sigma \approx \frac{B_\phi^2}{4\pi\gamma^2\rho_0 c^2} = \frac{U_f^2}{c^2} , \tag{7.19}$$

where $U_f = \sqrt{(B^2 - E^2)/(4\pi\rho_0)}$ is the phase speed of the fast magnetosonic waves.

When the flow crosses the fast magnetosonic surface ($\gamma V_p = U_f$) it is still Poynting-dominated, and thus $\mu = \gamma(1+\sigma) \approx \gamma\sigma$. Equation (7.19) then gives the value of the Lorentz factor: $\mu/\gamma_f \approx \gamma_f^2$, or, $\gamma_f \approx \mu^{1/3}$. (Indeed $\mu^{1/3} \ll \mu$ for relativistic flows with high μ.) The main part of the acceleration is realized in the super-fast magnetosonic part of the flow; for this reason it is enough to focus on the $x \gg 1$ regime.

7.4.1 The Bunching Function

Another useful approximate expression for the magnetization in the $x \gg 1$ regime is

$$\sigma \approx \frac{\sigma_{\rm M}}{\gamma V_p/c} S \,, \quad S = \frac{\varpi^2 B_p}{A} = \frac{\varpi |\nabla A|}{A} \,, \tag{7.20}$$

where S is the so-called bunching function (Vlahakis 2004a; Komissarov et al. 2009), a measure of the local poloidal magnetic field compared to the mean magnetic field A/ϖ^2. The electromagnetic part of the energy-to-mass flux ratio is $\gamma\sigma \approx \sigma_{\rm M}(c/V_p)S$. Since $V_p/c \approx \sqrt{1-1/\gamma^2}$, using the definition of $\sigma = (\mu-\gamma)/\gamma$ we find a relation between the Lorentz factor and the bunching function (Vlahakis 2004b)

$$\mu = \gamma + \frac{\sigma_{\rm M}}{V_p/c} S \Leftrightarrow \mu = \gamma + \frac{\sigma_{\rm M}}{\sqrt{1-1/\gamma^2}} S \,. \tag{7.21}$$

Differentiating the above equation we find that

$$\boldsymbol{V} \cdot \nabla \gamma = \frac{\sigma_{\rm M}}{\mu} \gamma(\gamma^2-1)(\mu-\gamma) \frac{\boldsymbol{V} \cdot \nabla S}{\mu - \gamma^3} \approx \sigma_{\rm M} \gamma^3 \frac{\boldsymbol{V} \cdot \nabla S}{\mu - \gamma^3} \,, \tag{7.22}$$

an expression which becomes $0/0$ at the fast magnetosonic point (finding again $\gamma_f = \mu^{1/3}$). For an accelerating flow, S should increase upstream from the fast magnetosonic point, and decrease downstream. The situation is similar to a De Lavan nozzle, with $1/S$ playing the role of the surface of the tube whose minimum determines the position where the flow becomes trans-fast magnetosonic, hence the term "magnetic nozzle" (Li et al. 1992).

There is also an interesting analogy with a particle motion in a velocity-dependent potential, see Appendix C in Komissarov et al. (2010). Equations (7.22) and (7.21) can be written as the Newton's law and its energy integral for a particle with rest energy $mc^2 = 1/\sigma_{\rm M}$ moving in a potential $\mathcal{V} = S/(V/c)$.

The proportionality between mass and magnetic flux gives an alternative expression for the bunching function $S \propto \varpi^2 \gamma \rho_0 V_p$. For two neighboring poloidal field/streamlines conservation of mass requires $\gamma\rho_0 V_p \delta S$, where δS is the area between them (a function of distance), see Fig. 7.3. Thus, $S \propto \varpi^2/\delta S \propto \varpi/\delta\ell_\perp$, meaning that the magnetic acceleration is at work whenever the flow expansion is such as the area between streamlines to increase faster than ϖ^2, or equivalently the distance between streamlines $\delta\ell_\perp$ to increase faster than the cylindrical distance ϖ.

Another way to see this condition is to think that energy is transferred from the field to the plasma if the Poynting flux $(c/4\pi)EB_\phi \propto B_\phi^2 \propto \varpi^2 B_p^2 \propto \varpi^2/(\delta S)^2$ (from magnetic flux conservation) decreases faster than the mass flux $\gamma\rho_0 V_p \propto 1/\delta S$.

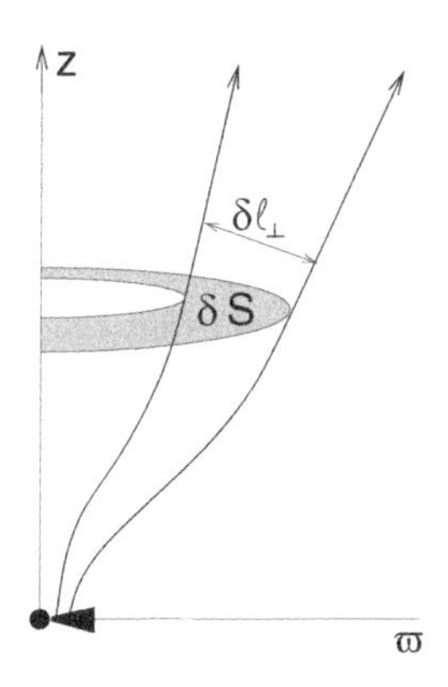

Fig. 7.3 The area between two poloidal field/streamlines $\delta S = 2\pi\varpi\,\delta\ell_\perp$ increases faster than ϖ^2 in efficiently accelerating outflows

The magnetic acceleration strongly depends on how S decreases with distance. Using Eq. (7.21) at the fast magnetosonic point where $\mu \gg \gamma_f \gg 1$ we find the value of the bunching function at this point $S_f \approx \mu/\sigma_{\rm M}$. The spatial decline of S is determined by the force balance in the transfield direction (although there are studies in the literature that assume a prescribed flow shape ignoring the transfield and concentrating only to the wind equation for given S function Michel 1969; Fendt and Ouyed 2004; Toma and Takahara 2013; Millas et al. 2014). Using its definition $S = \varpi|\nabla A|/A$ we can estimate its asymptotic value $S_\infty \sim 1$, corresponding to a quasi-monopolar flow shape in which $|\nabla A| \sim A/\varpi$. Thus the asymptotic Lorentz factor can be estimated from Eq. (7.21) to be (Vlahakis 2004a) $\gamma_\infty \sim \mu - \sigma_{\rm M}$.

In Michel's solution (1969) where radial flow shape was assumed, the bunching function is given as constant along the flow (δS scales exactly as ϖ^2). This is the reason why this solution corresponds to completely inefficient magnetic accelerators with $\gamma_\infty \approx \gamma_f \approx \mu^{1/3}$. In self-similar solutions on the other hand (Li et al. 1992; Vlahakis and Königl 2003a,b, 2004) the flow shape – that is self-consistently found by solving the transfield force balance equation – around the fast magnetosonic surface is parabolic $z \propto \varpi^2$, $A \propto \varpi^2/z$. This gives $S_f \approx 2$ resulting in $\sigma_{\rm M} \approx \mu/2$ and $\gamma_\infty \approx \mu/2$.

The same is true for a more general parabolic shape $z \propto \varpi^b$ for which $A \propto \varpi^2/z^{2/b}$ and $S = |2\hat{\varpi} - 2\varpi/(bz)\hat{z}| \approx 2$.

As will be discussed in the following, the environment of the outflow plays a significant role in defining the shape of the flow and thus the value of S_f that controls the acceleration efficiency. Numerical simulations (Komissarov et al. 2007, 2009) show that for values of μ up to a few tens, appropriate for AGN jets, even if the external pressure is very small resulting in radial outer jet boundary, the distribution of the magnetic flux inside the jet is such as to give high acceleration efficiencies. For higher μ though, appropriate for GRB jets, a radial jet boundary shape leads to radial geometry even inside the jet, and the acceleration efficiency is small except in a small region around the axis (Tchekhovskoy et al. 2009). For such high-μ flows (and potentially high Lorentz factor flows) the external pressure needs to be non-negligible for the magnetic acceleration to work in the whole jet volume. Practically jets are always ejected inside an environment, either an outer slower disk-wind or the intergalagtic/interstellar medium, providing the required external pressure. It is important to emphasize however that the self-collimation property of magnetic fields is crucial in the magnetic acceleration process. The expansion between neighboring field/streamlines needed for the magnetic acceleration to work is realized by a faster collimation of the inner parts of the jet compared to the outer ones, hence the term "collimation-acceleration" mechanism.

7.4.2 Spatial Scalings

From the preceding discussion it is evident that the transfield force-balance equation that controls the shape of the flow crucially affects the result of the magnetic acceleration.

The momentum equation (7.12) can be written as the sum of the following force densities (for simplicity we use hereafter the term force)

$$\boldsymbol{f}_G + \boldsymbol{f}_C + \boldsymbol{f}_I + \boldsymbol{f}_E + \boldsymbol{f}_B = 0\,, \tag{7.23}$$

where

$$\begin{aligned}
\boldsymbol{f}_G &= -\gamma\rho_0 \left(\boldsymbol{V}\cdot\nabla\gamma\right)\boldsymbol{V} && \\
\boldsymbol{f}_C &= \hat{\varpi}\gamma^2\rho_0 V_\phi^2/\varpi && :\text{centrifugal force}\\
\boldsymbol{f}_I &= -\gamma^2\rho_0 \left(\boldsymbol{V}\cdot\nabla\right)\boldsymbol{V} - \boldsymbol{f}_C && \\
\boldsymbol{f}_E &= \left(\nabla\cdot\boldsymbol{E}\right)\boldsymbol{E}/4\pi && :\text{electric force}\\
\boldsymbol{f}_B &= \left(\nabla\times\boldsymbol{B}\right)\times\boldsymbol{B}/4\pi && :\text{magnetic force}
\end{aligned}$$

The poloidal part of the $\boldsymbol{f}_I$ force is

$$-\gamma^2\rho_0\left(\boldsymbol{V}_p\cdot\nabla\right)\boldsymbol{V}_p = -\gamma^2\rho_0\left(V_p^2\frac{\partial\vartheta}{\partial\ell}\frac{\nabla A}{|\nabla A|} + \frac{\partial V_p}{\partial\ell}\boldsymbol{V}_p\right),$$

where ℓ is the arclength along the poloidal field line, ϑ is the angle between the poloidal magnetic field and the rotation axis ($\sin\vartheta = V_\varpi/V_p$), and the derivative $\partial/\partial\ell = \sin\vartheta\ \partial/\partial\varpi$ is taken keeping A constant. The radius of curvature of a poloidal field line is $\mathcal{R} = -(\partial\vartheta/\partial\ell)^{-1}$ (positive when the field line bends toward the axis, i.e., when ϑ decreases along the poloidal flow line).

The projection of Eq. (7.23) along $\hat{n} = \boldsymbol{E}/E = -\cos\vartheta\hat{\varpi} + \sin\vartheta\hat{z}$, gives the transfield force-balance equation

$$f_{C\perp} + f_{I\perp} + f_{E\perp} + f_{B\perp} = 0\,,$$

where the subscript $\perp$ denotes the vector component along $\hat{n}$. The various terms are:

- The azimuthal centrifugal term

$$f_{C\perp} = -\gamma^2 \rho_0 \frac{V_\phi^2}{\varpi} \cos\vartheta = -\frac{B_p^2}{4\pi\varpi}\left(\frac{MV_\phi}{V_p}\right)^2 \cos\vartheta \,.$$

- The rest of the inertial force along $\hat{n}$

$$f_{I\perp} = -\gamma^2 \rho_0 \hat{n} \cdot [(V \cdot \nabla) V] - f_{C\perp} = -\frac{B_p^2}{4\pi\mathcal{R}} M^2$$

  is the poloidal centrifugal term.
- The "electric field" force

$$f_{E\perp} = \frac{1}{8\pi\varpi^2}\hat{n}\cdot\nabla\left(\varpi^2 E^2\right) - \frac{E^2}{4\pi\mathcal{R}} \,.$$

- The "magnetic field" force along $\hat{n}$

$$f_{B\perp} = -\frac{1}{8\pi\varpi^2}\hat{n}\cdot\nabla\left(\varpi^2 B^2\right) + \frac{B_p^2}{4\pi\mathcal{R}} - \frac{B_p^2}{4\pi\varpi}\cos\vartheta \,.$$

The total electromagnetic force in the transfield direction $(f_{E\perp} + f_{B\perp})$ can be decomposed as

$$\underbrace{-\frac{1}{8\pi\varpi^2}\hat{n}\cdot\nabla\left[\varpi^2\left(B^2 - E^2\right)\right]}_{f_{EM1}} \underbrace{+\frac{B_p^2(1-x^2)}{4\pi\mathcal{R}}}_{f_{EM2}} \underbrace{-\frac{B_p^2}{4\pi\varpi}\cos\vartheta}_{f_{EM3}} \,. \tag{7.24}$$

Altogether, they give the following form of the transfield force-balance equation

$$\frac{B_p^2}{4\pi\mathcal{R}}\left(M^2 + x^2 - 1\right) = -\frac{1}{8\pi\varpi^2}\hat{n}\cdot\nabla\left[\varpi^2\left(B^2 - E^2\right)\right] - \frac{B_p^2}{4\pi\varpi}\cos\vartheta - \frac{B_p^2}{4\pi\varpi}\left(\frac{MV_\phi}{V_p}\right)^2\cos\vartheta \,. \tag{7.25}$$

(If we substitute the various physical quantities using the integrals of motion we find a second-order partial differential equation for A, the so-called Grad-Shafranov equation.)

The terms on the right-hand side of Eq. (7.25) are recognized as f_{EM1}, f_{EM3}, and $f_{C\perp}$, respectively. The left-hand side consist of the sum of $-f_{I\perp}$, and $-f_{EM2}$ terms, that both are proportional to the curvature of the poloidal field lines.

In the regime well beyond the light surface ($x \gg 1$) this equation can be simplified to

$$\frac{\gamma^2 \varpi}{\mathcal{R}} = \frac{\sigma}{1+\sigma} \frac{\varpi \nabla A \cdot \nabla}{|\nabla A|} \ln \frac{A\Omega\sigma}{\sigma_{\mathrm{M}}} - \frac{\gamma^2}{x^2} \frac{\hat{\varpi} \cdot \nabla A}{|\nabla A|}, \tag{7.26}$$

(see Vlahakis (2004b) for details; the most important points are that in the term f_{EM1} we approximate the comoving magnetic field as $\sqrt{B^2 - E^2} \approx |B_\phi|/\gamma = (A\Omega\sigma)/(c\sigma_{\mathrm{M}}\varpi)$, and eliminate M^2 using Eq. 7.18).
In the Poynting-dominated regime where σ is significantly larger than unity, order of magnitude estimation of the various terms (assuming that the characteristic spatial scale is the cylindrical distance $\nabla \sim 1/\varpi$) we further simplify the transfield equation as (Tchekhovskoy et al. 2008; Komissarov et al. 2009)

$$\frac{1}{\gamma^2} \approx \frac{\varpi}{\mathcal{R}} + \frac{1}{x^2}. \tag{7.27}$$

Depending on which term can be neglected we distinguish three different cases (see Komissarov et al. 2009 for more details).

- Linear acceleration regime: If the poloidal curvature term is negligible then

$$\gamma \approx x. \tag{7.28}$$

  This it the case for radial flows (which however do not continue to be accelerated for long; the efficiency is limited due to the constancy of the bunching function). It also applies to the initial phase of paraboloidal flows as long as the flow remains force-free (its inertia can be neglected compared to the electromagnetic forces), as well as to well collimated flows $z \propto \varpi^b$ with $b \geq 2$.
- Power-law acceleration regime: If the centrifugal term is negligible then $\gamma \approx \sqrt{\mathcal{R}/\varpi}$. This case applies to any paraboloidal flow shape $z \propto \varpi^b$ with $b < 2$. Estimating the curvature for paraboloidal flows

$$\frac{1}{\mathcal{R}} = -\left(\frac{B_z}{B_p}\right)^3 \frac{\partial^2 \varpi}{\partial z^2} \approx -\frac{\partial^2 \varpi}{\partial z^2} = \frac{b-1}{b^2} \frac{\varpi}{z^2}, \tag{7.29}$$

  we find that

$$\gamma \approx \frac{b}{\sqrt{b-1}} \frac{z}{\varpi} \propto \varpi^{b-1} \propto z^{(b-1)/b}. \tag{7.30}$$

- Ballistic regime: If the electromagnetic term is negligible then we have ballistic motion with constant γ.

7.4.3 The Role of the External Pressure

Pressure balance at the jet boundary requires that the internal jet pressure $(B^2 - E^2)/8\pi \approx B_\phi^2/(8\pi\gamma^2)$ equals $P_{\rm ext}$. The azimuthal component of the magnetic field drops as $B_\phi \propto 1/\varpi$. One way to see this is to use $B_\phi \propto \varpi B_p$ and $B_p \propto 1/\varpi^2$ from magnetic flux conservation; a second way is to think the poloidal current $I \propto \varpi B_\phi \propto \mu - \gamma$ which declines but slowly. Thus pressure balance gives the variation of Lorentz factor along the boundary $1/\gamma^2 \propto \varpi^2 P_{\rm ext}$. Assuming a profile $z^{-\alpha}$ for the external pressure we find $1/\gamma^2 = Cx^2 Z^{-\alpha}$ where $Z = z\Omega/c$ and C =const. Substituting this in Eq. (7.27) we find an equation specifying the shape of the boundary

$$\frac{d^2x}{dZ^2} + C\frac{x}{Z^\alpha} - \frac{1}{x^3} \approx 0\,. \tag{7.31}$$

We can solve this equation and the results are for various values of α (see Appendix A in Komissarov et al. 2009 for more details):

- $\alpha > 2$: The shape is conical asymptotically, $z \propto \varpi$ since the fast drop of pressure is not able to confine the flow.
- $\alpha = 2$: This is a degenerate case allowing paraboloidal shapes $z \propto \varpi^b$ with $1 < b \le 2$.
- $\alpha < 2$: In this case the shape is $z \propto \varpi^b$ with $b = 4/\alpha > 2$.

7.4.4 An Example Exact Solution

It is possible to find exact solutions of the relativistic MHD equations by assuming r self-similarity (Li et al. 1992; Contopoulos 1994; Vlahakis and Königl 2003a,b, 2004). These are the only possible exact steady-state solutions describing cold flows. They successfully capture the physics of MHD jets as was recently verified by means of numerical simulations. A solution like this was chosen as an example to have an opportunity to directly see the general characteristics of magnetized flows discussed in the previous sections. The presented solution is the second from Vlahakis and Königl (2004) (the one applied to the blazar 3C345). Figure 7.4 shows the poloidal field/streamline shape in logarithmic scale. The plotted curve holds for every poloidal field line, which is characterized by the value of $\varpi_{\rm A}$, the cylindrical distance of the Alfvén point in the particular line which practically coincides with the light surface ($\varpi_{\rm A} = c/(\Omega x_{\rm A})$ with $x_{\rm A} = 0.99$). In this model the light surface is conical, so as we move to lines away from the rotation axis the Alfvénic lever arm $\varpi_{\rm A}$ increases, so both ϖ and z scale accordingly.
In the main part of the flow the line-shape is parabolic $z \propto \varpi^2$, while asymptotically it becomes cylindrical (the slope $d \ln z/d \ln \varpi$ progressively increases as shown in Fig. 7.4).

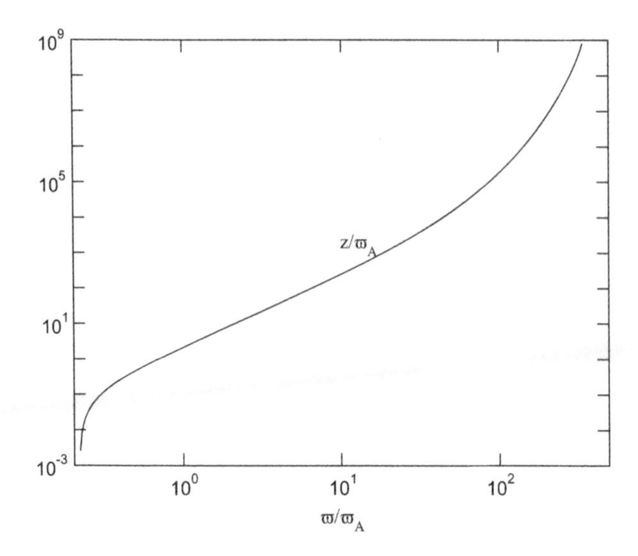

Fig. 7.4 Poloidal field/streamline shape in logarithmic scale

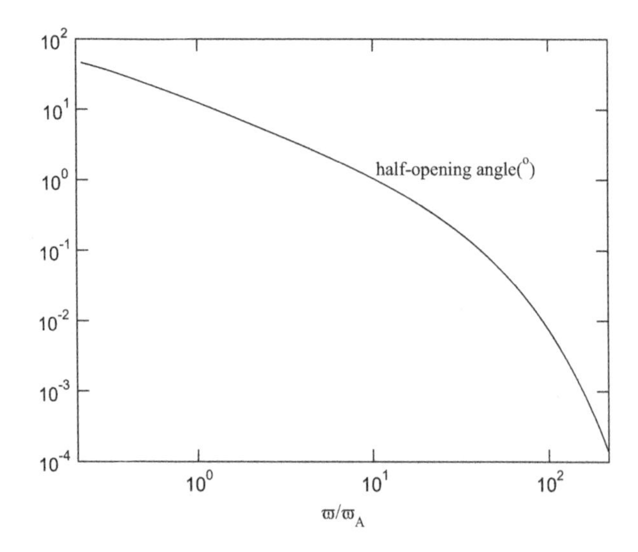

Fig. 7.5 Half-opening angle of the poloidal flow, $\vartheta = \arctan(V_\varpi / V_z)$

The half-opening angle of the outflow is shown in Fig. 7.5. The angle $\vartheta = \arctan(V_\varpi / V_z)$ that the poloidal flow makes with the rotation axis, initially has the value 45°, and decreases, for the main part of the flow, as $\vartheta \propto \varpi^{-1}$, a characteristic of the parabolic line-shape. At larger distances it decreases faster than $\propto \varpi^{-1}$, approaching the asymptotic value ≈ 0 when the flow becomes cylindrical.

Figure 7.6 shows $x = \varpi\Omega/c$ and $M = (\gamma V_p/B_p)\sqrt{4\pi\rho_0}$ (the "light cylinder" radius and the Alfvénic Mach number, respectively) as well as the fast-magnetosonic

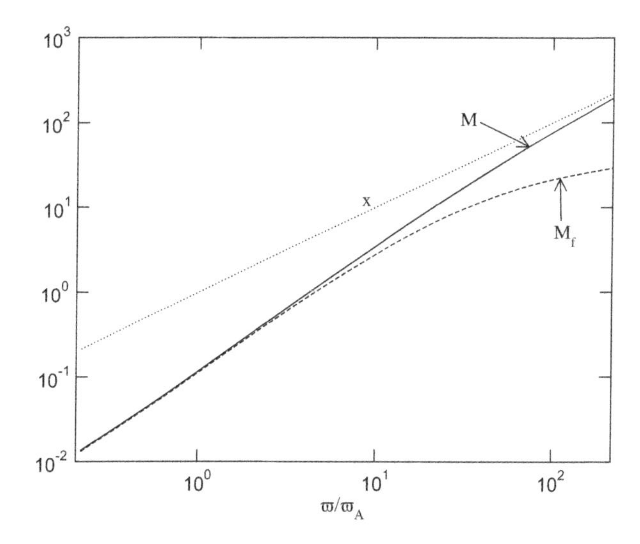

Fig. 7.6 Cylindrical distance in units of the "light surface" distance x, and the Mach numbers M and M_f

Mach number $M_f = \gamma V_p / U_f$. The point where $x = 1$ corresponds to the light surface, which practically coincides with the Alfvén surface $\varpi/\varpi_{\rm A} = 1$. At the Alfvén surface $x = x_{\rm A}$ and $M = (1 - x_{\rm A}^2)^{1/2}$. The point $M_f = 1$ (at $\varpi/\varpi_{\rm A} \approx 4.6$) is the fast-magnetosonic point.[4]

Figure 7.7 shows the two components of the magnetic field. It is seen that close to the base the poloidal part dominates, near the light surface the two parts become comparable, while at large distances the azimuthal part dominates. The two components scale as $B_p \propto 1/\varpi^2$ and $-B_\phi \propto 1/\varpi$. However, the small deviation of the product ϖB_ϕ from a constant is connected to the acceleration, since the Poynting-to-mass flux ratio is $-\varpi \Omega B_\phi / \Psi_A$, see Fig. 7.8. Also the deviation of $B_p \varpi^2$ is shown in Fig. 7.9, and plays an important role for the acceleration (since $-B_\phi \approx x B_p$, the quantities ϖB_ϕ and $B_p \varpi^2$ are proportional to each-other).

Figure 7.8 shows the two parts of the total energy-to-mass flux ratio, see Eq. (7.14). The matter part γ is increasing with increasing cylindrical distance in expense on the electromagnetic part $-\varpi \Omega B_\phi / \Psi_A c^2$. The small deviation of the ϖB_ϕ results in significant acceleration. A variation from $-\varpi \Omega B_\phi / \Psi_A c^2 = \mu$ near the base to $\sim\mu/2$ asymptotically, corresponds to acceleration up to $\gamma \sim \mu/2$, leading to equipartition between matter and electromagnetic energy fluxes at large distances. As we discussed in Sect. 7.4.1, the acceleration efficiency depends on

---

[4] The function M_f shows where the fast-magnetosonic surface is located. Besides that, M_f is related to the Mach-cone of the propagation of fast magnetosonic waves in the super-fast regime. This cone has half opening angle $\arctan(M_f^2 - 1)^{-1/2}$ (the cone's symmetry axis coincides with the poloidal field line).

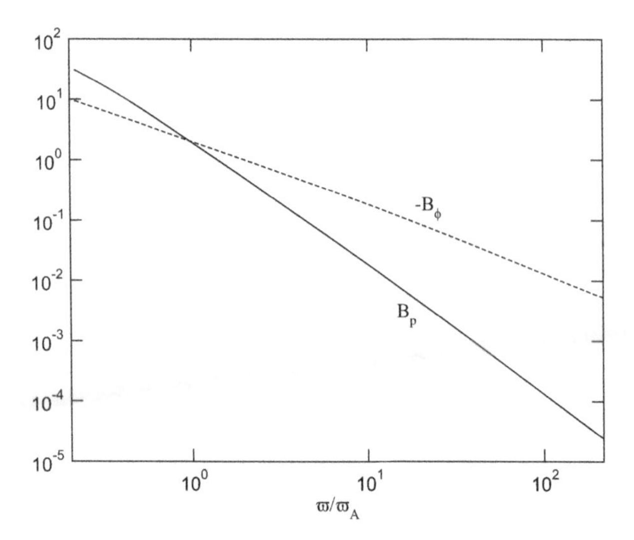

Fig. 7.7 Poloidal and azimuthal components of the magnetic field

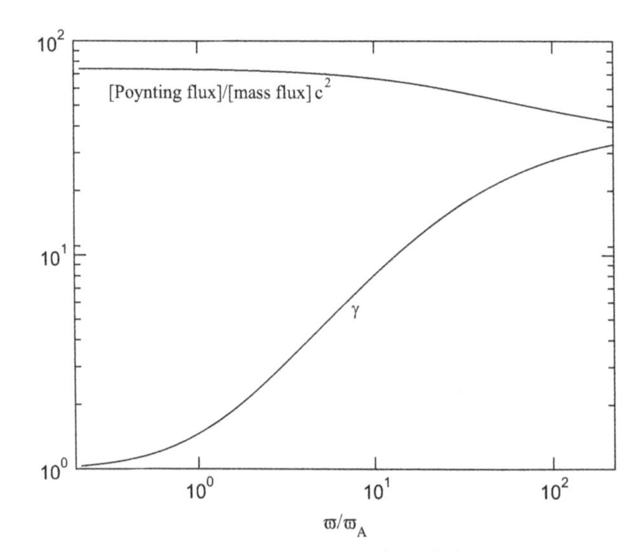

Fig. 7.8 The quantity $-\varpi \Omega B_\phi / \Psi_A c^2 = \mu - \gamma$, representing the Poynting-to-mass flux ratio (over c^2), and the Lorentz factor γ. These are the two parts of the total energy-to-mass flux ratio (over c^2); their sum equals the field line constant μ; their ratio is the magnetization function σ

the value of $B_p \varpi^2 / A$ at the fast magnetosonic surface. Since this value is 2 (see Fig. 7.9) we expect $(\mu - \gamma_\infty)/\mu \approx 1/2$, and this is indeed the case. Figure 7.9 also verifies the results of Sect. 7.4.1. In particular it shows that the analytic expression (7.21) is almost exact in the super-fast magnetosonic regime.

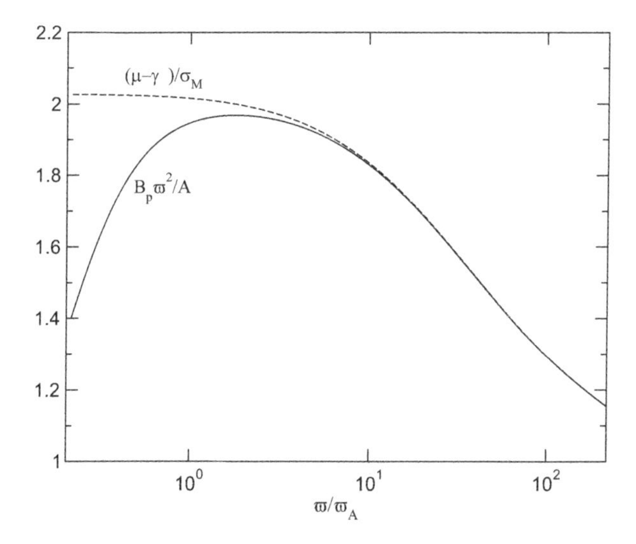

Fig. 7.9 The bunching function $S = B_p \varpi^2 / A$ and the electromagnetic part of the energy-to-mass flux ratio, over $\sigma_M c^2$. The two quantities become practically equal in the super-fast regime

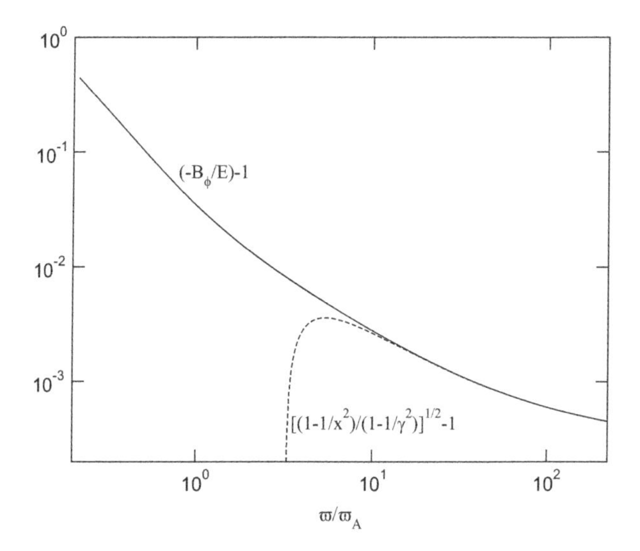

Fig. 7.10 The $-B_\phi/E$ ratio and its approximate expression

Figure 7.10 shows that the electric field is slightly smaller than the absolute value of the azimuthal magnetic field. To an almost perfect accuracy, their difference is given by the approximate expression $-B_\phi/E \approx \sqrt{(1-1/x^2)/(1-1/\gamma^2)}$, valid in the super-fast regime (see equation 13 in Vlahakis 2004b).

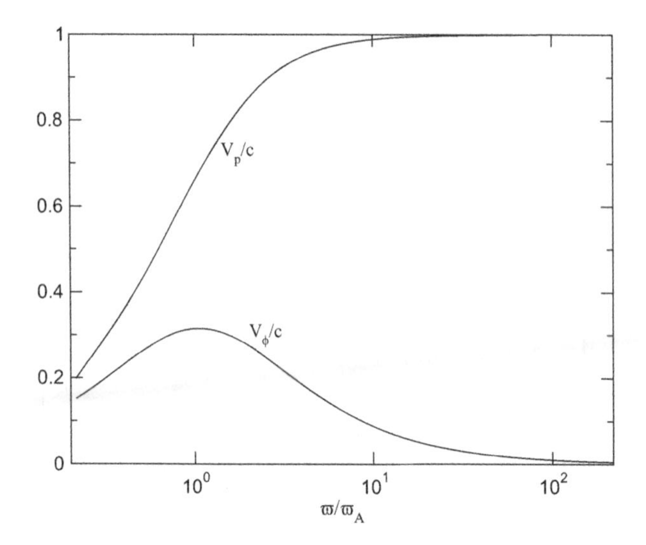

Fig. 7.11 Poloidal and azimuthal velocity components

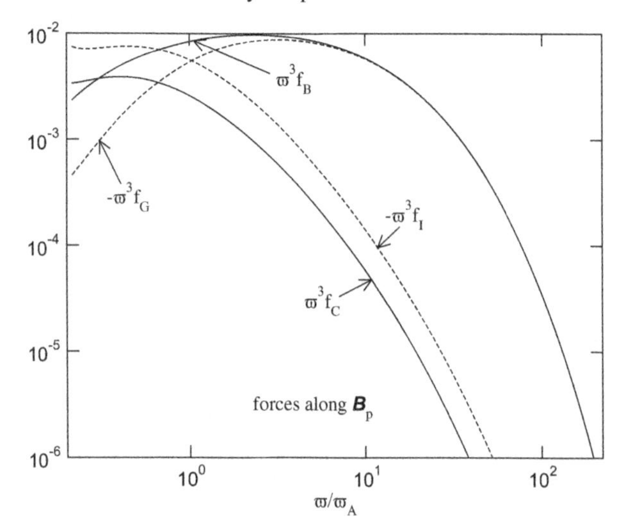

Fig. 7.12 Force density components along the poloidal flow (multiplied with ϖ^3, in arbitrary units)

The poloidal and azimuthal parts of the flow velocity are shown in Fig. (7.11). The azimuthal part has an initial value of the order of $\varpi_i \Omega$ and keeps increasing inside the roughly corotating sub-Alfvénic regime, while at large distances decreases as $\propto 1/\varpi\gamma$ from angular momentum conservation.

Figure 7.12 shows the various force components along the poloidal flow. It is seen that the magneto-centrifugal term $f_{C\parallel}$ is important near the base of the flow, but the magnetic force $f_{B\parallel}$ soon takes over. In response to the accelerating forces,

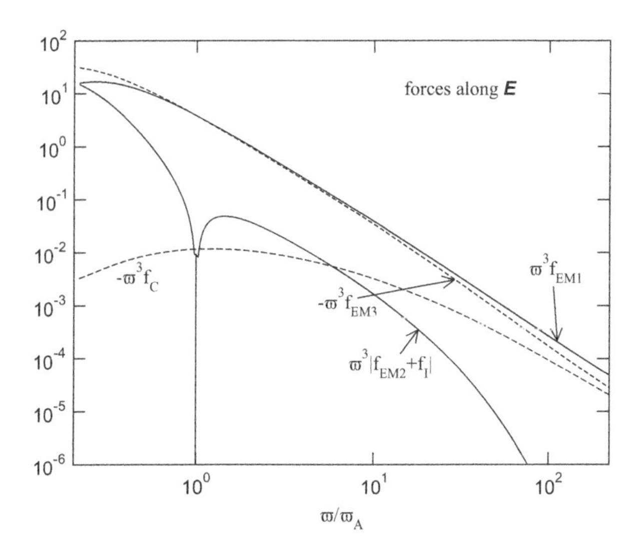

Fig. 7.13 Force density components in the transfield direction (along the electric field) (multiplied with ϖ^3, in arbitrary units)

the inertia terms $-f_{I\parallel} = \gamma^2 \rho_0 V_p \partial V_p / \partial \ell$ and $-f_{G\parallel} = \gamma \rho_0 V_p^2 \partial \gamma / \partial \ell$ are positive, resulting in an accelerating flow. As long as the flow remains mildly-relativistic the term $-f_{I\parallel}$ dominates, while at large distances the $-f_{G\parallel}$ is the main inertial force in the $\boldsymbol{V}_p$ direction.

The force components in the transfield direction on the poloidal plane (i.e. along the direction of the electric field) are shown in Fig. 7.13. The force-balance in this direction determines the poloidal field line shape, and indirectly the acceleration, as we discussed in the previous sections. In the sub-Afvénic regime the force $-f_{EM3}$ dominates, while in the super-Afvénic part of the flow the force f_{EM1} takes over. However, the $-f_{EM3}$ part remains comparable to the f_{EM1}; thus, it is not a correct approximation to ignore the poloidal field terms even at large distances. At small distances the centrifugal force is negligible. As a result, the difference between the f_{EM1} and $-f_{EM3}$ terms determines the poloidal curvature (this difference equals the $|f_{EM2} + f_{I\perp}|$ term). At large distances however, the centrifugal term $-f_{C\perp}$ is non-negligible, and equals the difference between the f_{EM1} and $-f_{EM3}$ terms (the curvature term $|f_{EM2} + f_{I\perp}|$ is much smaller, a characteristic of cylindrical asymptotics).

Figure 7.14 shows the curvature of the poloidal field lines. As expected from the analysis of Sect. 7.4.2, $\varpi/\mathcal{R} < 1/\gamma^2$. However, the approximation $\varpi/\mathcal{R} \sim 1/\gamma^2$ does not hold, because the force f_{EM3} related to the poloidal field is non-negligible as we discussed in the previous paragraph.

The solution that we analyzed above was presented by Vlahakis and Königl (2004) in the context of AGN jets. Similar solutions can be found in Li et al. (1992) and Contopoulos (1994). One could also find solutions of the same model

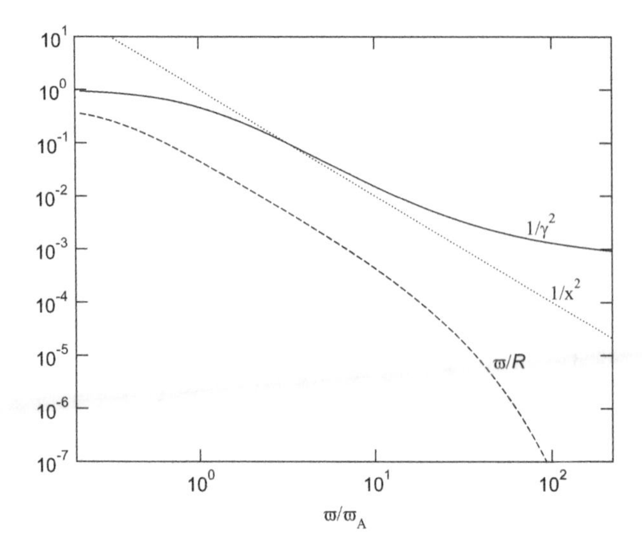

Fig. 7.14 The quantity $\varpi/\mathcal{R}$, where $\mathcal{R}$ is the curvature of the poloidal field lines, and the $1/x^2$, $1/\gamma^2$

in the context of GRB outflows in Vlahakis and Königl (2003a,b) (including also thermal and radiation pressure effects, which however do not significantly modify the conclusions for the super-fast magnetosonic regime). In addition, numerical simulations (e.g. Komissarov et al. 2007, 2009) offer another way to analyse the MHD problem; the steady-steady that they reach is in agreement with the preceding analysis.

7.5 Rarefaction Acceleration

From the analysis of Sect. 7.4.1 it is clear than sufficient expansion of the magnetic field leads to acceleration. In the collimation-acceleration mechanism this is achieved by differential collimation, i.e., with the streamlines that are closer to the symmetry axis being collimated faster than the outer ones, leading to a sufficient opening of the streamlines. There is another way to achieve this, by loosing the support of the external medium after some distance.

Suppose for example that in the space $x < 0$, $z < 0$ we have a uniform magnetized flow with super-fast magnetosonic velocity $V_j\hat{z}$, supported by some external pressure of a medium that fills the space $x > 0$, $z < 0$. If the space $x > 0$, $z > 0$ is empty, or in the more general case filled by a medium with smaller pressure, then the pressure discontinuity at the line $x = z = 0$ will drive a rarefaction wave inside the flow. The associated expansion of streamlines will obviously lead to flow bulk acceleration. This "rareafaction acceleration" mechanism has been seen in numerical simulations of hydrodynamic and magnetohydrodynamic flows (Aloy and Rezzolla 2006; Mizuno et al. 2008; Zenitani et al. 2010; Matsumoto et al. 2012). It is a powerfull mechanism with important applications in collapsar GRB jets, in which the jet is initially propagated inside the progenitor star and when it breaks out from the star looses its external pressure support. Related simulations by Tchekhovskoy et al. (2010) and Komissarov et al. (2010) discuss this effect, and show that the efficiency of the magnetic acceleration (which was in effect inside the star leading to super-fast magnetosonic jet speed at breakout) is significantly increased. Subsequently Sapountzis and Vlahakis (2013) analyzed the effect using a steady-state, self-similar, relativistic MHD description of the rarefaction wave in planar geometry. This is to first order sufficient to study the effect which is important near the "corner" formed by the intersection of the jet and star surfaces. The related equations can be found in Sapountzis and Vlahakis (2013).[5] The picture is much simplified if we neglect the poloidal magnetic field and the azimuthal bulk velocity (replaced by the $\hat{y}$ component in the planar geometry), both of which are negligible in the super-fast magnetosonic regime. In that case, and for ultrarelativistic flows $V_j = \sqrt{1 - 1/\gamma_j^2}$ with $\gamma_j \gg 1$, we can find analytical relations describing the effect.

The initial flow characteristics in the regime $z < 0$, $x < 0$ are density ρ_j, velocity $\boldsymbol{V} = V_j \hat{z}$, electromagnetic field $\boldsymbol{B} = B_j \hat{y}$ and $\boldsymbol{E} = (V_j/c) B_j \hat{x}$, Lorentz factor γ_j and magnetization $\sigma_j = B_j^2/4\pi\rho_j$. The rarefaction wave speed is the fast magnetosonic speed and thus only the part of the flow located inside the fast Mach cone with its origin at $x = z = 0$ will be affected. A Lorentz transformation of the wave proper speed (which is $U_f = c\sigma^{1/2}$ in the comoving frame) gives the opening of the Mach cone which defines the head of the rarefaction, at polar angle $\theta_{\rm head} = -\sigma_j^{1/2}/\gamma_j$. Figure 7.15 shows the flow streamlines together with the characteristics. The head of the rarefaction corresponds to the characteristic that passes through the points $x = 0, z = 0$ and $x = -1, z \approx 45$.

The integrals of motion give the density, the Lorentz factor, and the magnetic field as functions of the magnetization

$$\rho = \rho_j \frac{\sigma}{\sigma_j}, \quad \gamma = \gamma_j \frac{1 + \sigma_j}{1 + \sigma}, \quad B = B_j \frac{\sigma}{\sigma_j}. \tag{7.32}$$

Integration of the momentum equation gives the flow inclination with respect to the z-axis $\vartheta \approx V_x/c$

[5]Actually the equations describing the effect in planar geometry can be found from the corresponding ones in cylindrical geometry by replacing $\varpi \to \varpi_0 + x$, $z \to z_0 + z$, $A/\varpi_0 \to A$, $\varpi_0 \Omega/c \to \chi$, $x_{\rm A} \to \chi_{\rm A}$, $L/\varpi_0 \to P = \mu c \chi_{\rm A}^2/\chi$, $\varpi_{\rm A}/\varpi_0 \to \chi_{\rm A}/\chi$, and then taking the limit $\varpi_0 \to \infty$.

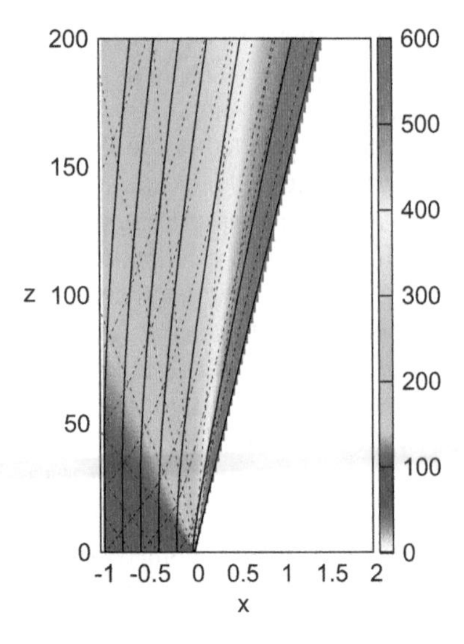

Fig. 7.15 Rarefaction solution corresponding to $\gamma_j = 100$, $\sigma_j = 5$. A few streamlines are shown (*solid lines*) together with the characteristics (*dashed lines*) and the Lorentz factor (*color*)

$$\vartheta = \frac{V_x}{c} = 2\frac{\sigma_j^{1/2} - \sigma^{1/2}}{\gamma_j(1+\sigma_j)} \,. \tag{7.33}$$

If the space outside the jet is empty, then the tail of the rarefaction corresponds to $\sigma = 0$ and its polar angle is $\theta_{\rm tail} = 2\sigma_j^{1/2}/\gamma_j(1+\sigma_j)$.

The momentum equation also gives the magnetization as a function of the polar angle $\theta = x/z$

$$\theta = \frac{x}{z} = \frac{2\sigma_j^{1/2} - 3\sigma^{1/2} - \sigma^{3/2}}{\gamma_j(1+\sigma_j)} \Leftrightarrow$$

$$\sigma = 4\sinh^2\left[\frac{1}{3}\text{arcsinh}\left(\sigma_j^{1/2} - \gamma_j\frac{1+\sigma_j}{2}\frac{x}{z}\right)\right] . \tag{7.34}$$

Figures 7.15 and 7.16 show the resulting rarefied flow, a solution from Sapountzis and Vlahakis (2013).

The flow is self-similar and thus the degree of acceleration which quickly happens in the part of the flow near the corner needs long time to happen in the parts that are initially at large $|x_i|$. The left panel of Fig. 7.17 shows the distribution of Lorentz factors across the jet at various heights z. Knowing the initial width of the jet we can estimate the acceleration efficiency of the whole flow using this figure.

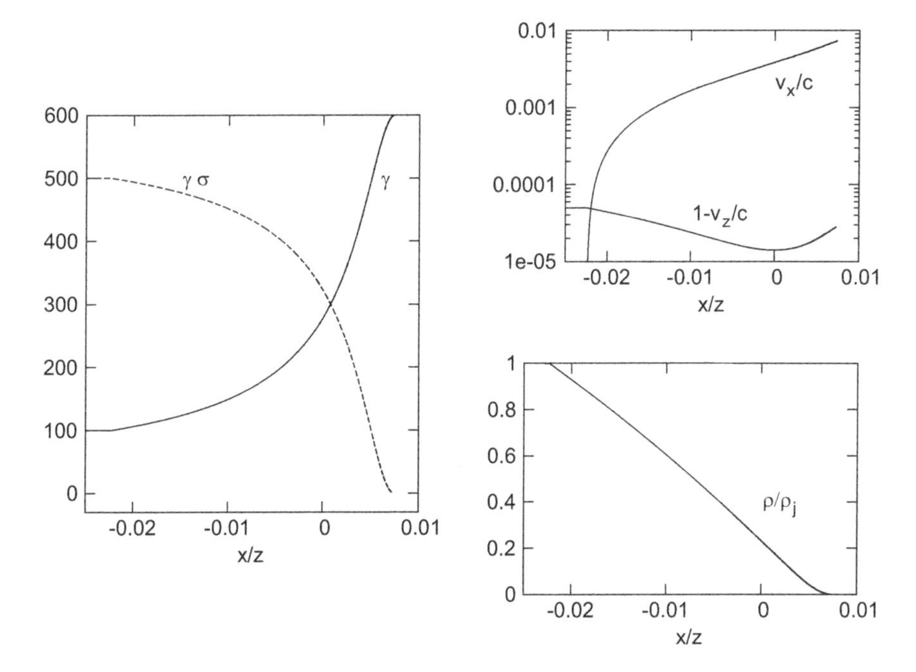

Fig. 7.16 Rarefied flow quantities. The figure on the *left* shows the two parts of the energy-to-mass flux ratio. A complete transformation of Poynting to mater kinetic energy characterizes the tail of the rarefaction at $x/z \approx 7 \times 10^{-3}$. The other two figures on the *right* show the velocity components (with V_x/c being also the opening ϑ of the flow), and the normalized density (which equals the normalized magnetization $\rho/\rho_j = \sigma/\sigma_j$)

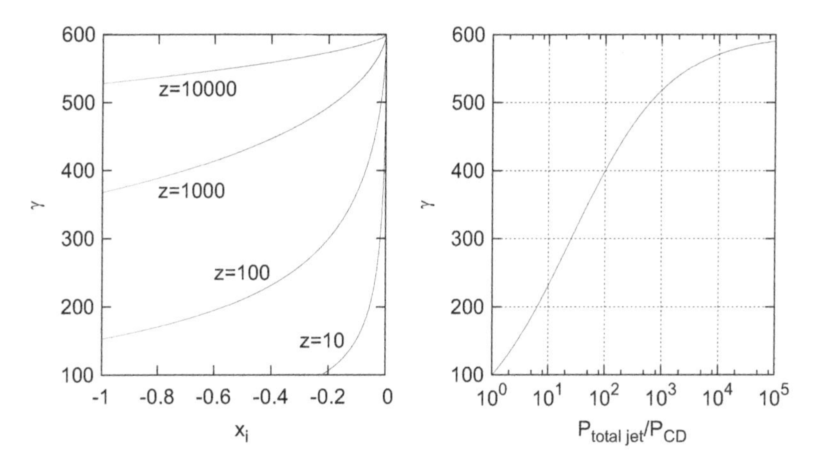

Fig. 7.17 *Left panel*: The Lorentz factor as a function of the starting position x_i of each part of the flow before the rarefaction starts, for various heights z. *Right panel*: Terminal Lorentz factor of the jet as a function of the ratio of the initial jet pressure over the pressure at the contact discontinuity

The described solution to its full extend applies in cases where the jet environment in the $z > 0$ regime is vacuum. In any other case the fluid of the environment will be compressed, a shock will likely be formed, and a pressure equilibrium will be reached between the jet and its environment at a contact discontinuity. The terminal Lorentz factor of the jet and the acceleration efficiency depends on the contrast of the initial jet pressure and the pressure at the contact discontinuity $P_{\rm CD}$. This is shown in the right panel of Fig. 7.17. A study of the shock wave propagating inside the environment will give the value of $P_{\rm CD}$ for given initial states of the jet and the environment. The outcome will depend on the density contrast as well. External densities much lower than the jet, as in the GRB collapsar model, will help to keep the picture almost the same as if the environment was empty space. In AGN jets on the other hand, the external densities are higher than the jet densities, something that may lead to high values of $P_{\rm CD}$ and low acceleration efficiencies.

Other factors that need to be taken into account include the effect of the finite jet width (the rarefaction wave will be reflected from the rotation axis and the collision of the two waves will likely create a shock inside the jet), the possible effects of cylindrical symmetry at distances far away from the "corner", and a generalization in three dimensions (which may affect the reflection from the axis).

7.6 Impulsive Acceleration

A similar mechanism has been suggested for the acceleration of AGN and GRB jets fronts (Granot et al. 2011), based however on the time dependent rarefaction wave that propagates in an initially static magnetized fluid, triggered by a pressure difference between the jet and its environment. Suppose at $t = 0$ a magnetized plasma fills the space $x < 0$ (in the following discussion this is the "left" state) while the space $x > 0$ is empty (this is the "right" state). The rarefaction can be described as a simple wave, i.e., all quantities are functions of the self-similarity variable x/ct. The equations describing the flow (Riemann invariants along the characteristics) can be found e.g., in Marti and Muller (1994).

Figure 7.18 shows an example solution. The wave evolution is similar to the steady-state rarefaction caused by sideways expansion (Sect. 7.5). There are however some important differences. As seen in Fig. 7.18 the mean value of the Lorentz factor in the body of the flow is quite low, only in a small width at the tail of the wave the proper speed becomes high; it equals $2\sigma_j + 1$. The mean value of γ as estimated in Granot et al. (2011) is only $\sim\sigma_j^{1/3}$.

The reflection of the wave from a conducting wall that ends the left state is considered in Granot et al. (2011). The interaction of the two waves strongly modifies the profile of the Lorentz factor in the plasma.

The profile is also strongly affected if the right state is not vacuum, but some medium with nonzero density, even if its pressure is zero. The full Riemann problem consist of the rarefaction wave propagating to the left and a shock wave propagating to the right inside the right state, with some velocity V_s.

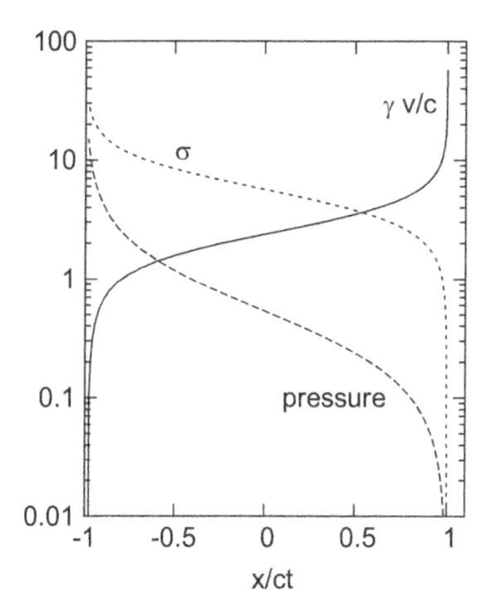

Fig. 7.18 Impulsive acceleration of a magnetized plasma at rest, with initial magnetization $\sigma_j = 30$. The evolution of σ, proper speed $\gamma V/c$ and magnetic pressure (in units of $\rho_j c^2$) are shown

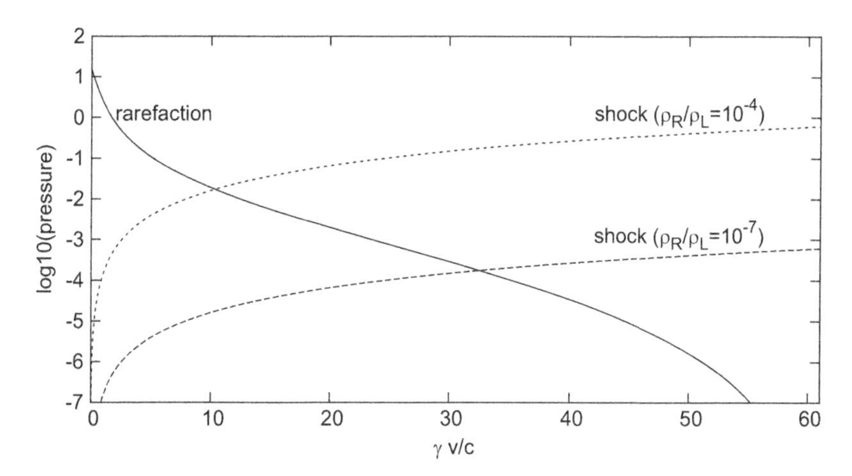

Fig. 7.19 Pressure-velocity curves for the rarefaction propagating to the left and for shock waves propagating to the right, for two cases of ρ_R/ρ_L

Following Marti and Muller (1994) (a similar study has been done in Lyutikov 2010) we can solve the equations for the rarefaction propagating to the left into the left state and find all possible pairs pressure-velocity that belong to the solution (decreasing curve in Fig. 7.19). We can also solve the jump conditions for a shock propagating to the right into the right state and find all the pairs pressure-velocity of the shocked region that correspond to various shock velocities. This is shown in the increasing curves of Fig. 7.19, for two values of the external (right) density. The intersection of the rarefaction and shock curves gives the pressure and velocity of the contact discontinuity that will be formed between the two fluids. Figure 7.20 shows the resulting profile of the rarefied left state and the shocked right state. Even for

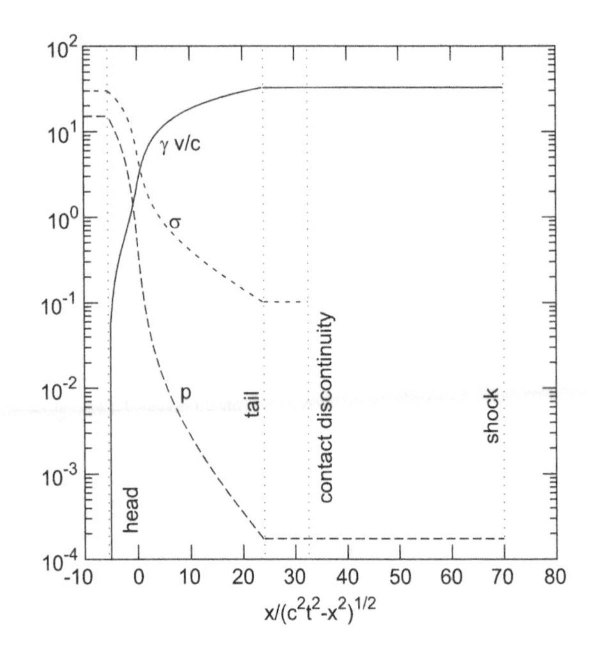

Fig. 7.20 The Riemann problem for a magnetized, cold, static left state with $\sigma_L = 30$ and a cold, static right state with $\rho_R/\rho_L = 10^{-7}$ and polytropic index $5/3$. The profile is shown as function of $(x/ct)/\sqrt{1-(x/ct)^2}$ in order to expand the region around the value $x/ct \approx 1$ where the tail, the contact discontinuity and the shock are located

$\rho_R/\rho_L = 10^{-7}$ the Lorentz factor at the tail of the rarefaction has been significantly reduced. More on the interaction of the impulsive flow with the external medium can be found in Lyutikov (2010) and Granot (2012).

References

Aloy, M.A., Rezzolla, L.: A powerful hydrodynamic booster for relativistic jets. ApJ **640**, L115 (2006). doi:10.1086/503608

Bekenstein, J.D., Oron, E.: New conservation laws in general-relativistic magnetohydrodynamics. Phys. Rev. D **18**, 1809 (1978)

Beskin, V.S.: MHD flows in compact astrophysical objects: accretion, winds and jets. Springer, Berlin/Heidelberg (2010)

Blandford, R.D., Payne, D.G.: Hydromagnetic flows from accretion discs and the production of radio jets. MNRAS **199**, 883 (1982)

Blandford, R.D., Znajek, R.L.: Electromagnetic extraction of energy from Kerr black holes. MNRAS **179**, 433 (1977)

Camenzind, M.: Centrifugally driven MHD-winds in active galactic nuclei. A&A **156**, 137 (1986)

Contopoulos, J.: Magnetically driven relativistic jets and winds: exact solutions. ApJ **432**, 508 (1994). doi:10.1086/174590

Fendt, C., Ouyed, R.: Ultrarelativistic magnetohydrodynamic jets in the context of gamma-ray bursts. ApJ **608**, 378 (2004). doi:10.1086/386363

Granot, J.: Interaction of a highly magnetized impulsive relativistic flow with an external medium. MNRAS **421**, 2442 (2012). doi:10.1111/j.1365-2966.2012.20473.x

Granot, J., Komissarov, S.S., Spitkovsky, A.: Impulsive acceleration of strongly magnetized relativistic flows. MNRAS **411**, 1323 (2011). doi:10.1111/j.1365-2966.2010.17770.x

Komissarov, S.S., Barkov, M.V., Vlahakis, N., Königl, A.: Magnetic acceleration of relativistic active galactic nucleus jets. MNRAS **380**, 51 (2007). doi:10.1111/j.1365-2966.2007.12050.x

Komissarov, S.S., Vlahakis, N., Königl, A., Barkov, M.V.: Magnetic acceleration of ultrarelativistic jets in gamma-ray burst sources. MNRAS **394**, 1182 (2009). doi:10.1111/j.1365-2966.2009.14410.x

Komissarov, S.S., Vlahakis, N., Königl, A.: Rarefaction acceleration of ultrarelativistic magnetized jets in gamma-ray burst sources. MNRAS **407**, 17 (2010). doi:10.1111/j.1365-2966.2010.16779.x

Li, Z.Y., Chiueh, T., Begelman, M.C.: Electromagnetically driven relativistic jets - a class of self-similar solutions. ApJ **394**, 459 (1992). doi:10.1086/171597

Lovelace, R.V.E., Mehanian, C., Mobarry, C.M., Sulkanen, M.E.: Theory of axisymmetric magnetohydrodynamic flows - disks. ApJS **62**, 1 (1986). doi:10.1086/191132

Lyutikov, M.: Simple waves in relativistic fluids. Phys. Rev. E **82**(5), 056305 (2010). doi:10.1103/PhysRevE.82.056305

Marti, J.M., Muller, E.: Analytical solution of the Riemann problem in relativistic hydrodynamics. J. Fluid Mech. **258**, 317 (1994). doi:10.1017/S0022112094003344

Matsumoto, J., Masada, Y., Shibata, K.: Effect of interacting rarefaction waves on relativistically hot jets. ApJ **751**, 140 (2012). doi:10.1088/0004-637X/751/2/140

Michel, F.C.: Relativistic stellar-wind torques. ApJ **158**, 727 (1969)

Millas, D., Katsoulakos, G., Lingri, D., Karampelas, K., Vlahakis, N.: Solutions of the wind equation in relativistic magnetized jets. Int. J. Mod. Phys. Conf. Ser. **28**, 1460200 (2014). doi:10.1142/S2010194514602002

Mizuno, Y., Hardee, P., Hartmann, D.H., Nishikawa, K.I., Zhang, B.: A magnetohydrodynamic boost for relativistic jets. ApJ **672**, 72 (2008). doi:10.1086/523625

Okamoto, I.: Relativistic centrifugal winds. MNRAS **185**, 69 (1978)

Sapountzis, K., Vlahakis, N.: Rarefaction acceleration in magnetized gamma-ray burst jets. MNRAS **434**, 1779 (2013). doi:10.1093/mnras/stt1142

Tchekhovskoy, A., McKinney, J.C., Narayan, R.: Simulations of ultrarelativistic magnetodynamic jets from gamma-ray burst engines. MNRAS **388**, 551 (2008). doi:10.1111/j.1365-2966.2008.13425.x

Tchekhovskoy, A., McKinney, J.C., Narayan, R.: Efficiency of magnetic to kinetic energy conversion in a monopole magnetosphere. ApJ **699**, 1789 (2009). doi:10.1088/0004-637X/699/2/1789

Tchekhovskoy, A., Narayan, R., McKinney, J.C.: Magnetohydrodynamic simulations of gamma-ray burst jets: beyond the progenitor star. NewA **15**, 749 (2010). doi:10.1016/j.newast.2010.03.001

Tchekhovskoy, A., Narayan, R., McKinney, J.C.: Efficient generation of jets from magnetically arrested accretion on a rapidly spinning black hole. MNRAS **418**, L79 (2011). doi:10.1111/j.1745-3933.2011.01147.x

Toma, K., Takahara, F.: Efficient acceleration of relativistic magnetohydrodynamic jets. Prog. Theor. Exp. Phys. **2013**(8), 083E02 (2013). doi:10.1093/ptep/ptt058

Vlahakis, N.: The efficiency of the magnetic acceleration in relativistic jets. Ap&SS **293**, 67 (2004a). doi:10.1023/B:ASTR.0000044653.77654.fb

Vlahakis, N.: Ideal magnetohydrodynamic solution to the σ problem in crab-like pulsar winds and general asymptotic analysis of magnetized outflows. ApJ **600**, 324 (2004b). doi:10.1086/379701

Vlahakis, N., Königl, A.: Relativistic magnetohydrodynamics with application to gamma-ray burst outflows. I. theory and semianalytic trans-alfvénic solutions. ApJ **596**, 1080 (2003a). doi:10.1086/378226

Vlahakis, N., Königl, A.: Relativistic magnetohydrodynamics with application to gamma-ray burst outflows. II. semianalytic super-alfvénic solutions. ApJ **596**, 1104 (2003b). doi:10.1086/378227

Vlahakis, N., Königl, A.: Magnetic driving of relativistic outflows in active galactic nuclei. I. interpretation of parsec-scale accelerations. ApJ **605**, 656 (2004). doi:10.1086/382670

Zenitani, S., Hesse, M., Klimas, A.: Scaling of the anomalous boost in relativistic jet boundary layer. ApJ **712**, 951 (2010). doi:10.1088/0004-637X/712/2/951

Chapter 8
X-Ray Binary Phenomenology and Their Accretion Disk Structure

Demosthenes Kazanas

Abstract We propose a scheme that accounts for the broader spectral and temporal properties of galactic black hole X-ray transients. The fundamental notion behind this proposal is that the mass accretion rate, $\dot{M}$, of the disks of these systems depends on the radius, as it has been proposed for ADIOS. We propose that, because of this dependence of $\dot{M}$ on radius, an accretion disk which is geometrically thin and cool at large radii converts into a geometrically thick, advection dominated, hot disk interior to a transition radius at which the local accretion rate drops below the square of the viscosity parameter, a condition for the existence of advection dominated flows. We argue also that such a transition requires in addition that the vertical disk support be provided by magnetic fields. As discussed in other chapters of this book, the origin of these fields is local to the disk by the Poynting Robertson battery, thereby providing a complete self-contained picture for the spectra and evolution of these systems.

8.1 Introduction

Accretion disks are the generic structures forming in the process of accretion of plasma onto compact objects, assumed in the context of the present article to be black holes. Viscous processes within these structures convert the kinetic energy of accreting plasma into radiation of specific spectra, readily observable on the Earth. As such they have been the subject of innumerable works that have modeled their structure, evolution and spectra at various levels of detail which were then implemented to account for the phenomenology, both of galactic X-ray binary sources (hereafter XRB) and active galactic nuclei (AGN). The present article is not a review of accretion disks (the interested reader can consult a number of such reviews e.g. Abramowicz and Fragile 2013); instead, following the spirit of the entire monograph, it aims to present alternatives to more conventional accretion disk theory, in an effort to provide an account of their broadest spectral and temporal

D. Kazanas (✉)
NASA, Goddard Space Flight Center, Code 663, Greenbelt, MD, USA
e-mail: demos.kazanas@nasa.gov

I. Contopoulos et al. (eds.), *The Formation and Disruption of Black Hole Jets*,
Astrophysics and Space Science Library 414,
DOI 10.1007/978-3-319-10356-3_8

features in ways that would lead to novel, fruitful insights in the structure of these systems. As such, it omits quoting many of the important works on the subject, while presenting some views that many people in the field would object against. This is precisely its aim! Were the more conventional works able to account for the established disk phenomenology, the entire subject would have been delegated to textbooks. The fact that it is not, implies that rather significant observational facts and correlations remain inaccessible to the more conventional models, thereby inviting introduction of more unorthodox views, in the hope that these may lead to modifications that will present an edifice that can accommodate, if not all, the majority of accretion power phenomenology.

Much of the work on accretion powered sources has been based on and influenced by the seminal work on the subject, namely that of Shakura and Sunyaev (hereafter SS73) (Shakura and Sunyaev 1973). These authors assumed the disks to be steady–state, roughly Keplerian ($V_r \ll V_\phi$), thin ($h \ll R$) and in hydrostatic equilibrium in the vertical direction. Then they solved for their thermal balance (local energy dissipation equals the energy radiated) and the transfer of angular momentum that allows for the accretion of matter, assuming that the viscous stresses that achieve this are proportional to the local gas pressure P ($t_{r\phi} \simeq \alpha P$; α is an unknown parameter to be determined by observation, or as is more recently the case, by numerical simulation). The usual assumption, enunciated by SS73 is that all energy transferred by the viscous torques is dissipated locally. At this point one should note that since angular momentum cannot be destroyed and since essentially all the disk kinetic energy is stored in its angular motion, none of that energy should be dissipated. However, viscous stresses can dissipate circulation and presumably it is the dissipation of circulation that powers the observed radiation of accretion disks. Because in axisymmetry circulation has the same form as the non-dissipative angular momentum, one can speculate that the observed presence of dissipation in accreting objects involves non-axisymmetric fluid modes.

A further simplifying assumption generally made in modeling accretion disks is that the dissipated energy is thermalized, i.e. that the disk particles achieve a thermal distribution of temperature determined so that it radiates away the locally dissipated power. The spectrum of the resulting radiation is thermal, of black body form if absorption is the dominant opacity process, or modified black body, if scattering opacity provides a significant contribution. The local opacity in turn depends on the vertical disk structure which (by angular momentum conservation) determines its local density and column, while at the same time the value of the local temperature feedbacks on the vertical disk structure. The same assumption of local dissipation of all energy transferred by the viscous stresses, determines also the disk temperature radial profile. In steady state disks (mass flux $\dot{M}$ independent of time) the local energy dissipation per unit area is proportional to $3GM\dot{M}/8r^3$; assuming this to be equal to σT^4 (black body emission), implies that the disk temperature should decrease like $T \propto r^{-3/4}$.

The innermost disk temperature can be estimated assuming that the emitting surface is roughly πr_{ISCO}^2 (ISCO is short for the innermost stable circular orbit; for a Schwarzschild black hole $r_{ISCO} = 3R_S \simeq 10^6\, M_0\,\text{cm}$, where $R_S = 2GM/c^2$

is the Schwarzschild radius and $M_n = M/10^n M_\odot$ is the black hole mass in 10^n solar masses). An estimate of the disk highest temperature then requires a value of the disk luminosity. Because the maximum luminosity of an accretion powered source is generally considered to be the object's Eddington luminosity, $L_{\rm Edd} \simeq 1.3\,10^{38}\,M_0$ erg/s, the corresponding maximum disk temperature is $T \simeq 10^{7.5}\,M_0^{-1/4}\ K = 10^{5.5}\,M_8^{-1/4}$ K. The fact that quasi-thermal features at the corresponding temperatures were observed at approximately these energies in galactic XRBs and AGNs have established the SS73 disk as a ubiquitous structure in both these classes of sources.

While quasi-thermal spectral components at temperatures in agreement with these estimates are observed in the spectra of accreting black holes, it was established, by even the earliest X-ray observations, that, in addition to these thermal components, a large fraction of these objects' luminosity is shared by an X-ray spectral component of power law form that extends to energies $E \simeq 100$ keV, clearly inconsistent with that of thermal radiation by an accretion disk. In analogy with the Sun, this component was then attributed to the presence of a hot ($T \sim 10^8 - 10^9$ K) corona, overlaying the accretion disk. Again, in analogy with the solar corona, this was proposed to be powered by magnetic fields that thread the accretion disk and dissipate part of its energy in this hot corona (Galeev et al. 1979), which Compton-scatters the disk thermal radiation to produce the observed high energy photons.

The advent of observations and data accumulation, then, established that the ratio of luminosities between the thermal accretion disk and the power law high energy X-ray emission is not random but varies in a systematic way, generally with the contribution of the quasi-thermal disk component increasing with the objects' luminosity and that of the harder, power law one waning. There are in fact rather abrupt transitions between states dominated by each of the spectral components outlined above as an object's luminosity varies. These variations are most clearly discernible in galactic XRB sources, in which their luminosity can go through low-high-low cycles of significant amplitude over intervals of a few months, thereby establishing the accretion rate $\dot{M}$ as the primary varying parameter in these systems. It is therefore important that the models be able to provide not only the steady state attributes of an accretion disk, but also its sequence of states with the variation of the accretion rate. Because the spectra of these objects change with $\dot{M}$, they are represented as trajectories in a Hardness-Luminosity Diagram (hereafter HLD). An important aspect of these trajectories is that they are not reversible in time but present a hysteresis, namely, in returning to the low hard state the sources spend some time in a low intensity soft state before they return to their hard one (see contribution by Kylafis and Belloni in this volume).

In AGN the thermal disk feature is in the optical – UV band and as such it is better discerned in their spectra. In fact it constitutes the dominant SED component in bright QSOs and a significant component in lower luminosity Seyferts; it is fit well by a multicolor disk whose innermost radius is that of ISCO and it is generally referred to as the Big Blue Bump (BBB). Its relation to the also ubiquitous X-ray emission is not at present clear. To begin with, the relative luminosity of these two components is not set by theory. Nonetheless, it was found that the logarithmic flux slope between the UV (2,500 Å) and 2 keV X-rays, the so-called α_{OX} parameter, increases with increasing source luminosity (Steffen et al. 2006) (see their fig. 5). The reason for the observed correlation is not clear, however, it is not unreasonable to consider that this correlation is similar to that between the XRB multicolor disk component and their power law X-rays (see Done et al. 2007 for a review). Another open issue is the geometry of the X-ray emitting plasma relative to the BBB component. Earlier models based on emission by a magnetically powered corona, assumed that the X-ray emitting plasma was overlying the BBB emitting thin accretion disk. However, issues concerning the cooling of the corona electrons (Haardt and Maraschi 1993), and the corresponding variability due to reprocessing of the X-rays on the thin disk (Berkeley et al. 2000), appear to require refinement of this geometric arrangement.

Accretion disk theory got a big boost in the 1990s with two significant developments: (i) Balbus and Hawley (1991) showed that a fluid rotating with angular velocity Ω and threaded by poloidal magnetic field is stable if $d\Omega^2/dr > 0$ (in disagreement with Rayleigh criterion for unmagnetized flows that demands $d(\Omega^2 r^4)/dr > 0$). According to this criterion therefore, Keplerian accretion disks are unstable, with the magnetic field acting as the agent that helps mediate the transfer of angular momentum necessary for accretion to take place (for a qualitative discussion of this and other instabilities of magnetized, rotating fluids and why it does not reduce to the Rayleigh criterion as the magnetic field goes to zero see Christodoulou et al. 1996). (ii) Narayan and Yi (1994; 1995) produced models of accretion disks that are optically thin and geometrically thick ($h \simeq R$), supported at each radius by the pressure of ions. The rotational velocity of these disks is sub-Keplerian with $V_r \simeq V_\phi \lesssim V_K$.

The reason the disks of Narayan and Yi are thick is that the proton cooling time through Coulomb collisions with the cooler electrons (assuming they achieve at some radius r their virial temperature, $kT_p \simeq GM/r$, heated both by the dissipation of the azimuthal motions and the pdV work of accretion), is longer than the local viscous time scale, which for $h \simeq r$ is only α times longer than the free fall time $t_{ff} \simeq r/V_r$. Such thick flows, are therefore possible only for $t_{cool} > t_{visc}$. Considering that $t_{cool} \propto 1/n(x)\sigma_T c$ (omitting here some additional dependence on the electron temperature), with $n(x) \propto \alpha^{-1}(\dot{m}/\sigma_T R_S)x^{-3/2}$ ($x = r/R_S$ is the radius normalized to the Schwarzschild radius $R_S \simeq 3 \times 10^5 M_0$ cm and $\dot{m}$ the accretion rate normalized to the Eddington accretion rate, $\dot{M}_{\rm Edd} = L_{\rm Edd}/c^2$ and σ_T the Thomson cross section), while $t_{visc} \propto r^2/\nu \simeq \alpha^{-1}(R_S/c)x^{3/2}$, with the viscosity coefficient ν given by $\nu \simeq \alpha r V_r \simeq \alpha h c_s$, the condition $t_{cool} > t_{visc}$ implies simply that $\dot{m} < \alpha^2$. It is interesting to note that this condition involves neither the mass of the object M nor the radius of the flow x (see however Narayan and Yi 1995 for a weak $x-$dependence). It imposes only a *global condition* on the dimensionless accretion rate; low accretion rate flows, if they start hot (i.e. with virialized protons), they will remain so all the way to the horizon of the accreting body with their azimuthal and radial velocities a fraction of the Keplerian one ($V_\phi \simeq V_r \sim 0.7 V_K$).

Therefore, since for these flows $h \simeq r$, they resemble spherical accretion; also, because the cooling time of the accreting gas is longer than the advection time onto the accreting object, their efficiency is reduced with much of the thermal energy being advected onto the black hole; as such they are referred to as either ADAF (Advection Dominated Accretion Flows) or RIAF (Radiatively Inefficient Accretion Flows). In a RIAF, the accretion luminosity L is not any longer simply proportional to the normalized accretion rate $\dot{m}$ but to $\dot{m}^2$, the extra power of $\dot{m}$ coming from the ratio t_{cool}/t_{visc}.

The great advantage that the ADAF paradigm offers to the broader view of accretion is that it provides the hot electrons demanded by the hard X-ray observations of (galactic and extragalactic) black holes as a result of the general accretion flow dynamics, rather than as a (putative) corona, unrestricted by the dynamics of accretion and introduced so that it would accommodate the observations. The notion of ADAF, then, led to the following novel picture for the spectral decomposition and evolution of galactic binary X-ray sources (Esin et al. 1997): At low luminosities, the thin SS73 disk terminates at some large radius r_{tr} and transitions to an ADAF in its interior region. As the luminosity (and the accretion rate increases), the inner radius of the SS73 disk decreases and so does the extent of the ADAF component, with the ADAF component disappearing at the largest luminosities. Presumably, as the luminosity decreases the scheme reverses itself.

Appealing as it appears, this picture leaves out some points not addressed by a well defined physical process: There is no compelling process detailing how a flow that starts as a thin SS73 disk at large radii (as implied by the presence of the thermal, multi-color disk component in the spectra) would puff-up at a given radius. It is also not clear why the inner edge of this flow would advance to smaller radii with increasing $\dot{m}$. Despite the absence of an account of these points, the combination of an inner geometrically thick, optically thin ADAF component and a cooler, geometrically thin, optically thick SS73 component with an inner edge that depends on luminosity provides good fits to the data and it is broadly accepted as a means of accounting for the correlated flux-spectral evolution of galactic XRBs. Besides the above issue, there is an additional one that makes ADAF somewhat problematic: Their Bernoulli integral, *Be*, the sum of their kinetic, thermal and potential energies per unit mass is positive, a fact noted in the original references on the subject (Narayan and Yi 1994, 1995); therefore, these flows provide for the potential escape of their fluid to infinity, raising thus the possibility of producing the outflows observed in accretion powered sources (this is not the case in SS disks whose internal thermal energy is assumed to be radiated away promptly resulting in $Be < 0$).

The reason for the positivity of *Be* was elucidated in a (very important to this author's opinion) paper by Blandford and Begelman (1999): It is due to two facts: (i) The viscous stresses that allow matter to sink toward the gravitating object, transfer outward besides angular momentum also (low entropy) mechanical energy, which adds to that released locally by gravity (see Sect. 8.2.1) (ii) The local cooling time is longer than the viscous time, so this energy is stored in the fluid rather than being radiated away (as it is assumed for SS73 disks), leading to the positivity of *Be*.

The resolution to this issue offered by Blandford and Begelman (1999) is that the excess energy and angular momentum are expelled off the disk in the form of a wind. This leaves the disk with less matter to accrete but with matter that it is now bound gravitationally, i.e. with $Be < 0$. As such, the (normalized) accretion rate $\dot{m}$ is not any longer constant, but depends on the (normalized) radius $x = r/R_S$, i.e. $\dot{m} = \dot{m}(x) \propto x^p$ and $1 > p > 0$. The positive value of p implies that these flows eject to a wind most of the mass available for accretion at their outer edge, named for this reason ADIOS (advection dominated inflow outflow solutions).

There is mounting evidence for the presence of such winds in the spectra of AGN and XRBs. These manifest themselves by absorption features in their X-ray and UV spectra. From the plethora of X-ray transitions one can estimate the absorption column N_H per decade of the ionization parameter ξ ($\xi = L/nr^2$ is the ratio of ionizing photon flux to electron densities), referred to as the absorption measure distribution (AMD) (Behar 2009), which was found only weakly dependent on ξ, namely $AMD \equiv dN_H/d\log\xi \propto \xi^{\alpha},\ 0 \lesssim \alpha \lesssim 0.2$. This relation can be inverted to infer the wind density as a function of r (or x) and then the mass flux rate as a function of (normalized) radius. Thus it is found that $\dot{m}(x) \propto x^p$ implies $p = (1-\alpha)/(2\alpha+2)$, leading to accretion rates $\dot{m}(x) \propto x^{1/2}$ (Fukumura et al. 2010a) or $\dot{m}(x) \propto x^{1/3}$ for $\alpha = 0$ and $\alpha = 0.2$ respectively. With these values of p, assuming the outer edge of the disk to be at $x_M \sim 10^6$, the fraction of the available mass that accretes onto the compact object is 0.001 and 0.01 respectively for $\alpha = 0$ and $\alpha = 0.2$, the remaining being lost in a wind that is launched over the entire extent of the accretion disk. The mass flux thus estimated is found to be much higher than the one needed to power the observed accretion luminosity, in agreement with the arguments given above. Also, detailed photoionization modeling of such winds (Fukumura et al. 2010a,b) provide very good agreement with the warm absorber observations of AGN.

As it will be described in some detail in Sect. 8.2.3, the radius – dependent accretion rate, $\dot{m}(x)$, of ADIOS offers a natural resolution to the observed behavior of XRBs, namely the presence of a hot, ADAF-type flow at small radii, and a much cooler, thin, SS73-type disk at larger ones, with the transition radius, r_{tr}, decreasing with increasing luminosity. We further propose there that the transition from a geometrically thin, "cool" disk to a geometrically thick, ADAF-type flow is facilitated if the hydrostatic disk balance is maintained by magnetic fields. This then divorces the disk height from the temperature of the emitting plasma; the disk is much less dense than an SS73 one of the same temperature; this allows the local cooling time to be sufficiently long to convert the flow into a geometrically thick one.

A final aspect of the correlated luminosity – spectral evolution of these systems is that the sequence of states followed by these systems upon decrease of the global value of $\dot{m}$, is not a time reversed version of the increase in $\dot{m}$. We have argued elsewhere (Kylafis et al. 2012) (as it is also discussed in this volume by Kylafis and Belloni), that this behavior could be explained in terms of the magnetic flux confined near the black hole horizon and created during the accretion process by the Poynting-Robertson battery (Contopoulos and Kazanas 1999).

The structure of this article is the following: After an introduction of the basic accretion disk notions, including those of ADAF and ADIOS and magnetically supported disks (Sect. 8.2), we discuss in detail our proposal for the spectral state transitions with varying disk luminosity and the corresponding variation in r_{tr}. Then we compute the ratio R of the disk luminosities for $r > r_{tr}$ (quasi-thermal) and $r < r_{tr}$ (hard X-rays), which we plot as a function of r_{tr} (Sect. 8.3); this ratio we compare to the data of GRO 1655-40 at its different spectral states, while we also provide arguments concerning the corresponding time variability Power Spectral Density (PSD). We conclude in Sect. 8.4 with a summary of our results and a discussion.

8.2 The Dynamics of Accretion Disks

8.2.1 General Accretion Disk Structure

The structure of accretion disks is given by the transfer of angular momentum, which determines the radial distribution of stresses responsible for this action, the hydrostatic equilibrium in the direction perpendicular to the disk plane and the mass flux conservation. Finally, the disk spectrum is computed assuming that all energy released locally is dissipated and emitted in black body form.

(i) The hydrostatic equilibrium assumption of a thin disk implies

$$\frac{dP}{dz} = -\rho \frac{GM}{r^2}\frac{z}{r} \quad \text{or} \quad \frac{P}{h} \simeq \rho \frac{GM}{r^2}\frac{h}{r} \tag{8.1}$$

upon setting $\Delta P \approx P$ and $\Delta z \approx h \approx z$ we have

$$\frac{P}{\rho} \simeq c_s^2 \simeq \frac{GM}{r}\frac{h^2}{r^2} = v_K^2 \frac{h^2}{r^2} \simeq \Omega^2 h^2 \tag{8.2}$$

where Ω, v_K are the disk Keplerian frequency and velocity; then the disk height read $h \simeq c_s/\Omega$, where $c_s \simeq (P/\rho)^{1/2}$ is the sound speed in the disk.

(ii) Angular Momentum Conservation: If $\dot{\ell} = \dot{M}(GMr)^{1/2}$ is the rate at which angular momentum is transported inward at radius r by the accretion of matter at accretion rate $\dot{M}$, and $\dot{\ell}_I = \dot{M}(GMr_I)^{1/2}$ the rate at which angular momentum accreted onto the black hole at its innermost, stable, circular orbit radius (ISCO) $r_I = 3R_S$ (R_S is the black hole Schwarzschild radius), their difference implies the presence of a torque $\mathscr{T} = t_{r\phi}(2h \cdot 2\pi r) \times r = \dot{\ell} - \dot{\ell}_I$, which transfers their difference outward. In this expression $t_{r\phi}$ is the viscous stress (i.e. the force per unit area in the ϕ–direction) and $2h$ the total thickness of the disk. It is generally assumed (Shakura and Sunyaev 1973) that the viscous stress is proportional to the local pressure P, so that $t_{r\phi} = \alpha\, P$ with $\alpha < 1$.

Angular momentum conservation then leads to the following expression for the stress $t_{r\phi}$ in terms of the disk parameters

$$2ht_{r\phi} = \frac{\dot{M}}{2\pi r^2}(GMr)^{1/2}\left[1 - \zeta\left(\frac{r_I}{r}\right)^{1/2}\right] = \frac{\dot{M}}{2\pi r^2}(GMr)^{1/2}J(r) \tag{8.3}$$

The parameter ζ in the square brackets determines the value and sign of the torques at $r = r_I$, i.e. the radius at which matter free-falls onto the black hole; for $\zeta = 1$ the torque is also zero at the same radius.

With the above expression for the stresses, one can now compute the heat generation rate *per disk unit area*, $2Q$ (Q is the emission from each side of the disk), considering that the energy so generated *per unit volume* is $\dot{\epsilon} = 2t_{r\phi}\sigma_{r\phi}$, where $\sigma_{r\phi}$ is the disk shear; on substituting $2ht_{r\phi}$ from Eq. (8.3) above, with $\sigma_{r\phi} = (3/4)\Omega = (3/4)(GM/r^3)^{1/2}$ Novikov and Thorn (1972) we obtain

$$2h\dot{\epsilon} = (2ht_{r\phi})(2\sigma_{r\phi}) = 2Q = \frac{3\dot{M}}{4\pi r^2}\frac{GM}{r}J(r) \tag{8.4}$$

Therefore, the heat generated from a radius $r_1 (\gg r_I)$ to infinity is

$$2Q_{tot}(r_1) = \int_{r_1}^{\infty} 2h\dot{\epsilon}\; 2\pi r\, dr \simeq \frac{3}{2}\dot{M}\frac{GM}{r_1} \tag{8.5}$$

One should note that gravitational energy V from infinity to r_1 is released at a rate $GM\dot{M}/r_1$, but because of the virial theorem, $2T + V = 0$, only half of this can be converted to heat, with the rest remaining as orbital energy ($T = -V/2$). Thus the rate at which gravitational energy is converted into heat is $GM\dot{M}/2r_1$. The remainder needed to make up the difference with Eq. (8.5), is provided by the viscous stresses which transport outward not only angular momentum but energy at a rate

$$\dot{E} = \Omega\mathscr{T} = \Omega 2\pi r^2\, 2ht_{r\phi} = \frac{GM\dot{M}}{r_1}J(r_1) \simeq \frac{GM\dot{M}}{r_1}\;(r \gg r_I)\,. \tag{8.6}$$

As noted in Blandford and Begelman (1999), this is an important issue because in cases that the energy transferred by viscous stresses cannot be radiated away, it leads to positive Bernoulli integral. The increasing outward mass flux of the winds invoked to resolve this issue has significant observational consequences that appear consistent with observations (Fukumura et al. 2010a,b). We propose that this last fact is responsible for much of the AGN and XRB phenomenology, as it will be discussed below.

8.2.2 General Accretion Disk Scalings

It is instructive to present the accretion disk equations in dimensionless form, as this makes apparent the dependence of their properties relative to their natural units, most notably their accretion rate in terms of the Eddington accretion rate and the Schwarzschild radius. Normalizing the disk radius r by R_S, i.e. setting $r = xR_S$, and its accretion rate by the Eddington accretion rate, i.e. setting $\dot{M} = \dot{m}\dot{M}_{\rm Edd}$, with $\dot{M}_{\rm Edd} = L_{\rm Edd}/c^2 = 2\pi\, m_p\, cR_S/\sigma_T$, the hydrostatic equilibrium and angular momentum transfer equations (Eqs. 8.2 and 8.3) can be both solved for the disk pressure P to obtain (bearing in mind the prescription $t_{r\phi} = \alpha P$)

$$P = \rho\frac{GM}{r^2}\frac{h}{r} = \frac{m_p c^2}{2} n(x) x^{-1}\left(\frac{h}{r}\right)^2 \tag{8.7}$$

$$P = \frac{\dot{M}}{4\pi h\alpha}\frac{(GMr)^{1/2}}{r^2}J(r) = \frac{m_p c^2}{\sigma_T R_S}\dot{m}(x)x^{-5/2}\left(\frac{h}{r}\right)^{-1}\frac{J(x)}{2\sqrt{2}\alpha} \tag{8.8}$$

with $J(x) = 1 - (3/x)^{1/2}$, and following the arguments given in the introduction we assume that the accretion rate $\dot{m}$ is also a function of the radius x.

From the above equations one can obtain an expression for the disk density $n(x)$

$$n(x) = \frac{\dot{m}(x)}{\sigma_T R_S}x^{-3/2}\left(\frac{r}{h}\right)^3\frac{J(x)}{2^{3/2}\alpha} \tag{8.9}$$

and for the energy emitted per unit disk area (from one of its sides)

$$Q = \frac{3}{8\pi}\frac{GM\dot{M}(r)}{r^3}J(r) = \frac{3}{4}\frac{m_p c^3}{\sigma_T R_S}x^{-3}\dot{m}(x)J(x) \tag{8.10}$$

The total luminosity can be obtained by integrating the above expression over the surface of the disk $2\pi r\, dr$ from r_I to infinity, i.e. (one must use $2Q$ to take into account both sides of the disk)

$$L = \int_{r_I}^{\infty}(2Q)\,2\pi r\, dr = \frac{GM\dot{M}}{r_I}\left(\frac{3}{2}-\zeta\right) = \frac{\pi\, m_p c^3}{\sigma_T}\frac{R_S\dot{m}(x)}{x_I}\left(\frac{3}{2}-\zeta\right) \tag{8.11}$$

It is generally assumed that the pressure P is given either by the sum of gas and radiation pressures, each becoming dominant at different radii of the disk and for different values of the accretion rate $\dot{m}(x)$. Each such approximation leads to different run of the disk parameters with the radius x (see Svensson and Zdziarski 1994 for a detailed study). At present we will include also magnetic pressure. For simplicity, herein we will assume only gas and magnetic pressures P_g, P_B, with the magnetic pressure being dominant (see also Pariev et al. 2003) and the stresses being again proportional to $\alpha(P_g + P_B)$. The presence of three components of the magnetic stresses makes the problem necessarily more complicated, however we will consider here only their contribution to the vertical disk structure. The inclusion of magnetic field contribution in the disk vertical structure divorces the disk height from the local plasma temperature, thereby allowing transitions between hot and cool states as implied by observations and detailed in the next subsection.

Even at this simplified approach, the system likely entails far more detail than presented herein (e.g. field annihilation on an equatorial current sheet). Our simplified model averages all that over the disk height. Then the hydrostatic equilibrium equation reads

$$P = \frac{B_\phi^2(x)}{4\pi} + n(x)m_p c_s^2 = \rho(x)\frac{GM}{r^3}h(x)^2 \tag{8.12}$$

where $c_s(x)$ is the sound speed of the gas. Considering that $B_\phi^2/4\pi\rho(x) = V_A^2$ with V_A the Alfvén velocity, and that $GM/r = V_K^2$, the disk Keplerian velocity, the hydrostatic equilibrium condition reads

$$\frac{h^2}{r^2} = \frac{B_\phi^2}{4\pi\rho V_K^2}[1 + \beta(x)] = \frac{V_A^2}{V_K^2}[1 + \beta(x)] \tag{8.13}$$

where $\beta(x)$ is the usual gas-to-magnetic pressures parameter of the plasma.

From mass conservation (bearing in mind that, as discussed above the accretion rate depends on the radius r),

$$\dot{M}(x) = 2\pi hr\, n(x)m_p V_r \tag{8.14}$$

and employing Eq. (8.9), we obtain an expression for the radial flow velocity V_r

$$\frac{V_r}{c} = \frac{1}{x^{1/2}}\left(\frac{h}{r}\right)^2 \frac{\alpha}{J(x)} \quad \text{or} \quad \frac{V_r}{V_K} \simeq \left(\frac{h}{r}\right)^2 \frac{\alpha}{J(x)} \tag{8.15}$$

Considering that the viscous time scale

$$t_{visc} \simeq \frac{r^2}{\nu} \simeq \frac{r}{V_r} \simeq \frac{R_S}{c}x^{3/2}\left(\frac{r}{h}\right)^2 \frac{J(x)}{\alpha} \tag{8.16}$$

(with ν the coefficient of viscosity), the above expression implies $\nu \simeq rV_r \simeq hV_A[1 + \beta(x)]^{1/2}$, which if V_A is replaced by the disk thermal velocity we obtained the standard result of $\nu \simeq hc_s$.

Finally, assuming proportionality of the magnetic field with the pressure P one can employ the equation of hydrostatic equilibrium and the expression of Eq. (8.9), to obtain the scaling of the magnetic field B_ϕ, i.e.

$$\frac{B_\phi^2}{4\pi} \simeq \frac{n(x)m_p c^2}{2}\frac{1}{x} = \frac{m_p c^2}{\sigma_T R_S}\frac{\dot{m}(x)}{2}x^{-5/2}\left(\frac{r}{h}\right)\frac{J(x)}{\alpha} \tag{8.17}$$

and from that an expression for the Alfvén velocity; then one can easily see that the scaling of the rate at which magnetic flux is annihilated at the disk plane, $\propto B_\phi^2 V_A$, is essentially identical to that of Eq. (8.10); this suggests that dissipation through magnetic field annihilation at an equatorial current sheet can produce similar scaling relations to those of the standard accretion disks; however, one should bear in mind that the proportionality constant could be less than one, implying storing the excess energy in the disk plasma.

8.2.3 On ADAF, ADIOS and the Black Hole Spectral States

The introduction of Advection Dominated Accretion Flows (Narayan and Yi 1994, 1995) brought a totally different view of accretion flows than that of the standard Shakura-Sunyaev (Shakura and Sunyaev 1973) picture. Accretion flows are now allowed to be thick ($h \simeq r$), with their ion temperatures close to virial values $T_i \propto m_p(GM/r)$.[1] The radial force balance is now effected not only by the centrifugal force $\Omega^2 r$, but also by the gradient of the radial pressure, so that the value of Ω is below its Keplerian value Ω_K. At the same time, the gas is also heated,in addition to its viscous stress heating, also by the radial compression of the flow; this alone leads to a temperature $T \propto n(x)^{(\gamma-1)}$, which for adiabatic index $\gamma = 5/3$ and density $n(x) \propto x^{-3/2}$, leads to $T \propto 1/r$; the thermal energy gradient can therefore provide the radial momentum balance against gravity without the need of rotation, leading to $\Omega^2 \propto (5\gamma - 3)/r^3 \rightarrow 0$ as $\gamma \rightarrow 5/3$, thereby precluding the rotation of an accreting flow with $\gamma = 5/3$ (Blandford and Begelman 1999). This issue is circumvented by assuming the presence of magnetic fields in the flow which reduce γ below the value 5/3 at the expense of increasing the value of *Be* and making the need for the disk winds discussed above ever more pressing.

As noted in Narayan and Yi (1994), the presence of an ADAF, i.e. a flow with $h \simeq r$ implies that the plasma remains hot, close to its virial temperature on times scales longer than it takes it to accrete through the flow. Therefore, its cooling time through Coulomb collisions, $t_{cool} \sim 1/n(x)\sigma_T c$ should be longer than its viscous (or flow) time scale of Eq. (8.16). This then implies

$$\frac{R_S}{c}x^{3/2}\left(\frac{r}{h}\right)^2\frac{J(x)}{\alpha} < \frac{R_S}{c}\frac{1}{\dot{m}(x)}x^{3/2}\left(\frac{h}{r}\right)^3\frac{\alpha}{J(x)} \quad \text{or} \tag{8.18}$$

$$\dot{m}(x) < \alpha^2 \tag{8.19}$$

[1] The electron temperatures are determined from the balance between Coulomb heating by the ions and the cooling processes that produce the observed X-ray emission; for sufficiently large radii with $T_i < 10^9$ K, $T_e = T_i$; however, the fast cooling of electrons with $T_e \sim 10^9$ K, limits their temperatures at the smallest flow radii to roughly this value, despite the much higher values of T_i.

assuming in the last step that $h \simeq r$ and $J(x) \simeq 1$, i.e. $x \gg 3$. Thus the existence or not of an ADAF depends on a *single, global* condition on the *normalized* accretion $\dot{m}$ and it is independent of the mass of the accreting object M and weakly dependent on the flow radius x (Narayan and Yi 1995). This provides a great economy of assumptions in fitting the spectra of accretion powered sources with realistic accretion flow models, a feature of wide applicability because of the mass independence of this criterion. At this point one must also bear in mind that the luminosity released in the accretion is smaller by a factor $\dot{m}(x)$, to take into account the fact that fraction of the internal energy of the disk is not radiated away but advected into the black hole. So the expression for the luminosity of Eq. (8.11) should be $L \propto R_S \dot{m}(x)^2$.

Indeed, ADAF have been employed to model the X-ray spectra of AGN and XRB. The great advantage of such an approach is that the sources' high energy (X-ray) emission emerges naturally as part of the dynamics of the accretion flow. With ADAF it is no longer necessary to invoke a corona whose origin, geometry and share of the total accretion energy budget must be parameterized or simply assumed, in order for the models to confront successfully the combined, optical, UV and X-ray observations.

Of the plethora of such models we will refer to that of Esin et al. (1997) whose basic premises are widely used in the literature: As noted in the introduction and elsewhere in this volume, the XRB (and AGN) spectra, besides the power law X-ray continua, attributed to Comptonization by the hot ADAF electrons, include also quasithermal features attributed to emission by a geometrically thin, optically thick accretion disk. More importantly, the relative importance of this component with respect to the harder, power-law like X-ray emission appears to increase with the bolometric luminosity of the system. The authors of Esin et al. (1997) provide an account of this behavior by proposing that the ADAF comprises only the inner section of the entire accretion flow, with its outer section, at $r > r_{tr}$ consisting of a geometrically thin, SS73–type disk, which produces the quasi-thermal spectral component. Furthermore, in order to account for the increasing dominance of the thermal component with luminosity, they propose that the transition radius, r_{tr}, decreases with increasing luminosity. While successful in fitting the data, such models vitiate the global character and advantages of ADAF by adding components as demanded by observation rather than as imposed by the internal consistency of the theory. A well known additional difficulty is the transition from a geometrically thin to a thick disk, as there no reason for the cooled plasma of a thin disk to heat up again and form the ADAF demanded at $r < r_{tr}$.

We believe that this specific behavior can be accounted for within the framework of broader accretion flow models, namely those of ADIOS (Blandford and Begelman 1999), precisely because they advocate an accretion rate that increases with radius, i.e. $\dot{m}(x) = \dot{m}_0 x^p$:

It is now likely that there will be a radius $x_{tr} = r_{tr}/R_S$ such that for $x < x_{tr}$, $\dot{m}(x) < \alpha^2$ (assuming α to be roughly constant across the entire disk). It is hence reasonable to assume that the flow in this region is of an ADAF type (i.e. hot, geometrically thick), while for $x > x_{tr}$, where $\dot{m}(x) > \alpha^2$, the flow will be geometrically thin and much cooler, similar to an SS73 disk (but supported vertically by magnetic pressure). It is now reasonable to consider that as the global accretion rate increases, the value of x_{tr} decreases, to be pushed close to the ISCO as the mass flux through this radius $\dot{m}_0$ gets close to α^2.

We propose that in its "hot/thick" state the flow is supported by the thermal pressure of ions as discussed in Narayan and Yi (1994), while in its "cool/thin" its vertical support is effected by magnetic fields as discussed above. Vertical support by magnetic fields rather than gas, implies that when the local value of $\dot{m}$ drops below α^2, the disk height will naturally increase, decreasing the local density and thus lead to $t_{cool} > t_{visc}$; this will then restore the hot ion dominance of its vertical support i.e. return (this particular section of) the disk to its ADAF-like structure. It appears, to this author, that without the magnetic disk support, a thin, steady-state disk will remain thin even at a small $\dot{m}$ because its density is too high to make the cooling time longer than the viscous one. In a standard, gas pressure supported Shakura-Sunyaev disk (Svensson and Zdziarski 1994), $(h/r)^2 \propto \dot{m}^{2/5}$ implying $t_{visc} \propto \dot{m}^{-2/5}$, while $t_{cool} \propto 1/n(x) \propto \dot{m}^{-2/5}$; hence a decrease in $\dot{m}$ does not appear to lead to a reduction in the ratio t_{visc}/t_{cool}, needed to convert the flow from a "cool/thin" to a "hot/thick" state.

8.3 Accretion Disk Spectral States and Variability

The above analysis set the groundwork for a more detailed discussion of the observed black hole states, their variability and their relation to the accretion flows of AGN. The most important issue, in this author's personal view, is that of the relative luminosities between the quasi-thermal and the hard/Comptonized radiation components as well as the shape of the overall spectrum, as these provide clues about the geometry and relative position of the hot advection flow and the cooler geometrically thin one that constitute the major components of the spectra of sources powered by accretion onto black holes.

Following Esin et al. (1997), we assume that the entire luminosity of the "hard" (i.e. $E \gtrsim 3\,\text{keV}$), power-law section of the spectrum $L(x < x_{tr})$ is, by and large, the result of thermal Comptonization by the plasma of the "hot", advection dominated, section of the accretion flow interior to r_{tr}, while the luminosity of the quasi-thermal component, which provides most of the Comptonization seed photons, $L(x > x_{tr})$, from the "cooler" flow section at $r > r_{tr}$. The luminosities of each of these components can now be computed within this broader framework, by assuming a specific dependence of the mass flow on r (Blandford and Begelman 1999) and bearing in mind that the luminosity of the advection dominated, hot flow is proportional to $\dot{m}(x)^2$, while that of its $r > r_{tr}$ geometrically thin disk section is simply proportional to $\dot{m}(x)$.

To be specific and consistent with the work of Fukumura et al. (2010a,b) on accretion disk wind absorbers, we will assume that $p = 1/2$ or that $\dot{m}(x) = \dot{m}_0\, x^{1/2}$, bearing in mind that different values of p may present better fits to the data;

it may also be that different ranges in r require different values of p, especially if $\zeta \neq 0$, as in this case the no-torque condition at ISCO does not hold and there may be transfer of energy to the flow via torques from a rotating black hole. The location of the ISCO, r_I, is also of importance (another issue related to the black hole spin), especially for the value of the luminosity of the advection dominated section of the flow $L(x < x_{tr})$.

With the above qualifications and assumptions, the transition radius x_{tr} given a value of the accretion rate onto the black hole $\dot{m}_0$ and of the index p (= 1/2 at present) is given by the expression

$$\dot{m}(x_{tr}) = \dot{m}_0 \, x_{tr}^{1/2} \simeq \alpha^2 \tag{8.20}$$

Then from Eq. (8.11) we obtain the following expressions of $L(x > x_{tr})$ and $L(x < x_{tr})$

$$L(x > x_{tr}) \propto \int_{x_{tr}}^{\infty} \frac{\dot{m}(x)}{x} J(x) dlnx \sim \frac{\dot{m}_0}{x_{tr}^{1/2}} \left[1 - \frac{\zeta}{2} \left(\frac{x_I}{x_{tr}} \right)^{1/2} \right] \tag{8.21}$$

$$L(x < x_{tr}) \propto \int_{x_I}^{x_{tr}} \frac{\dot{m}(x)^2}{x} J(x) dlnx \sim \dot{m}_0^2 \left[ln \left(\frac{x_{tr}}{x_I} \right) - 2\zeta + 2\zeta \left(\frac{x_I}{x_{tr}} \right)^{1/2} \right] \tag{8.22}$$

One should note that the ratio R of these two luminosities is proportional to $1/\alpha^2$:

$$R = \frac{L(x > x_{tr})}{L(x < x_{tr})} = \frac{1}{\dot{m}_0 \, x_{tr}^{1/2}} \left(\frac{P1}{P2} \right) = \frac{1}{\alpha^2} \left(\frac{P1}{P2} \right) \tag{8.23}$$

with the last equality because of Eq. (8.20) and with $P1$, $P2$ the values of the square brackets in Eqs. (8.21) and (8.22) respectively.

8.3.1 Application to the Spectra of GRO 1655-40

Figure 8.1 depicts the ratio R as a function of the transition radius x_{tr}, for $x_I = 1.5, \zeta = 1$ and $\alpha = 1/3$.[2] Of interest in this figure is the value of $x_{tr} \simeq 100$ for which $R \simeq 1$, because this sets the size of the disk at which the cool/thin disk quasi-thermal luminosity matches that of the harder radiation produced in the advection dominated accretion flow section. This will then correspond to the blue,

[2] While this value may appear too large, one should note that Pariev et al. (2003), argue that for disks at which the pressure is dominated by magnetic fields $\alpha = 1/\sqrt{3}$, while Bai and Stone (2013) in their study of shear-box simulations with a net magnetic flux find $\alpha \simeq 1$ for disks with gas to (poloidal) magnetic pressure, greater than 0.01.

hard spectrum of Fig. 8.2, obtained from observations of GRO 1655-40 (Done et al. 2007), while the general shape of the entire flow in this state depicted by the cartoon of the same color in the last panel of this figure. Equation (8.20), then, implies that at this point, $\dot{m}_0 \simeq \alpha^2/(100)^{1/2} \simeq 1/30$ and since the hard component luminosity is $L(x < x_{tr}) \simeq \dot{m}_0^2 P2 \simeq \dot{m}_0^2 \simeq 10^{-3} L_{\rm Edd} \simeq 10^{36}$ erg/s, assuming the black hole mass of GRO 1655-40 is about ten solar masses (the value of $P2$, the square bracket of Eq. (8.22), is approximately equal to 1 for $x_{tr} \simeq 100$).

For $x_{tr} \simeq 5$, Fig. 8.1 implies $R \simeq 10$ and by Eq. (8.20) $\dot{m}_0 \simeq 1/8$; the corresponding spectra then are those given by the red curves (US,HS) of the first panel of Fig. 8.2, while the corresponding bolometric luminosity $L \gtrsim L_{Edd}/20$, in good agreement with the data, for the assumed value of the black hole mass. The red color cartoon on the last panel of Fig. 8.2 depicts the geometry of this flow indicating that the extent of the advection dominated component is now smaller, and for the same reason, so is its luminosity relative to the thermal component (and it should go to zero as $x_{tr} \rightarrow x_I$). It should be noted that both in the blue

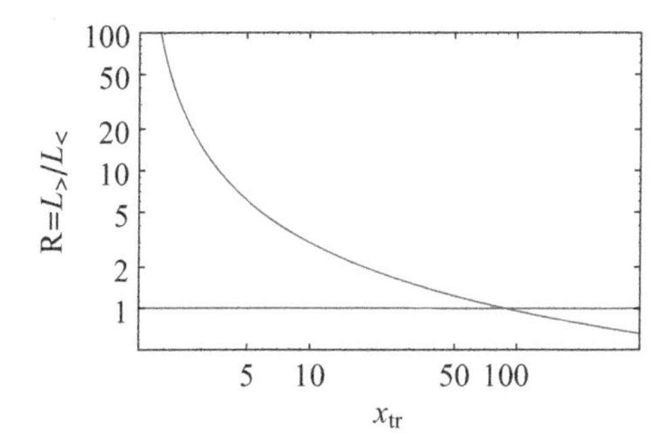

Fig. 8.1 The value of the ratio R of disk luminosities at $x > x_{tr}$ and $x < x_{tr}$, as a function of the transition radius x_{tr}, given by Eq. (8.20) in terms of the mass accretion rate onto the black hole $\dot{m}_0$ and for $\alpha = 1/\sqrt{3}$, $p = 1/2$, $\zeta = 1$ and $x_I = 1.5$. $R \sim 1$ for $x_{tr} \sim 100$ and increases with the value of $\dot{m}_0$ as described in the text

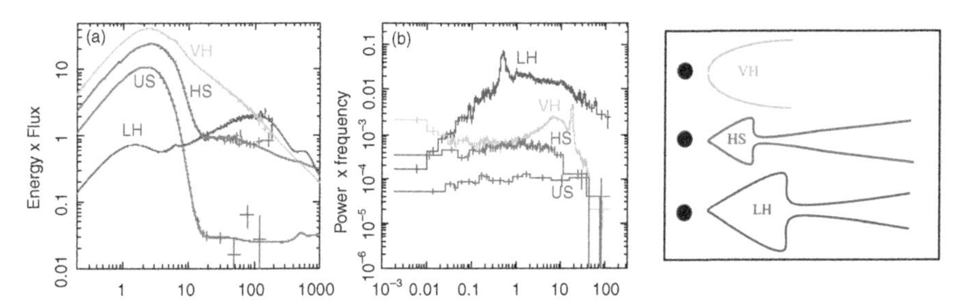

Fig. 8.2 *Left:* The νF_ν spectra of GRO 1655-40 at its different states: *Blue* Low Hard, *red* High Soft, *green* Very High. *Middle:* The corresponding timing properties corresponding the these energy spectra, color coded for correspondence. *Right:* Schematic depiction of the disk structure corresponding to the spectral states of the same color as the figure to the left

(LH) and the red (US,HS) spectra, the high energy component does not connect to the peak of the quasi-thermal component, implying that the electron source covers only a fraction of the soft photon source solid angle, in agreement with the cartoon figures of these flows. On the other hand, in the spectrum VH (green) of Fig. 8.2, the Comptonization power-law component joins to the peak of the quasi-thermal spectrum, suggesting that the majority of the soft photons of this source get up-Comptonized. In this case, whose luminosity is close to Eddington, either the high energy emission originates in a corona that overlies the entire inner disk, or/and it is the result of bulk Comptonization by the plasma that free falls onto the black hole interior to ISCO (Titarchuk et al. 1997). In this case, the advection component is absent but the disk is thick (slim, see Abramowicz et al. 1988) due to the dominant contribution of radiation pressure.

8.3.2 *The Timing Properties of GRO 1655-40*

The present model, as outlined above, provides also some insights into the timing properties of these systems. An example of these is given in the middle panel of Fig. 8.2 for the black hole transient source GRO 1655-40, color coded to the corresponding spectral states of the left panel of the same figure. It is generally considered that while details vary from source to source, the overall behavior of their combined spectral/timing properties are similar to those shown in this figure.

The fundamental premise of the model is the junction of geometrically thin and thick flows at $r = r_{tr}$, with a continuous mass flow between them. Since the flow (viscous) time at $r \lesssim r_{tr}$ is shorter than that at $r \gtrsim r_{tr}$ by virtue of its smaller disk height, the mass flux across this surface will not be continuous but it will appear as discrete, random, accretion events for the thick flow at $r \lesssim r_{tr}$. As such, the X-ray PSD should exhibit a white noise spectrum at time scales $t \lesssim t_{visc}(r_{tr})$; this is in agreement with the blue PSD of middle panel of Fig. 8.2, which rises linearly in the νP_ν vs. ν coordinates employed.

Associating this frequency with the viscous time scale at $x_{tr} \simeq 100$, the value employed to account for the spectra of this state, and assuming $R_S \simeq 3 \cdot 10^6$ cm, the free-fall time from this distance is $t_{ff} \simeq (R_S/c)x_{tr}^{3/2} \simeq 10^{-1}$ s and the corresponding frequency $\nu \simeq 10$ Hz. Since $\nu_{visc} \simeq 1/t_{visc} \simeq (1/t_{ff})(1/\alpha)(h/r)^2$, the observed break at $\nu \simeq 0.5$ Hz, implies $(1/\alpha)(h/r)^2 \simeq 0.05$ and for $\alpha \simeq 1/\sqrt{3}$, one obtains $h/r \sim 0.3$, in line with earlier discussion favoring magnetic vertical support of these disks and h/r values larger than those implied by the plasma temperature of the disk quasi-thermal spectral component. It is worth noting that similar values of h/r are inferred by similar time scale associations in Gilfanov and Arefiev (2005).

The QPO at the same frequency (0.5 Hz) has no apparent account within these arguments; its relatively high–Q value, implies a rather coherent process at approximately the same frequency, it is however similar to c_s/h, where c_s the proton sound speed of the quasi-thermal spectral component and $h \sim R/3$ (see below), i.e. to slow magnetosonic oscillations. An increase in the break frequency with increasing $\dot{m}$ is in agreement with the arguments put forward in Sect. 8.2. The concomitant increase in the QPO frequency argues for the association of this process with the decrease of r_{tr} with increasing $\dot{m}$, but we are not willing to speculate at this point.

With the increase of $\dot{m}$ and transition to the source soft state, the PSD acquires a form $\propto \nu^{-1}$ attributed to fluctuations of α in the entire thin disk that dominates now the X-ray emission. What is most puzzling is a break of the PSDs of all spectral states at a frequency ν_b $(10 < \nu_b < 100)$ Hz. This frequency seems to depend on $\dot{m}$ far more weakly than that associated with the variation of r_{tr} discussed above. This cut-off frequency persists at the VS state too, one in which the radiation pressure is assumed to dominate the disk pressure. The radiation pressure dominance of this state makes all the more intriguing (at least to this author's mind) the presence of the very high–Q QPO at $\nu \simeq 20$ Hz. We refrain from speculations on the nature of this frequency at this point, as these issues will be treated at a different contribution of this volume (Kylafis and Belloni).

8.4 Discussion

In the sections above we have put forward a picture of the structure of accretion disks that produce the observed radiation in XRBs whose compact object is a black hole rather than a neutron star. This restricted view avoids the complications of the radiation emitted by the boundary layer and makes for an easier interpretation of the spectra. Furthermore, these considerations should be applicable to the accretion disks of AGN.

There are two central points at which our treatment differs from the standard ones in the literature: (i) We propose that the disk vertical structure is effected by magnetic rather than thermal plasma pressure. This implies that the disk height can now be larger than inferred by assuming that the supporting pressure is that of their quasi-thermal spectral component. (ii) Because of the magnetic disk support, it is also not necessary that the energy transferred by the viscous stresses (through the action of the magnetic fields) be dissipated locally. As a result, one is likely facing a situation similar to that of ADIOS (Blandford and Begelman 1999), in which this excess energy is given off in winds launched off these disks. This then results to accretion flows with radius dependent mass flux $\dot{m}$ (under steady state conditions), more specifically with $\dot{m}$ increasing with radius. The variation of the disk $\dot{m}$ with x then allows for transitions from geometrically thin to geometrically thick disks, when the local cooling time becomes less that the viscous one, i.e. when the local (normalized) accretion rate becomes smaller than the square of the local viscosity parameter. Then the disks "puff-up" to assume the geometry of ADAF interior to the specific transition radius for which $\dot{m} < \alpha^2$, as proposed in the literature.

This approach makes transparent why the transition radius decreases with increasing $\dot{m}$, a fact consistent with observations but without a straightforward interpretation within the standard disk theory. We showed that the observed spectra, more specifically the relative importance of the quasi-thermal to the harder Comptonization produced X-ray components can be easily interpreted through these considerations. We also showed, that insisting on such an interpretation provides estimates of the disk's viscosity parameter $\alpha \sim 1$; we believe that similar considerations can be further refined to be used at probes of the disks' viscosity. This higher value of α can be traced to the presence of significant poloidal magnetic flux threading the disk, generated in the black hole vicinity, one of the general themes of the present volume. It becomes apparent from the above, that accretion onto a black hole, is accompanied not only by the outflow of photons but also the outflow of mechanical energy in the form of winds and also of magnetic flux as proposed in Contopoulos and Kazanas (1999), Contopoulos et al. (2006), Christodoulou et al. (2008), and Kylafis et al. (2012) and elsewhere in this volume.

Finally, we argued that the scheme proposed above provides an account for several of the timing properties of these flows, with characteristic frequency associations consistent with its general premises. It can in principle account for the broader PSD forms, however, the presence of QPOs remains vexing; nonetheless, the model provides some new ideas on the origin of several of these oscillations which may prove fruitful in future studies.

References

Abramowicz, M.A., Fragile, P.C.: Foundations of black hole accretion disk theory. Living Rev. Relativ. **16**, 1–88 (2013)

Abramowicz, M.A., Czerny, B., Lasota, J.P., Szuszkiewicz, E.: Slim accretion disks. Astrophys. J. **332**, 646–658 (1988)

Bai, X-N., Stone, J.M.: Local study of accretion disks with a strong vertical magnetic field: magnetorotational instability and disk outflow. Astrophys. J. **767**, 30–48 (2013)

Balbus, S., Hawley, J.: A powerful local shear instability in weakly magnetized disks. Astrophys. J. **376**, 214–233 (1991)

Behar, E.: Density profiles in Seyfert outflows. Astrophys. J. **703**, 1346–1351 (2009)

Berkeley, A.J., Kazanas, D., Ozik, J.: Modeling the X-ray-ultraviolet correlations in NGC 7469. Astrophys. J. **535**, 712–720 (2000)

Blandford, R.D., Begelman, M.C.: On the fate of gas accreting at a low rate on to a black hole. Mon. Not. R. Astron. Soc. **211**, P1–P5 (1999)

Christodoulou, D., Contopoulos, I., Kazanas, D.: Interchange method in incompressible magnetized couette flow: structural and magnetorotational instabilities. Astrophys. J. **462**, 865–873 (1996)

Christodoulou, D.M., Contopoulos, I., Kazanas, D.: Simulations of the poynting-robertson cosmic battery in resistive accretion disks. Astrophys. J. **674**, 388–407 (2008)

Contopoulos, I., Kazanas, D.: A cosmic battery. Astrophys. J. **508**, 859–863 (1999)

Contopoulos, I., Kazanas, D., Christodoulou, D.M.: The cosmic battery revisited. Astrophys. J. **652**, 1451–1456 (2006)

Done, C., Gierlinski, M., Kubota, A.: Modelling the behaviour of accretion flows in X-ray binaries. Astron. Astrophys. Rev. **15**, 1–66 (2007)

Esin, A.A., McClintock, J.E., Narayan, R.: Advection-dominated accretion and the spectral states of black hole X-ray binaries: application to Nova Muscae 1991. Astrophys. J. **489**, 865–889 (1997)

Fukumura, K., Kazanas, D., Contopoulos, I., Behar, E.: MHD accretion disk winds as X-ray absorbers in active galactic nuclei. Astrophys. J. **715**, 636–650 (2010a)

Fukumura, K., Kazanas, D., Contopoulos, I., Behar, E.: Modeling high-velocity QSO absorbers with photoionized MHD disk winds. Astrophys. J. **723**, L228–L232 (2010b)

Galeev, A.A., Rosner, R., Vaiana, G.S.: Structured coronae of accretion disks. Astrophys. J. **229**, 318–326 (1979)

Gilfanov, M., Arefiev, V.: X-ray Variability, Viscous Time Scale and Lindblad Resonances in LMXBs. arXiv:astro-ph/0501215 (2005)

Haardt, F., Maraschi, L.: X-Ray spectra from two-phase accretion disks. Astrophys. J. **413**, 507–517 (1993)

Kylafis, N.D., Contopoulos, I., Kazanas, D., Christodoulou, D.M.: Formation and destruction of jets in X-ray binaries. Astron. Astrophys. **538**, A5–A9 (2012)

Narayan, R., Yi, I.: Advection-dominated accretion: a self-similar solution. Astrophys. J. **428**, L13–L16 (1994)

Narayan, R., Yi, I.: Advection-dominated accretion: self-similarity and bipolar outflows. Astrophys. J. **444**, 231–243 (1995)

Novikov, I.D., Thorn, K.S.: Astrophysics of black holes. In: DeWitt, DeWitt (eds.) Black Holes, p. 343–450. Gordon and Breach, New York (1972)

Pariev, V.I., Blackman, E.G., Boldyrev, S.A.: Extending the Shakura-Sunyaev approach to a strongly magnetized accretion disk model. Astron. Astrophys. **407**, 403–421 (2003)

Shakura, N., Sunyaev, R.: Black holes in binary systems. Obs. Appear. Astron. Astrophys. **24**, 337–355 (1973)

Steffen, A.T., et al.: The X-ray-to-optical properties of optically selected AGN over wide luminosity and redshift ranges. Astron. J. **131**, 2826–2842 (2006)

Svensson, R., Zdziarski, A.A.: Black hole accretion disks with coronae. Astrophys. J. **436**, 599–606 (1994)

Titarchuk, L.G., Mastichiadis, A., Kylafis, N.: X-ray spectral formation in a converging fluid flow: spherical accretion into black holes. Astrophys. J. **487**, 834–846 (1997)

Chapter 9
A Cosmic Battery around Black Holes

Ioannis Contopoulos

Abstract Energetic astrophysical jets have always been associated with relatively strong large-scale magnetic fields that extract energy from the rotation of a central black hole and its surrounding accretion disk. This natural association does not answer the fundamental question what is the origin of the large-scale magnetic field.

When the disk is non-diffusive, standard MHD advection can bring the field in from large distances, as attested by the multitude of ideal MHD simulations of magnetized astrophysical accretion disks. When this is not the case, however, the large-scale dipolar magnetic field expected at the origin of astrophysical jets is generated naturally by the anisotropic radiation pressure around the central black hole. In this Chapter, we discuss the scenario where the innermost part of the accretion disk generates and holds one polarity of the magnetic field, while the return polarity diffuses outward through the outer diffusive part of the disk. This is the Cosmic Battery first proposed by Contopoulos and Kazanas in 1998.

9.1 Magnetic Fields Around Black Holes

The most energetic astrophysical jets have always been associated with black holes surrounded by rotating disks of ionized matter (plasma), the two immersed in relatively strong large scale magnetic fields (e.g. Blandford 2001). The large scale magnetic field is held from escaping to infinity by the ionized matter, while the matter is held by the gravity of the central compact object. Obviously, there is a maximum magnetic field pressure $B^2/4\pi$ that a certain amount of matter with ram pressure ρv^2 or thermal pressure P can hold. This yields magnetic field estimates that do not exceed 10^7-10^8 G around stellar mass black holes, or 10^3-10^4 G around super-massive black holes in the centers of active galaxies. The latter maximum magnetic field values correspond to what is termed as *equipartition* between the field and the matter (or MADness; see Chapter 3). The field is also assumed to thread the surrounding accretion disk and to decrease radially as some power of the

I. Contopoulos (✉)
Research Center for Astronomy and Applied Mathematics, Academy of Athens, Athens, Greece
e-mail: icontop@academyofathens.gr

I. Contopoulos et al. (eds.), *The Formation and Disruption of Black Hole Jets*,
Astrophysics and Space Science Library 414,
DOI 10.1007/978-3-319-10356-3_9

radial distance r. Such large scale magnetic fields extract rotational energy from both the black hole and the surrounding disk. The details of the energy extraction involve an electromagnetic Penrose effect for the black hole (Blandford and Znajek 1977; Komissarov 2008; Lasota et al. 2014; Nathanail and Contopoulos 2014), and a centrifugal 'slingshot' (Blandford and Payne 1982; Contopoulos and Lovelace 1994; Contopoulos 1994) or magnetic 'spring' (Contopoulos 1995; Lynden-Bell 1996) for the accretion disk. Whatever the specific energy extraction mechanism may be, however, the above mental picture has the basic elements of a Faraday disk (Lovelace 1976; Blandford 1976), and that is why it has almost always been associated with astrophysical jets. Unfortunately, this 'natural association' does not answer the fundamental question

> What is the origin of astrophysical magnetic fields around black holes?

nor the more philosophical question

> Why both stellar and super-massive black hole systems contain large scale magnetic fields?

The origin of large scale cosmic magnetic fields remains one of the open questions of astrophysics. Their origin has been sought in the entire history of the universe as far back as the Planck time. However, the consideration that in its initial state the universe was to a high degree homogeneous and isotropic precludes the presence of large-scale magnetic fields. Most theories attempt to produce the observed magnetic fields by starting with some 'seed' field and following its amplification by means of a dynamo mechanism, i.e. a mechanism that can convert the kinetic energy of the conducting fluid (into which the magnetic field is supposedly frozen) to magnetic field energy. Unfortunately, when one considers the back-reaction of the stretched and folded magnetic field on the dynamics of the conducting fluid, dynamo action seems to be very ineffective: as the field at small diffusive scales reaches equipartition, the large-scale component remains several orders of magnitude weaker than astrophysical observed magnetic fields (Vainshtein and Cattaneo 1992; Zrake and MacFadyen 2012). In addition to these problems with the effectiveness of the dynamo theory, the origin of the seed magnetic field that is needed for amplification is also an issue. The seed field is considered to be generated by some 'battery mechanism', like the Biermann battery: the thermal pressure acting on the plasma electrons imparts a differential velocity between the electrons and the ions which leads to an electric current and a consequent magnetic field (see Eq. 9.17 below).

In spite of the above difficulties, jets seem to be present everywhere wherever there is rotation and accretion around compact astrophysical objects in the Universe, which as we said, according to the current belief (or 'paradigm'...), implies that equipartition-level large scale magnetic fields are omnipresent whenever they are needed. This is rather strange because as we said above

> Dynamo mechanisms in the turbulent accretion disk generate equipartition level turbulent, *not* large scale, magnetic fields.

Back in the 1990s, it seemed natural to put the large scale magnetic field in place as an outer boundary condition of the problem (e.g. Contopoulos and Lovelace 1994). What was the outcome? Astrophysical jets and winds naturally! Some people dared to address the issue of how large scale magnetic fields were held in place by the accreting flow since this involved consideration of the magnetic diffusivity in the disk (Ferreira and Pelletier 1995; Li 1995; Contopoulos 1996). Unfortunately, this is a formidable problem that can only be addressed phenomenologically under extremely simplifying considerations. There is still strong debate in the astrophysics community on whether accretion can bring in a large scale magnetic field presumably generated somewhere far away from the central compact object (van Ballegooijen 1989; Lubow et al. 1994; Lovelace et al. 1994). Recent supercomputer simulations (e.g. Tchekhovskoy et al. 2011; McKinney et al. 2012) consider an ideal MHD accretion flow (something that is certainly *not* justified by the physics of thin accretion disks; Shakura and Sunyaev 1973) inside an infinite reservoir of uniform magnetic flux. What did we get out of it? An accretion flow filled with a large scale magnetic field naturally! The inconvenient physical reality, though, is that accretion disks are inherently turbulent, thus also diffusive, and therefore, the bulk of their material *cannot* support the inward advection of the hoped-for magnetic field. There have been proposals that only the surface layers of the disk that are ionized by cosmic radiation support the inward field advection but that proposal too fails to account for the magnitude of the accreted field since the field can only reach equipartition with the surface layers (and not the bulk) of the disk material (Lovelace et al. 2009). In summary, more and more researchers agree that

> Magnetic diffusivity does not allow for the inward advection of large scale magnetic fields.

In the late 1990s, we were not satisfied with the answers given to the above questions, so we set out to find a natural source for the large scale magnetic field that inundates the central black hole inside the inner edge of the accretion disk. We then realized that it is precisely in that region where radiation pressure is maximal (a significant fraction of the Eddington radiation pressure), plasma velocities are maximal (close to the speed of light), and radiation is maximally misaligned with the plasma motion (90°) that a battery mechanism works most effectively. In other words,

> The radiation pressure at the inner edge of an accretion disk around a central astrophysical compact object imparts a relative azimuthal velocity on the electrons with respect to the ions, thus generating an azimuthal electric current that becomes the origin of the sought for large scale magnetic field that inundates the central object and closes through the surrounding disk

This effect is so powerful that it manages to generate the sought for magnetic field on astrophysically relevant timescales. Not only that. It also naturally accounts for the large scale magnetic field that threads the accretion disk, and for the observed disk winds surrounding the central black hole jet. We have named it the *Cosmic Battery* (Contopoulos and Kazanas 1998).

9.2 Radiation Dynamics

We will give here a simple description of how radiation acts on a plasma. Radiation forces are usually introduced in the equation of motion for the plasma, written in non-relativistic form as

$$\rho \frac{\mathrm{d}\mathbf{v}}{\mathrm{d}t} = \rho \mathbf{g} + \frac{1}{c}\mathbf{J} \times \mathbf{B} - \nabla P + \frac{\rho}{m_{\mathrm{i}}}\mathbf{f}_{\mathrm{rad}} , \tag{9.1}$$

where, ρ, P and $\mathbf{v}$ are the plasma matter density, thermal pressure and velocity respectively, $\mathbf{g}$ is the acceleration of gravity, $\mathbf{J}$ is the electric current density, $\mathbf{B}$ is the magnetic field, and $\mathbf{f}_{\mathrm{rad}}$ is the radiation force per ion. m_{i} is the average ion particle mass, and c is the speed of light.

Equation (9.1) leads to the concept of the Eddington luminosity L_{Edd} defined as the luminosity of a central isotropic radiation source (e.g. a star) that imparts a force strong enough to hold ions against gravity. In other words,

$$f^r_{\mathrm{rad}_{\mathrm{Edd}}} \equiv \frac{L_{\mathrm{Edd}}\sigma_T}{4\pi r^2 c} = \frac{GMm_{\mathrm{i}}}{r^2} . \tag{9.2}$$

Here, M is the mass of the central compact object. What is to be noticed here is that radiation and pressure forces are felt only by the plasma electrons (what enters the above equation is the Thomson cross-section σ_T for the electrons; the Thomson cross-section for the ions is several million times smaller), yet the ions and thus the full plasma are influenced by the radiation and pressure forces. Why this is so becomes clear when we consider the equations of motion for the plasma electrons and ions in a cold plasma independently:

$$m_{\mathrm{e}} \frac{\mathrm{d}\mathbf{v}_{\mathrm{e}}}{\mathrm{d}t} = m_{\mathrm{e}}\mathbf{g}_{\mathrm{grav}} - \frac{\mathrm{e}}{\mathrm{c}}\mathbf{v}_{\mathrm{e}} \times \mathbf{B} - \frac{m_{\mathrm{i}}\nabla P}{\rho} + \mathbf{f}_{\mathrm{rad}} - \mathrm{e}\mathbf{E} , \text{ and} \tag{9.3}$$

$$m_{\mathrm{i}} \frac{\mathrm{d}\mathbf{v}_{\mathrm{i}}}{\mathrm{d}t} = m_{\mathrm{i}}\mathbf{g}_{\mathrm{grav}} + \frac{\mathrm{e}}{\mathrm{c}}\mathbf{v}_{\mathrm{i}} \times \mathbf{B} + \mathrm{e}\mathbf{E} . \tag{9.4}$$

$\mathbf{E}$ is the local electric field that develops due to the various non-gravitational forces in the plasma, m_{e} is the mass of the electron, and e is the proton charge. Since, $m_{\mathrm{e}} \ll m_{\mathrm{i}}$, one can ignore the m_{e}-terms in Eq. (9.3) and write

$$\mathbf{E} \approx -\frac{1}{\mathbf{c}}\mathbf{v}_{\mathrm{e}} \times \mathbf{B} - \frac{m_{\mathrm{i}}\nabla P}{\mathrm{e}\rho} + \frac{\mathbf{f}_{\mathrm{rad}}}{\mathrm{e}} . \tag{9.5}$$

Putting this back in Eq. (9.4) we obtain

$$\frac{\mathrm{d}\mathbf{v}_{\mathrm{i}}}{\mathrm{d}t} = \mathbf{g} + \frac{\mathrm{e}}{cm_{\mathrm{i}}}(\mathbf{v}_{\mathrm{i}} - \mathbf{v}_{\mathrm{e}}) \times \mathbf{B} - \frac{\nabla P}{\rho} + \frac{\mathbf{f}_{\mathrm{rad}}}{m_{\mathrm{i}}} , \tag{9.6}$$

which is just Eq. (9.1) under the approximation that $\mathbf{v} \approx \mathbf{v}_\mathrm{i} \approx \mathbf{v}_\mathrm{e}$, and $\mathbf{J} = e\rho(\mathbf{v}_\mathrm{i} - \mathbf{v}_\mathrm{e})/m_\mathrm{i}$. In other words, the ions feel the radiation force through an electric field that develops in the plasma. This important effect is often ignored by the younger generation of researchers.

What is the radiation field in the vicinity of an astrophysical source of X-rays? Most studies, as in the calculation of the Eddington luminosity in Eq. (9.2), assume an isotropic source of radiation at the center. This is a valid approximation only when we view astrophysical sources of luminosity L from a great distance. In that limit, radiation introduces a radial force component per ion, given by Eq. (9.2), which obviously modifies the dynamics of the surrounding plasma. This is not the only force component in an isotropic radiation field though. Matter is in orbit around the center, and therefore, one needs to also take into account the Poynting-Robertson effect on the plasma electrons (Poynting 1903; Robertson 1937; Bini et al. 2009, 2011). The main previous application of this effect has been in the study of orbits of dust grains around the Sun. Without radiation drag, the orbits are circular or elliptical. In the presence of radiation, the grains scatter photons emitted by the central star, lose angular momentum, and inspiral toward the star. The question that arises, is how can a force perpendicular to the motion remove angular momentum. The answer is obvious when we view the radiation field in the frame of the moving particle. In that frame, the radiation field is aberrated, namely slightly concentrated or beamed in the direction of motion. This abberation results in an extra force opposite to the target's motion that slows it down. In flat spacetime, the Poynting-Robertson azimuthal drag force per ion f^{ϕ}_{PR} can be directly calculated as

$$f^{\phi}_{\mathrm{PR}} = -f^{r}_{\mathrm{rad}}\frac{\mathrm{v}^{\phi}}{c} = -\frac{L\sigma_T \mathrm{v}^{\phi}}{4\pi r^2 c^2}, \tag{9.7}$$

where, v^{ϕ} is the plasma azimuthal velocity.

Obviously, the astrophysical setting of an accreting rotating black hole is far more complex. As material is pulled from its surroundings (companion star, stellar wind, interstellar medium, etc.) it forms a rotating disk around the center. This material slowly loses angular momentum through a number of different mechanisms of varying efficiency (magnetorotational instability-MRI, magnetic braking by disk winds/jets, Poynting-Robertson drag, viscosity, turbulence, tidal forces, gravitational waves, etc.) and inspirals towards the center. As this material accretes, the temperature rises to values on the order of 10^7 K or even higher. At these temperatures, matter radiates in the X-ray part of the spectrum, as in an X-ray binary (in the case of a stellar mass black hole), or an AGN (in the case of a supermassive black hole). At this point the disk speed is a considerable fraction of the speed of light and the radiation luminosity a considerable fraction of its Eddington value, we therefore expect that the dynamical effects of radiation will be significant.

It is interesting that the first realistic calculation of the accretion disk radiation field in the vicinity of a spinning black hole was performed only very recently by Koutsantoniou and Contopoulos (2014). In Kerr spacetime, photon orbits are curved and the photon frequency varies along the photon's trajectory. The fact that now the source of radiation is also rotating, further complicates the problem by introducing an extra Doppler shift. If the central object is not a black hole but a slowly rotating compact spherical star, it is somewhat easier to obtain a solution (e.g. Abramowicz et al. 1990; Miller and Lamb 1993, 1996; Lamb and Miller 1995). On the other hand, astrophysical sources associated with accretion disks generate a complex radiation field with ensuing complex dynamics around the central black hole. Therefore, in order to study the radiation field in the vicinity of a black hole surrounded by an astrophysical accretion disk, one is compelled to perform fully relativistic ray tracing that takes into account the spatial extent and the rotation of its source, the accretion disk itself.

9.3 Calculation of the Radiation Field

The calculation of the radiation force in the immediate environment of an accreting rotating black hole must be performed in the Kerr spacetime metric. A Kerr black hole is characterized by its mass M and its angular momentum J, or equivalently its spin parameter $a = J/M$. a takes values between zero (for a non-rotating Schwarzschild black hole) and M (for a maximally rotating black hole). Physical quantities are measured by Zero Angular Momentum Observers (ZAMOs; also known as local Fiducial Observers or Fidos) in their Locally Non-Rotating Frame (LNRF). In Boyer-Lindquist (t, r, θ, ϕ) coordinates, the Kerr metric reads

$$ds^2 = -\alpha^2 dt^2 + \varpi^2 (d\phi - \omega dt)^2 + \frac{\Sigma}{\Delta} dr^2 + \Sigma d\theta^2 \,, \tag{9.8}$$

where,

$$\begin{aligned}
\alpha &= (\Delta\Sigma/A)^{1/2}, \\
\omega &= 2aMr/A \,, \\
\varpi &= (A/\Sigma)^{1/2} \sin\theta \,. \\
\Sigma &= r^2 + a^2 \cos^2\theta \,, \\
\Delta &= r^2 - 2Mr + a^2 \,, \\
A &= (r^2 + a^2)^2 - a^2 \Delta \sin^2\theta
\end{aligned}$$

Here α is the lapse function, ω is the angular velocity of ZAMOs, and ϖ is the cylindrical radius. We have assumed here geometrical units in which $c = G = 1$. Latin/Greek indices denote space/spacetime components respectively.

The calculation of the radiation force is performed in the LNRF, mainly because it is in that frame where MacDonald and Thorne (1982) formulated black hole electrodynamics. The radiation field is experienced by target electrons orbiting the central black hole inside a certain depth from the surface of the accretion disk. Our present preliminary calculation considers only electrons at the inner edge of the disk at the position of the Innermost Stable Circular Orbit-ISCO. A more general calculation will also take into acount the optical depth of the disk (Koutsantoniou & Contopoulos in preparation). The radiation force per ion in the accretion disk plasma, $f^i_{\rm rad}$, is a non-gravitational term that enters the equation of motion Eq. (9.1), and is connected to the radiation flux F^i through the formula

$$f^i_{\rm rad} = \sigma_T F^i \ . \tag{9.9}$$

For simplicity, we take the Thomson cross-section σ_T to be independent of the radiation frequency and scattering angle. Following Miller and Lamb (1996), the radiation flux components F^i are given by

$$F^i = h^i_\nu T^{\kappa\nu} u_\kappa \ , \tag{9.10}$$

where $T^{\kappa\nu}$ is the radiation stress-energy tensor given by

$$\begin{aligned} T^{\mu\nu} &= \iint I_\nu(r,\theta,a,b;\nu)\,{\rm d}\nu\; n^\mu n^\nu\; {\rm d}\Omega \ , \\ &= \int I(r,\theta,a,b)\; n^\mu n^\nu\; {\rm d}\Omega \ , \end{aligned} \tag{9.11}$$

where, I_ν and I are the frequency dependent and integrated specific intensities respectively in the LNRF at the position of the target electron. Here, u^μ is the four-velocity of the disk at the position of the target electron at the ISCO, $h^\mu_\nu = -\delta^\mu_\nu - u^\mu u_\nu$ is a tensor that projects orthogonally to the target electron four-velocity. Radiation photons originate on the surface of the accretion disk, travel to the position of the target, and reach the target from a certain direction n^i that corresponds to a solid angle element ${\rm d}\Omega = \sin a\ {\rm d}a\ {\rm d}b$. We define angles a, b in which

$$n^r = \cos a,\ n^\theta = \sin a \cos b,\ n^\phi = \sin a \sin b \ . \tag{9.12}$$

The integral in Eq. (9.11) has contributions *only* from those directions that correspond to photon trajectories that originate on the radiation source, in our case the surface of the hot innermost accretion disk. Therefore, the calculation of the radiation field requires the *backward integration* of photon trajectories from the position of the target to their origin on the surface of the disk along all directions (a, b) in the sky of the target particle. For each such photon trajectory, the specific intensity I_ν that appears in Eq. (9.11) is *different* from the source specific intensity $I_{\nu,\rm source}$. In order to obtain I_ν we take advantage of the fact that, along the path of a light ray, $I_\nu/\nu^3 = I_{\nu,\rm source}/\nu^3_{\rm source}$ for the frequency dependent specific intensity, or equivalently

$$I = \left(\frac{\nu}{\nu_{\rm source}}\right)^4 I_{\rm source} \tag{9.13}$$

for the frequency integrated intensities. Notice that the ratio ν/ν_{source} expresses a frequency shift that is *independent* of the frequency itself and depends only on the form of the spacetime and the angle of emission. It accounts for three different phenomena: the gravitational redshift caused by gravitational time dilation, the Doppler shift caused by the motion of the emitting surface, and the frame dragging shift caused by the 'differential rotation' of spacetime. The first two shifts can be encountered in any spacetime, while the latter only in rotating spacetimes. It is straightforward to show (see Koutsantoniou and Contopoulos 2014 for details) that the total shift between the emitted and received frequencies ν_{source} and ν is

$$\frac{\nu}{\nu_{\text{source}}} = \frac{\alpha_{\text{source}}\left[1+\omega\frac{p_\phi}{p_t}\right]}{\gamma\alpha(1-\text{v}^\phi\cos\psi)\left[1+\omega_{\text{source}}\frac{p_\phi}{p_t}\right]}\,. \tag{9.14}$$

Here, P_θ, P_t are the photon momentum components, and γ is the Lorentz factor of the flow in the disk. From Eqs. (9.13) and (9.14), the specific intensity we seek is given by

$$I = \frac{\alpha^4_{\text{source}}}{\gamma^4\alpha^4[1-\text{v}^\phi\cos\psi]^4}\left(\frac{1+\omega\frac{p_\phi}{p_t}}{1+\omega_{\text{source}}\frac{p_\phi}{p_t}}\right)^4 I_{\text{source}} \tag{9.15}$$

along each photon trajectory (Fig. 9.1).

As explained before, the complex astrophysical setup that we are considering (accretion disk + rotating black hole) requires backward ray tracing along every direction (a, b) in the sky of the target. The photon trajectories that we obtain are divided into three categories, depending on their point of origin: those that cross the event horizon, those that originate from a point at 'infinity' or the 'outer disk' and those that originate from a point of the inner disk. Trajectories from the first two categories do not give any significant contribution to the radiation field. Radiation pressure is due to photons of the third category that originate from the innermost hotter part of the disk. For those 'allowed photon trajectories' we calculate their point of origin, their direction of emission, and the resulting frequency shifts

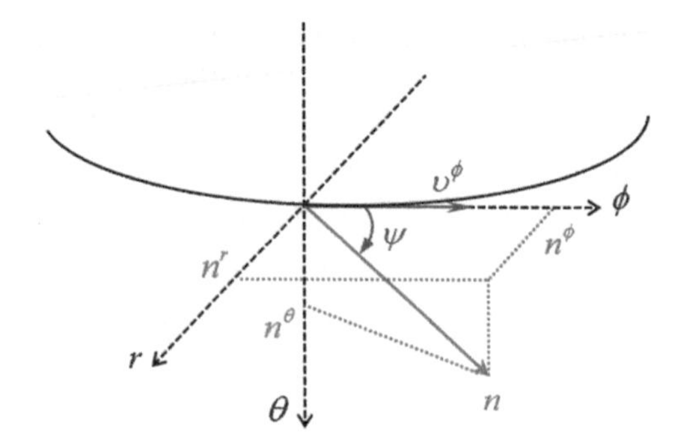

Fig. 9.1 The photon emission along the direction n and the disk motion with velocity $\text{v} \equiv \text{v}^\phi$ at the point of emission. The arc represents a part of the circular orbit of the disk element that emitted the photon (taken from Koutsantoniou and Contopoulos 2014)

between the emission and destination points. For those photons we may assume that the surface of the disk radiates as a black body with temperature T that varies as a known function of distance $r_{\rm source}$ on the disk (e.g. $T \propto r_{\rm source}^{-3/4}$ as in Shakura and Sunyaev 1973). In that case,

$$I_{\rm source} = \frac{\sigma_B}{\pi} T^4(r_{\rm source}) \tag{9.16}$$

where σ_B is the Stefan-Boltzmann constant (Rybicki and Lightman 1986).

In Fig. 9.2, we present sky maps, namely the sky as seen in the LNRF at the position of the moving target particle at the ISCO, for various values of the black hole spin parameter $j \equiv a/M$ and for various accretion disk configurations. In each frame, the black hole direction is at the center and the direction at 90° to that is at the circumference. The circle of radiation around the black hole horizon that appears in most of the images is similar to an Einstein ring. We have performed these calculations only for prograde accretion disks. Notice the difference in radiation intensity from left to right due to the following two reasons:

1. The material on the right is moving toward the target, whereas the material on the left is moving away from it.
2. For faster and faster spinning black holes, photons reach the target primarily along the direction of spacetime rotation.

In Tables 9.1 and 9.2, we present numerical results for the r- and ϕ-components of the radiation force per ion (normalized to the canonical gravitational force per proton $GMm_{\rm p}/r_{\rm ISCO}^2$) for various black hole spin parameters j and for various accretion disk geometries. $r_{\rm ISCO}$ is expressed in units of M. We have assumed here for simplicity that the accretion disk is in prograde Keplerian rotation (Bardeen et al. 1972) with $v^r \ll v^\phi$, and $T = 10^7\ {\rm K}(r_{\rm ISCO}/r)^{-3/4}$ along the disk. What we found is most interesting and rather unexpected:

1. In the theoretical case of an infinitely thin disk, radiation hits the electrons at the ISCO from all directions. It is thus expected that the radiation field is almost isotropic, and therefore, the abberation due to the orbital motion of the target electrons at the ISCO results in radiation drag. This is the general relativistic generalization of the PR drag for thin astrophysical accretion disks. A similar effect is expected to take place when the accretion disk is geometrically thick but optically thin, as is the case of an ADAF disk (Narayan and Yi 1994; Koutsantoniou and Contopoulos in preparation).
2. As the inner edge of the disk thickens, photons can reach the ISCO electrons only from the half space facing the black hole (we have assumed zero optical depth in the disk interior). The important new element is that the radiation that originates on the ISCO contributes more and more to the total radiation pressure, and the radiation field is distorted by the disk and spacetime rotations. As a result, the radiation force acts *along the direction of rotation*, and the azimuthal force component changes sign from negative to positive. In other words, *the azimuthal effect of radiation changes from drag to acceleration*. Considering the analogy with the classical Poynting-Robertson effect, this result was rather unexpected.

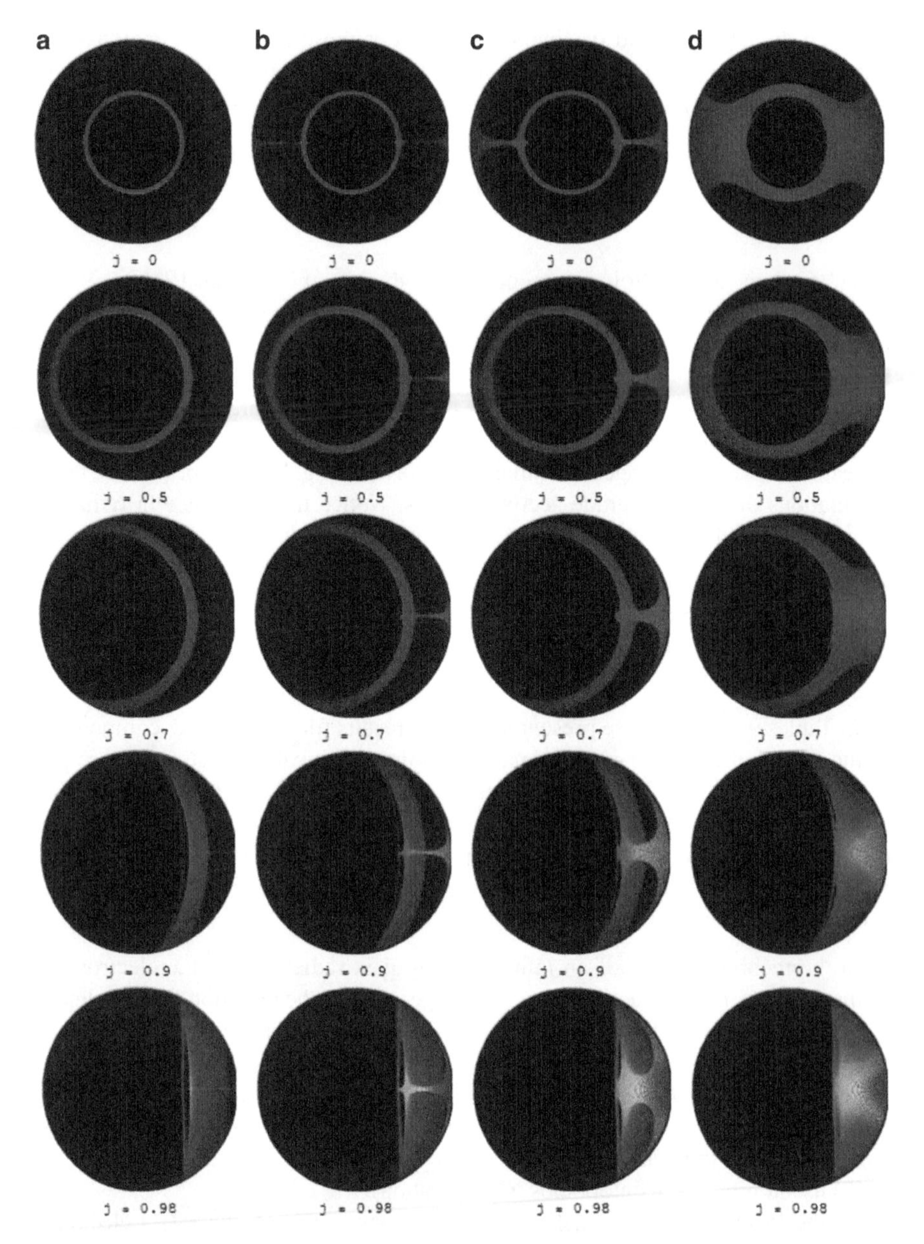

Fig. 9.2 The sky as seen in the LNRF at the position of the moving target electron at the ISCO, for various values of the black hole spin parameter $j = a/M$ and for various disk configurations (**a**: infinity thin, **b**: thin, **c**: thick, **d**: torus). In each frame, the black hole direction is at the center, and the direction at 90° to that is on the circumference. The black hole horizon is the distorted black circle. The circle of radiation around the black hole horizon is an Einstein ring generated by the upper and lower surface of the hot inner disk. Notice the difference in radiation intensity from left to right due to the disk and spacetime rotations. We only considered here prograde accretion disk rotation

Table 9.1 The radial normalized radiation force per electron

j	r_{ISCO}	$f^r_{\text{rad}}/\frac{GMm_p}{r^2_{\text{ISCO}}}$			
		Inf. disk	Thin disk	Thick disk	Torus
0	6.000	0.017	0.021	0.038	0.140
0.1	5.669	0.018	0.022	0.040	0.136
0.2	5.329	0.018	0.023	0.041	0.132
0.3	4.979	0.018	0.023	0.042	0.128
0.4	4.614	0.018	0.024	0.044	0.124
0.5	4.233	0.018	0.024	0.046	0.120
0.6	3.829	0.018	0.024	0.047	0.115
0.7	3.393	0.016	0.024	0.049	0.109
0.8	2.907	0.013	0.025	0.052	0.102
0.9	2.321	0.004	0.025	0.055	0.090
0.92	2.180	0.001	0.025	0.056	0.085
0.94	2.024	−0.004	0.025	0.056	0.080
0.96	1.843	−0.013	0.025	0.055	0.073
0.98	1.614	−0.030	0.024	0.052	0.061

Table 9.2 The azimuthal normalized radiation force per electron

j	r_{ISCO}	$f^\phi_{\text{rad}}/\frac{GMm_p}{r^2_{\text{ISCO}}}$			
		Inf. disk	Thin disk	Thick disk	Torus
0	6.000	−0.007	0.003	0.037	0.111
0.1	5.669	−0.008	0.003	0.040	0.118
0.2	5.329	−0.010	0.003	0.043	0.126
0.3	4.979	−0.012	0.002	0.047	0.134
0.4	4.614	−0.015	0.001	0.052	0.145
0.5	4.233	−0.018	0.001	0.059	0.157
0.6	3.829	−0.023	0.002	0.071	0.175
0.7	3.393	−0.028	0.021	0.098	0.198
0.8	2.907	−0.039	0.036	0.124	0.214
0.9	2.321	−0.061	0.054	0.163	0.230
0.92	2.180	−0.070	0.060	0.174	0.232
0.94	2.024	−0.084	0.067	0.187	0.235
0.96	1.843	−0.106	0.075	0.203	0.236
0.98	1.614	−0.158	0.089	0.227	0.233

9.4 The Cosmic Battery

Once the radiation field in the vicinity of the ISCO around a rotating black hole is known, one can investigate its role in the electrodynamics of the central black hole through the induction equation $\partial \mathbf{B}/\partial t = -c\nabla \times (\alpha \mathbf{E})$ in the LNRF (MacDonald and Thorne 1982). Using Eq. (9.5) this yields,

$$\frac{\partial \mathbf{B}}{\partial t} \approx -c\alpha\nabla \times \left(-\mathrm{v} \times \mathbf{B} + \frac{\mathbf{f}_{\text{rad}}}{\mathrm{e}}\right) + \frac{c a m_{\mathrm{i}} \nabla P \times \nabla \rho}{\mathrm{e}\rho^2}\,. \tag{9.17}$$

In most astrophysical cases of interest, P is a function of the density ρ, and therefore, the last term in the above equation vanishes. In cases where $P \neq P(\rho)$, this term yields a non-zero source-term in the induction equation. This is precisely the thermoelectric effect of the Biermann battery.

In order to investigate the effect of radiation, we ignore the pressure gradient term in Eq. (9.17), add magnetic diffusivity η, and consider for simplicity the integrated form of that equation in axisymmetry on the equatorial plane,

$$\frac{\partial \Psi}{\partial t} = 2\pi\alpha r c \left([\mathbf{v} \times \mathbf{B}]^{\phi} - \frac{f_{\rm rad}^{\phi}}{e} + \eta [\nabla \times \mathbf{B}]^{\phi} \right) , \tag{9.18}$$

where, $\Psi = \pi r^2 \mathcal{B}$ is the magnetic flux contained inside radius r ($\mathcal{B}$ is the average magnetic field inside r).

The second term in the r.h.s. of Eq. (9.18) generates an axial magnetic field which is along the direction of the angular velocity vector in the disk if $f_{\rm rad}^{\phi}$ is negative (i.e. if radiation results in an azimuthal drag force in the LNRF at the inner edge of the disk), or opposite to that if $f_{\rm rad}^{\phi}$ is positive (i.e. if radiation results in an accelerating azimuthal force in the LNRF at the inner edge of the disk). The magnetic flux that builds up obviously closes further out through the accretion disk where $f_{\rm rad}^{\phi}$ drops to zero. The axial magnetic field will be carried by the accretion flow (first term in the r.h.s. of Eq. 9.18), assuming ideal MHD conditions around and inside the ISCO, and the flux accumulated inside the inner edge of the disk will keep growing. The growth would cease, and the mechanism would saturate if the flow begins to also carry inward the return polarity of the magnetic field (Contopoulos and Kazanas 1998; Bisnovatyi-Kogan et al. 2002; Contopoulos et al. 2006). Nevertheless, the return polarity lies in a region where magnetic diffusivity is significant and the third term in the r.h.s. of Eq. (9.18) dominates over the first one (van Ballegooijen 1989; Lubow et al. 1994; Lovelace et al. 1994; see also, however, Lovelace et al. 2009). Therefore,

> The return magnetic field *diffuses outward through the disk*, the mechanism does not saturate, and the Cosmic Battery increases the magnetic field to equipartition values

(see Figs. 9.3, 9.4). Note that this latter point was missed by Bisnovatyi-Kogan et al. (2002). This natural scenario where the innermost part of the accretion disk generates and holds one polarity of the magnetic field, while the return polarity diffuses outward through the outer diffusive part of the disk is a central element of the *Cosmic Battery* (Contopoulos and Kazanas 1998).

A naive approach to understanding the Cosmic Battery mechanism is that, since only the plasma electrons feel the radiation force, they are the ones that slow down or accelerate in the azimuthal direction gradually building up a relative velocity between electrons and ions (see also Chapter 1 in this volume). This is equivalent to an azimuthal electric current which gives rise to the poloidal magnetic field through the ISCO. Obviously, one cannot just 'turn on' an electric current in a plasma, and this is why the investigation of the Cosmic Battery must proceed through the induction equation (Eq. 9.18). This investigation showed that the azimuthal

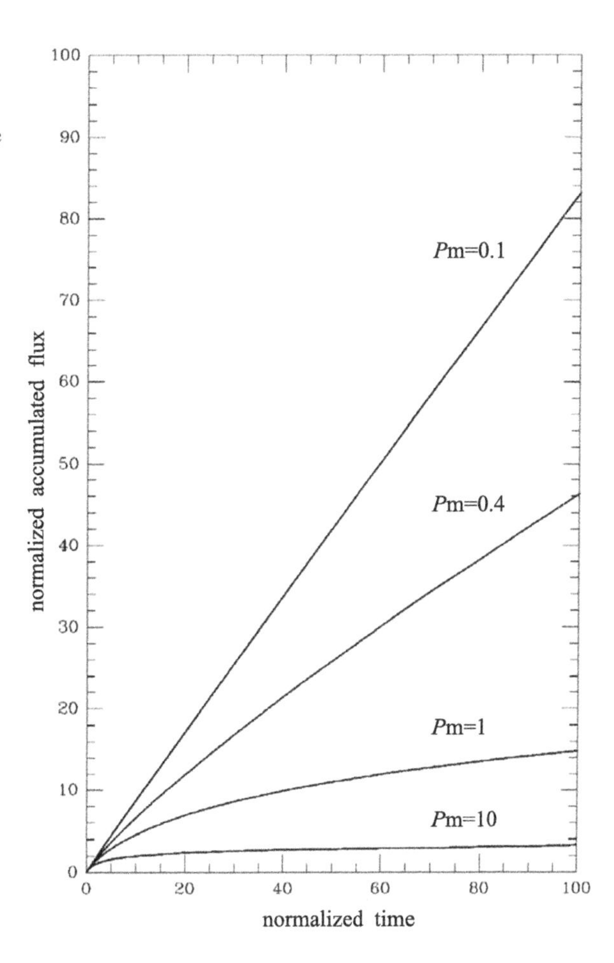

Fig. 9.3 Growth of the magnetic flux contained inside the inner edge of the disk for different values of the magnetic Prandtl number $\mathcal{P}_m$. Time normalized to the time for accretion from the inner edge. Flux normalized to the flux generated by the Cosmic Battery during one accretion time. Notice the field saturation when $\mathcal{P}_m > 0.5$ (conductive disk), and the linear field growth when $\mathcal{P}_m < 0.5$ (diffusive disk). Obviously, the linear field growth will saturate as we approach equipartition (adapted from Contopoulos and Kazanas 1998)

radiation force works similarly to a Biermann battery in its most extreme (maximum radiation pressure, maximum plasma velocities, and maximum radiation-plasma motion misalignment) to grow astrophysically significant magnetic fields of order $\mathcal{B}_o$ inside the ISCO over timescales roughly equal to

$$t_{\rm CB} \sim \frac{{\rm e}\mathcal{B}_o r_{\rm ISCO}}{\alpha_{\rm ISCO} f^{\phi}_{\rm rad} c} \,. \tag{9.19}$$

The latter estimate is obtained from dimensional analysis of Eq. (9.18). The values of $t_{\rm CB}$ that correspond to $\mathcal{B}_o = 10^7\,$G for a $5M_\odot$ black hole for various black hole spin parameters and for various accretion disk models are shown in Table 9.3. The characteristic timescales that we obtain vary from a few hours (in the case of maximally rotating ones) to several days (in the case of slowly rotating black holes). These timescales scale roughly proportionally to $M^{3/2}$ with black hole mass, and therefore the corresponding times to reach equipartition range from about 10^9 years to about 10^{10} years for $10^8 M_\odot$ supermassive black holes.

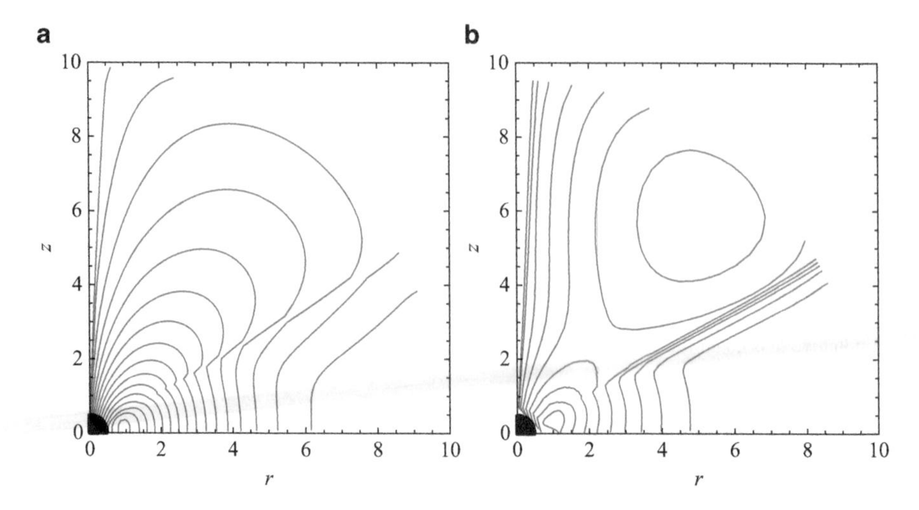

Fig. 9.4 Time sequence (left to right) of the magnetic field evolution inside and outside the accretion disk. The Cosmic Battery generates magnetic field loops around the position of the ISCO ($r = 1$ in this plot) where the azimuthal radiation force per electron is maximal. These loops are stretched by the differential rotation in the disk and open up to infinity. Their inner part is advected by the accretion flow toward the black hole horizon, whereas their outer part diffuses outward through the diffusive/turbulent accretion disk. As a result, the Cosmic Battery mechanism does not saturate, and the field grows to equipartition

Table 9.3 Timescales t_{CB} for the growth of equipartition magnetic fields of order $\mathcal{B}_o \sim 10^7$ G for various types of accretion disks around a $5M_\odot$ black hole for various spin parameters j (taken from Koutsantoniou and Contopoulos 2014)

j	r_{ISCO}	t_{CB} (h)			
		Inf. disk	Thin disk	Thick disk	Torus
0	6.000	625	1,532	115	39
0.1	5.669	440	1,282	92	31
0.2	5.329	301	1,130	72	25
0.3	4.979	213	1,348	55	19
0.4	4.614	143	1,610	40	15
0.5	4.233	93	2,796	29	11
0.6	3.829	58	537	19	8
0.7	3.393	34	46	10	5
0.8	2.907	18	19	5	3
0.9	2.321	7	8	3	2
0.92	2.180	5	6	2	2
0.94	2.024	4	5	2	1
0.96	1.843	3	4	1	1
0.98	1.614	2	3	1	1

One cannot fail to notice that when the outside magnetosphere supports electric currents (i.e., when there is no vacuum outside), the magnetic field geometry obtained in our numerical simulations leads naturally to the electromagnetic interaction between the central region and the extended disk. The reason for this is that the differential rotation 'twists' the newly formed field, generating magnetic torques on the central part. This magnetic field twisting leads to another very important effect, namely, the overall opening of the closed-loop geometry (Newman et al. 1992; Goodson et al. 1997). As one can see in the above references, the axial part of the expanding plasma + magnetic field remains well collimated. Under certain physical circumstances (Shibata et al. 1990; Matsumoto et al. 1996), this effect might also proceed explosively, leading to the expulsion of a significant fraction of the disk material in the form of collimated, fast, axially moving plasma outflows (the astrophysical plasma gun; see Contopoulos 1995; Matsumoto et al. 1996).

There have been several efforts to associate the predictions of the Cosmic Battery with astronomical observations of the Faraday rotation measure gradient across astrophysical jets (Contopoulos et al. 2009; Gabuzda et al. 2012; see also Chapter 5 in this volume). It has been shown that the toroidal component of the magnetic field across astrophysical jets is along a universal direction that corresponds to the axial electric current *outflowing* from the core to infinity along the jet (e.g. Kronberg et al. 2011). This observational result is consistent with our latest relativistic estimates of the direction of the azimuthal radiation force term in the induction equation. The results are not yet conclusive, and more observations are needed to confirm our theoretical estimates.

9.5 Astrophysical Applications

The Cosmic Battery mechanism could be important in accounting for the presence of magnetic fields in several other astrophysical systems (neutron stars, protostars, galaxies) as outlined in Contopoulos and Kazanas (1998). Around black holes, the mechanism generates a magnetic field of the right large-scale dipolar topology required for the Blandford- Znajek mechanism of electromagnetic energy extraction to work (Blandford and Znajek 1977; Blandford 2002). In this respect we would like to point the reader's attention to the intriguing possibility that these considerations could in fact apply to understanding the cyclic variability observed in the Galactic X-ray binaries and microquasars (e.g. microquasar GRS 1915 + 105; Pooley and Fender 1998; Ueda et al. 2002). A distinguishing characteristic of these sources is their repeating X-ray/IR/radio flares that lead to mildly relativistic outflows. In fact, Kylafis et al. (2012) showed that the Cosmic Battery naturally accounts for the rich phenomenology of the jet appearance, disappearance, and re-appearance as these sources follow a 'q'- shaped curve in the so-called Hardness-Intensity Diagram-HID (first introduced in the present context by Miyamoto et al. 1995; see also Fig. 2.1, Chapter 2, and Chapter 10) as follows:

In an HID, a steady jet exists even when the sources are in the quiescent state, where typically the X-ray luminosity is six to eight orders of magnitude smaller than the Eddington luminosity (Kong et al. 2002; Hameury et al. 2003; Remillard and McClintock 2006). In that state, the accretion disk is optically thin geometrically thick (ADAF). The sources spend most of their time in this state, and therefore, the Cosmic Battery has enough time to establish the magnetic field needed to sustain a jet. During an outburst, and as the sources trace the vertical part of the 'q'-shaped curve in the HID, the X-ray luminosity of the sources increases. As a consequence, the electric current produced in the inner ADAF and the magnetic field in the steady jet also increase. Thus, the radio emission also increases (e.g., Giannios 2005). It is not surprising then that the radio luminosity and the X-ray luminosity are correlated. As the sources reach the upper right part of the 'q'-shaped curve, the X-ray luminosity is at its maximum and the steady jet is at full strength. As the outburst evolves, the disk accretion rate apparently increases, and as the sources reach the so called 'jet line', the disk becomes geometrically thin and extends all the way to the ISCO. Whatever magnetic field existed inside the ISCO of the geometrically thick ADAF will now have trouble to be held there by the thin disk. The thin disk becomes unstable to non-axisymmetric magnetic 'Rayleigh-Taylor' instability modes, and the accumulated magnetic field escapes to the outer disk in the form of magnetized 'strands', as manifested in modern state-of-the-art numerical simulations of Magnetically Arrested Disks-MADs (Tchekhovskoy et al. 2011; Dexter et al. 2014). Such an instability explains naturally the flaring nature of the jets as the sources approach the jet line during the hard-to-soft transition. To the left of the jet line, the sources are in the soft state and the disk remains geometrically thin. In this state, the thickness of the disk is much smaller than before, the efficiency of the Cosmic Battery is significantly reduced, and the timescales for the growth of an equipartition magnetic field are significantly increased. Some sources, including the archetypal black-hole XRB GX 339-4, cross the jet line only once (Fender et al. 2009). While in the soft state, they may trace small closed loops (see, e.g., Fig. 7 of Fender et al. 2004). No jet has ever been detected when the sources trace such loops. On the other hand, in cases in which a source decides to cross the jet line leisurely several times, then (a) the inner accretion disk gradually thickens, the magnetic field is being built at an accelerated rate, the disk becomes stable to non-axisymmetric Raleigh-Taylor modes, and a steady jet is created as the hardness ratio increases (soft to hard state transition); (b) the inner disk becomes thin, the accumulated field makes the thinner disk unstable, and a flaring jet appears as the sources return to the jet line (hard to soft state transition); and (c) just as in the first crossing of the jet line, the jet disappears after the system returns to the soft state. This is exactly what is observed in GRS 1915+105 (Rushton et al. 2010; see also Brocksopp et al. 2002 for XTE J1859+226). This source is known to be in the soft state for long periods of time, although it makes repeated excursions to the hard state. Finally, when the sources cross the jet line at lower X-ray intensity, the inner part of the accretion disk has thickened substantially, the Cosmic Battery again operates very efficiently, and a steady jet is established in the low hard state. Once established, the steady jet persists as the sources transit to the quiescent state. This re-generation of the steady magnetic field and jet is the key element that is missing from other theoretical models for the HIS (e.g. Begelman and Armitage 2014).

We conclude that the Cosmic Battery mechanism enters as a new, previously unaccounted-for player in the ongoing study of the generation of astrophysical magnetically driven black hole jets.

Parts of this Chapter may be found in Contopoulos and Kazanas 1998; Kylafis et al. 2012, and Koutsantoniou and Contopoulos 2014. We acknowledge discussions with Demos Kazanas, Nick Kylafis, Leela Koutsantoniou, Denise Gabuzda, Dimitris Christodoulou, and Matthaios Katsanikas. This work was supported by the General Secretariat for Research and Technology of Greece and the European Social Fund in the framework of Action 'Excellence'.

References

Abramowicz, M.A., Ellis, G.F.R., Lanza, A.: ApJ **361**, 470 (1990)
Bardeen, J.M., Press, W.H., Teukolsky, S.A.: ApJ **178**, 347 (1972)
Begelman, M.C., Armitage, P.J.: ApJ **782**, 18 (2014)
Bini, D., Jantzen, R.T., Stella, L.: CQGrav **26**, 5009 (2009)
Bini, D., Geralico, A., Jantzen, R.T., Semerak, O., Stella, L.: CQGrav **28**, 5008 (2011)
Bisnovatyi-Kogan, G.S., Lovelace, R.V.E., Belinski, V.A.: ApJ **580**, 380 (2002)
Blandford, R.D.: MNRAS **176**, 465 (1976)
Blandford, R.D.: PThPhS **143**, 182 (2001)
Blandford, R.D.: To the Lighthouse. In: Gilfanov, M., Sunyaev, R., Churazov, E. (eds.) Lighthouses of the Universe, vol. 381. Springer, Berlin (2002)
Blandford, R.D., Payne, D.G.: MNRAS **199**, 883 (1982)
Blandford, R.D., Znajek, R.L.: MNRAS **179**, 433 (1977)
Brocksopp, C., et al.: MNRAS **331**, 765 (2002)
Contopoulos, J.: ApJ **432**, 508 (1994)
Contopoulos, J.: ApJ **450**, 616 (1995)
Contopoulos, J. ApJ **460**, 185 (1996)
Contopoulos, I., Kazanas, D.: ApJ **508**, 859 (1998)
Contopoulos, J., Lovelace, R.V.E.: ApJ **429**, 139 (1994)
Contopoulos, I., Kazanas, D., Christodoulou, D.M.: ApJ **652**, 1451 (2006)
Contopoulos, I., Christodoulou, D.M., Kazanas, D., Gabuzda, D.C.: ApJ **702**, L148 (2009)
Contopoulos, I., Nathanail, A., Katsanikas, M., Koutsantoniou, L.E.: (2014, in preparation)
Dexter, J., McKinney, J.C., Markoff, S., Tchekhovskoy, A.: MNRAS **440**, 2185 (2014)
Fender, R.P., Belloni, T.M., Gallo, E.: MNRAS **355**, 1105 (2004)
Fender, R.P., Homan, J., Belloni, T.M.: MNRAS **396**, 1370 (2009)
Ferreira, J., Pelletier, G.: A& A **295**, 807 (1995)
Gabuzda, D.C., Christodoulou, D.M., Contopoulos, I., Kazanas, D.: JPhCS **355**, 2019 (2012)
Giannios, D.: A&A **437**, 1007 (2005)
Goodson, A.P.,Winglee, R.M., Bohm, K.-H.: ApJ **489**, 199 (1997)
Hameury, J.-M., Barret, D., Lasota, J.-P., McClintock, J.E., Menou, K., Motch, C., Olive, J.-F., Webb, N.: A&A **399**, 631 (2003)
Komissarov, S.S.: JKPhS **54**, 2503 (2008)
Kong, A.K.H., McClintock, J.E., Garcia, M.R., Murray, S.S., Barret, D.: ApJ **570**, 277 (2002)
Koutsantoniou, L.E., Contopoulos, I.: ApJ **794**, 27 (2014)
Kronberg, P.P., Lovelace, R.V.E., Lapenta, G., Colgate, S.A.: ApJ **741**, 15 (2011)
Kylafis, N.D., Contopoulos, I., Kazanas, D., Christodoulou, D.M.: A&A **538**, 5 (2012)
Lamb, F.K., Miller, M.C.: ApJ **439**, 828 (1995)
Lasota, J.-P. et al.: PhysRevD **89**, 024041 (2014)

Li, Z.-Y.: ApJ **444**, 848 (1995)
Lovelace, R.V.E., Romanova, M.M., Newman, W.I.: ApJ **437**, 136 (1994)
Lovelace, R.V.E.: Nature **262**, 649 (1976)
Lovelace, R.V.E., Rothstein, D.M., Bisnovatyi-Kogan, G.S.: ApJ **701**, 885 (2009)
Lubow, S.H., Papaloizou, J.C.B., Pringle, J.E.: MNRAS **267**, 235 (1994)
Lynden-Bell, D.: MNRAS **369**, 1167 (1996)
MacDonald, D., Thorne, K.S.: MNRAS **198**, 345 (1982)
Matsumoto, R., Uchida, Y., Hirose, S., Shibata, K., Hayashi, M.R., Ferrari, A., Bodo, G., Norman, C.: ApJ **461**, 115 (1996)
McKinney, J.C., Tchekhovskoy, A., Blandford, R.D.: MNRAS **423**, 3083 (2012)
Miller, M.C., Lamb, F.K.: ApJ **413**, L43 (1993)
Miller, M.C., Lamb, F.K.: ApJ **470**, 1033 (1996)
Miyamoto, S., Kitamoto, S., Hayashida, K., Egoshi, W.: ApJ **442**, L13 (1995)
Narayan, R., Yi, I.: ApJ **428L**, 13 (1994)
Nathanail, A., Contopoulos, I.: ApJ **788**, 186 (2014)
Newman, W.I., Newman, A.L., Lovelace, R.V.E.: ApJ **392**, 699 (1992)
Pooley, G.G., Fender, R.P.: GRS 1915+105: Flares, QPOs and Other Events at 15 GHz. In: Zensus, J.A., Taylor, G.B., Wrobel, J.M. (eds.) IAU Colloq. 164, Radio Emission from Galactic and Extragalactic Compact Sources. ASP Conference Series, vol. 144, p. 333. ASP, San Francisco (1998)
Poynting, J.H.: MNRAS **64**, A1 (1903)
Remillard, R.A., McClintock, J.E.: ARA&A **44**, 49 (2006)
Robertson, H.P.: MNRAS **97**, 423 (1937)
Rushton, A., Spencer, R., Fender, R., Pooley, G.: A&A **524**, A29 (2010)
Rybicki, G.B., Lightman, A.P.: Radiative Processes in Astrophysics. Wiley, New York (1986)
Shakura, N.I., Sunyaev, R.A.: A&A **24**, 337 (1973)
Shibata, K., Tajima, T., Matsumoto, R.: ApJ **350**, 295 (1990)
Tchekhovskoy, A., Narayan, R., McKinney, J.C.: MNRAS **418**, 79 (2011)
Ueda, Y., et al.: ApJ **571**, 918 (2002)
Vainshtein, S.I., Cattaneo, F.: ApJ **393**, 165 (1992)
van Ballegooijen, A.A.: Magnetic fields in the accretion disks of cataclysmic variables. In: Belvedere, G. (ed.) Accretion Disks and Magnetic Fields in Astrophysics, p. 99. Dordrecht, Kluwer (1989)
Zrake, J., MacFadyen, A.: ApJ **744**, 32 (2012)

Chapter 10
Accretion and Ejection in Black-Hole X-Ray Transients

N.D. Kylafis and T.M. Belloni

Abstract A rich phenomenology has been accumulated over the years regarding accretion and ejection in black-hole X-ray transients (BHTs). Here we summarize the current observational picture of the outbursts of BHTs, based on the evolution traced in a hardness – luminosity diagram (HLD), and we offer a physical interpretation with two assumptions, easily justifiable. The first is that the mass-accretion rate to the black hole in a BHT outburst has a generic bell-shaped form. This is guaranteed by the observational fact that all BHTs start their outburst and end it at the quiescent state, i.e., at very low accretion rate. The second assumption is that at low accretion rates the accretion flow is geometrically thick, ADAF-like, while at high accretion rates it is geometrically thin. Both, at the beginning and the end of an outburst, a strong poloidal magnetic field develops locally in the ADAF-like part of the accretion flow, and this explains naturally why a jet is always present in the right part of the HLD. "Memory" of the system explains naturally why BHTs traverse the q-shaped curves in the HLD always in the counterclockwise direction and that no BHT is expected to ever traverse the entire curve in the clockwise direction. The only parameter in our picture is the accretion rate.

10.1 Introduction

The spectral evolution of black-hole X-ray transients (BHTs) has been presented using a hardness – luminosity diagram (HLD), where the sources often exhibit a q-shaped curve, complicated by small excursions (see Fig. 10.1, top curve; Homan et al. 2001; Belloni et al. 2005; Homan and Belloni 2005; Gierliński and Newton 2006; Remillard and McClintock 2006; Fender et al. 2009; Motta et al. 2009;

N.D. Kylafis (✉)
Physics Department, Institute of Theoretical and Computational Physics, University of Crete, 71003 Heraklion, Crete, Greece
e-mail: kylafis@physics.uoc.gr

T.M. Belloni
INAF-Osservatorio Astronomico di Brera, Via E. Bianchi 46, I-23807 Merate (LC), Italy
e-mail: tomaso.belloni@brera.inaf.it

I. Contopoulos et al. (eds.), *The Formation and Disruption of Black Hole Jets*,
Astrophysics and Space Science Library 414,
DOI 10.1007/978-3-319-10356-3_10

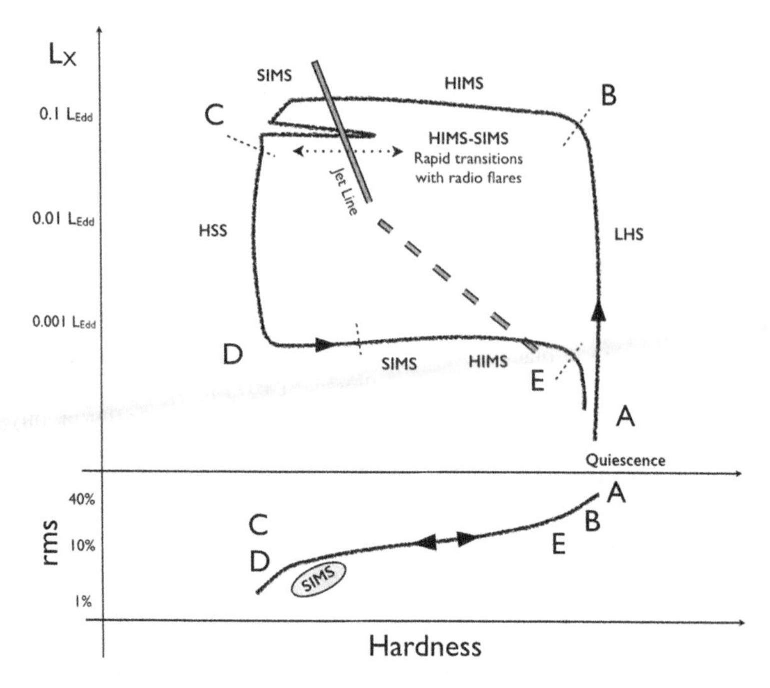

Fig. 10.1 Schematic representation of the q-shaped curve in a hardness-luminosity diagram (*top curve*) and a hardness – fractional-rms diagram (*bottom curve*) for black-hole X-ray binaries. The *dotted lines* mark the transition between states

Belloni 2010; Munoz-Darias et al. 2011a; Stiele et al. 2011). Such a hysteresis was first noticed by Miyamoto et al. (1995). Complementary to HLDs are the hardness – fractional-rms-variability diagrams (HRDs) (Fig. 10.1, bottom curve; Belloni et al. 2005) and the absolute-rms-variability – luminosity diagrams (RLDs) (Fig. 10.2; Munoz-Darias et al. 2011a,b). The last ones exhibit curves topologically similar to the curves in HLDs. These three types of diagrams, especially the HRDs and the RLDs, have shown that the BHTs exhibit some characteristic states (Belloni et al. 2005; Belloni 2010; Belloni et al. 2011). These are the quiescent state (QS), the hard state, which historically is called low/hard state (LHS) but the word "low" has been proven inappropriate, the hard intermediate state (HIMS), the soft intermediate state (SIMS), and the soft state, which historically is called high/soft state (HSS), but again the word "high" has been proven inappropriate. A different classification scheme, based on quantitative criteria regarding the energy spectrum and the variability, was introduced by Remillard and McClintock (2006) and McClintock et al. (2009). A comparison between the two classification schemes has been provided by Motta et al. (2009). For the purposes of this paper, we will use the first classification.

A simple, generic model (see, e.g., Gilfanov 2010) assumes that, at all times during an outburst, there is a hot corona near the black hole and further out an accretion disk, which is geometrically thin, optically thick, relatively cool, and well

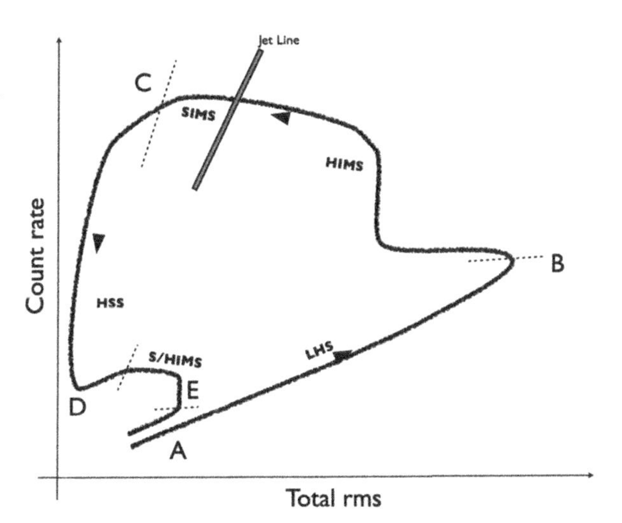

Fig. 10.2 Schematic representation of the curve traversed by black-hole X-ray binaries in a luminosity versus total-integrated-rms diagram. The *dotted lines* mark the transition between states

described by the work of Shakura and Sunyaev (1973). The relative sizes of the corona and the accretion disk determine the spectral state. In the HSS the corona is small and the accretion disk extends all the way to the inner stable circular orbit (ISCO). In the LHS, the corona is large and the accretion disk is restricted to the outer region.

A more realistic and quantitatively successful model for the various spectral states was provided by Esin et al. (1997); see also Narayan et al. (1996, 1997) and Esin et al. (1998, 2001), who identified the unphysical corona with a physical Advection Dominated Accretion flow (ADAF). For the interpretation of the q-shaped curve (Fig. 10.1, top panel) we make use of the picture of Esin et al. (1997), but we also improve it and expand it, taking into account the knowledge that has been accumulated in the intervening years.

For the traversal of the q-shaped curve, we want to make the following remark. Imagine taking a video of a BHT as it traverses the q-shaped curve (Fig. 10.1, top curve) from quiescence (below point A), vertically to high luminosity (point B), left to small hardness ratio at about constant luminosity (point C), vertically down to a significantly lower luminosity (point D), horizontally to large hardness ratio (point E), and finally vertically to quiescence (below point A). If the video is shown backwards, then the left turn occurs at low luminosity (point E) and the rest of the q-shaped curve is traversed in the clockwise direction. No law seems to be violated in this time-reversed video. Thus, some sources ought to be observed to traverse the q-shaped curve clockwise and others counterclockwise. However, all BHTs and in all their outbursts, if more than one outburst have been observed, traverse the q-shaped curve in the counterclockwise direction. It is therefore important to understand what breaks this time-reversal symmetry. In our interpretation of the q-shaped curve, we address this point. We remark that not all sources exhibit a nice q-shaped curve during their outbursts (see Belloni 2010). In this paper we will restrict ourselves to sources exhibiting a nice q-shaped curve in a HLD, with GX 339-4 as the prototype.

Our work is based on two assumptions: (1) The accretion rate into BHTs as a function of time has a generic bell-shaped curve. It starts from a very low accretion rate, it increases steadily up to an accretion rate comparable to the Eddington rate, and ends again with a very low accretion rate. This is justified by the fact that BHTs start and end their outbursts at very low luminosity. (2) At high accretion rates, the accretion disk is optically thick, geometrically thin, and it is described well by the model of Shakura and Sunyaev (1973). At low accretion rates, the accretion flow is ADAF (Narayan and Yi 1994, 1995; Abramowicz et al. 1995) or ADAF-like.

Both of these assumptions are well accepted, have been verified by numerical simulations (Ohsuga et al. 2009), and have been applied quantitatively to observations many times in the past.

As mentioned above, our knowledge has increased significantly in recent years. Thus, despite its success, the Esin et al. (1997) model has some shortcomings. In particular, the very high state in their picture is now identified with the HIMS and the SIMS transitions, which come *before* the HSS, not after. A second shortcoming is the fact that their picture is "one-dimensional", i.e., in a hardness luminosity diagram, the sources traverse the same curve forward and backward, they do not traverse a loop (see their Fig. 10.1). Finally, jet formation and destruction is not addressed. In our work, we extend the model of Esin et al. (1997) and subsequent work (for a comprehensive recent review see Done et al. 2007) to account for: (1) The *return path* of the q-shaped curve (see Fig. 10.1, top panel, from point C to points D and E). To our knowledge, the outburst decay has not been modeled theoretically, despite the fact that it has been studied observationally in detail (Kalemci et al. 2001, 2003, 2004, 2005, 2006). (2) Jet formation, destruction, and re-formation. By taking into account a novel mechanism for magnetic field generation in the ADAF, we offer an explanation for (a) the formation and evolution of a compact jet as the sources move from quiescence, to the hard, and then to the hard intermediate state, (b) the eruptive disappearance of the compact jet before the sources reach point C in Fig. 10.1 (HIMS to SIMS transition) and (c) the smooth re-appearance of the compact jet before the sources reach point E. As we will show in detail below, the only parameter in our model is the mass-accretion rate.

An inferred quantity, which is a function of the accretion rate, is the transition radius R_{tr} (Esin et al. 1997; see also the Chapter of Kazanas in this volume) between the outer thin disk and the inner ADAF. Work on R_{tr}, taking into account evaporation, has been done by Liu et al. (1999); see also Meyer et al. (2000), Meyer-Hofmeister and Meyer (2001), Qian et al. (2007), and Meyer-Hofmeister et al. (2009). According to this work, we show in Fig. 10.3 a schematic of R_{tr} versus mass-accretion rate $\dot{M}$. In quiescence, $\dot{M}$ is small and R_{tr} is large. In the LHS, say at points between A and B, $\dot{M}$ increases and R_{tr} decreases. In the HIMS and the SIMS, i.e., at points between B and C, $\dot{M}$ keeps increasing until it reaches its maximum value (here we have allowed for three different maximum values), while R_{tr} keeps decreasing until it reaches the innermost stable circular orbit at $R_{\rm ISCO}$. Then, $\dot{M}$ decreases, but R_{tr} remains at $R_{\rm ISCO}$ until point D is reached. There, the inner part

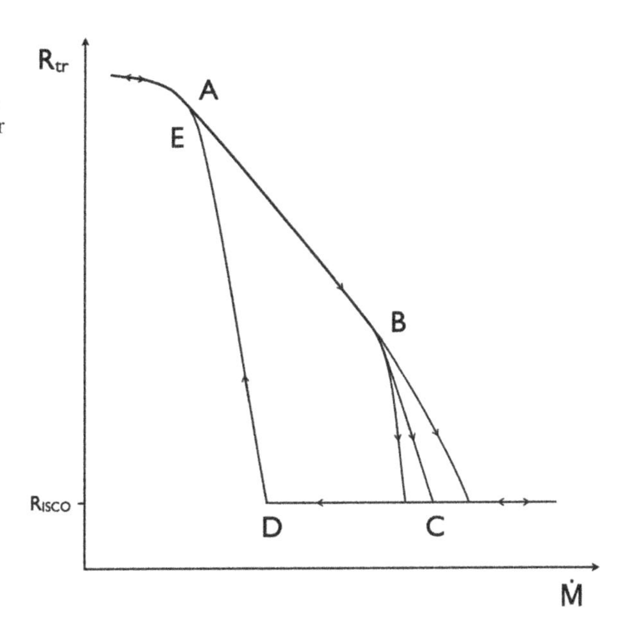

Fig. 10.3 Schematic representation of the transition radius R_{tr} between the inner ADAF and the outer thin disk as a function of mass-accretion rate $\dot{M}$. The points *A*, *B*, *C*, *D*, and *E* reflect the same points in Figs. 10.1 and 10.2

of the accretion disk becomes radiatively inefficient and puffs up (Das and Sharma 2013). As $\dot{M}$ keeps decreasing, an increasing part of the inner flow becomes ADAF-like and therefore R_{tr} increases. Finally, at point E the sources reach the LHS.

Our presentation is qualitative, but it relies on well-accepted theoretical ideas and calculations. We concentrate on the "zeroth order effects", while for the "first order ones" we either speculate and say so or leave them unexplained. In many respects, especially jet formation and destruction, our picture is applicable to neutron-star and white-dwarf X-ray binaries (Kylafis et al. 2012).

In Sect. 10.2 we describe our main ideas and explain the "zeroth order" phenomenology, in Sect. 10.3 we make some remarks, and in section "Conclusions" we present our conclusions.

10.2 Following BHTs Along an Outburst

BHTs spend most of their time in the QS (see Levine et al. 2006), where the X-ray luminosity is very low and the sources are usually undetected within short observing times. Then, a significant amount of matter begins accreting into the black hole and an outburst occurs. The fact that BHTs start their outbursts in the QS, increase their luminosity, and come back to the QS means that the mass-accretion rate has, to "zeroth order", the form of a bell-shaped curve, beginning and ending at the accretion rate of the QS. This does not mean a Gaussian, but rather any curve that starts at low values, increases, reaches a maximum or a plateau or multiple local maxima and minima, and returns to low values.

Using multi-wavelength observations, it has been demonstrated (McClintock et al. 2003; see also Marsh et al. 1994; Orosz et al. 1994) that in quiescence the matter that leaves the companion star forms a thin disk at large radii ($r \gtrsim 2\times10^4 R_g$), where $R_g = GM/c^2$ is the gravitational radius and M is the mass of the black hole. At radii interior to this disk, and for the relatively small amount of matter that accretes into the black hole in quiescence, the flow is according to our second assumption ADAF-like.

It is generally considered that, matter accumulates in the above thin disk and, through the ionization instability (Meyer and Meyer-Hofmeister 1981; Smak 1984; for a review see Lasota 2001), matter is released, falls into the black hole, and powers the X-ray outburst.

Before proceeding to a discussion of the outburst, we will comment on the formation of the hard and the soft X-ray spectra in BHTs and on their observed time variability.

10.2.1 X-Ray Spectra of BHTs

The soft X-ray spectra of BHTs (i.e., below a few keV) are generally believed to be multi-temperature blackbody spectra emitted by a geometrically thin and optically thick accretion disk, of the type proposed by Shakura and Sunyaev (1973).

For the hard X-ray power-law spectra of BHTs, observed in the LHS, there is no general agreement for their formation, though a consensus is forming (see below). In models where a corona is assumed (see Gilfanov 2010 for a review), the hard X-ray spectra are produced by inverse Comptonization of soft seed photons by thermal or non-thermal electrons in the corona. In view of the fact that simple corona models cannot explain the time-reversal asymmetry, discussed in the Introduction, we will not consider them further.

Since all BHTs exhibit a compact radio jet when they are in the LHS or in the HIMS, and the jet originates in an advection dominated accretion flow (ADAF; Narayan and Yi 1994, 1995; Abramowicz et al. 1995), it is natural to assume that the jet and/or the ADAF play a significant role in the formation of the hard X-ray spectrum. The ADAF always plays a role. On the other hand, the jet clearly has the potential to play a role. It has been demonstrated (Reig et al. 2003) that a simple jet model can easily explain the hard X-ray spectra of BHTs in the LHS and the HIMS. The same is true for ADAF models (for a review see Done et al. 2007). Thus, there is no need to invoke a thermal or non-thermal corona, when an ADAF and a jet are present, which can explain observational facts that corona models do not ever address (Giannios et al. 2004; Giannios 2005; Kylafis et al. 2008; Ingram et al. 2009; Ingram and Done 2011, 2012).

Markoff et al. (2005) proposed that the upscattering of soft photons, to create the hard spectrum, takes place *at the base* of the jet. However, if the seed photons enter the base of the jet, there is nothing to prevent them from "exploring" the entire body of the compact jet. Indeed, Compton upscattering of soft seed photons *in the entire body of the jet* fits nicely the observed spectra from radio to hard X-rays and explains in a natural way many other observational facts (Giannios et al. 2004; Giannios 2005). Thus, we will assume that the hard X-ray spectrum in the LHS and the HIMS is produced by inverse Compton scattering of soft seed photons mainly in the ADAF and possibly in the compact jet. After all, there is hardly any energetic difference if the Comptonization takes place in the ADAF or the jet, since it is the ADAF that creates the jet and one expects that, to "zeroth order", the mean electron energy in the jet and the ADAF is about the same.

Before closing, we remark that the steep power-law, high-energy spectra observed in the HSS of BHTs constitute only a small fraction of the total luminosity. Thus, a limited attention has been paid to them and a possible explanation is that these spectra are formed by inverse Compton scattering in non-thermal flares above and below the thin accretion disk (Poutanen et al. 1997; Gierliński et al. 1999). If a jet is ever detected in the HSS, then the above interpretation must be re-considered.

10.2.2 Time Variability of BHTs

The properties of fast (<1 s) time variability change regularly during an outburst, following a well repeated pattern (see Belloni 2010; Belloni et al. 2011). These variations can be best followed by producing in addition to the HLD, two more diagrams: the HRD (Fig. 10.1, bottom curve) and the RLD (Fig. 10.2). As already mentioned, the HRD shows the total fractional rms variability integrated over a broad frequency range, as a function of hardness. Here, to "zeroth order", one can notice the absence of hysteresis: all observations of a source lie along a line which goes from 40 % rms in the LHS, down to $\sim$1 % in the HSS, and back up to 40 % when the source returns to the LHS. The SIMS points occupy a separate "cloud" at lower % rms than the main line. The power density spectra (PDS) in the LHS show large variability in the form of band-limited noise components, usually fitted with broad Lorentzian models (see Belloni et al. 2002). As the source brightens and slightly softens, the total rms decreases and the characteristic frequencies of all Lorentzian components increase (see, e.g., Belloni et al. 2005). In the HIMS, when the source experiences a fast and strong softening (see below), the rms further decreases down to 10 %, while all Lorentzians continue increasing their frequency. In this state, quasi-periodic oscillations (QPOs) appear together with the noise. This type of QPO is called type-C (Casella et al. 2005). The transition to the SIMS is very fast, it can take place on time scales below 1 h (see Nespoli et al. (2003). Also, Casella et al. (2004), Belloni et al. (2005), and Motta et al. (2011) show fast transitions), and is characterized by a radical change in the timing properties: the broad-band noise disappears to be replaced by a weaker power-law component, while the type-C QPO is replaced by a so-called type-B QPO, whose properties are radically different (see Motta et al. 2011). Minor transitions between HIMS and SIMS are often observed later, and the timing properties change accordingly. As the source moves to the HSS, weak QPOs of yet another type

(type-A) (see Motta et al. 2011 and references therein) and reduced noise level in the form of a power law are observed. In the bottom branch, the sequence is reversed and there is little evidence for hysteresis in the HRD.

Complementary to the HRD is the RLD (Fig. 10.2), where the horizontal axis shows the total (*not the fractional*) integrated rms, and the vertical axis the luminosity. All transitions described above are clearly visible in this diagram, which has the advantage of not requiring any spectral resolution. Minor transitions between HIMS and SIMS can be seen in such a diagram also (see Munoz-Darias et al. 2011a).

10.2.3 Quiescent State

In the HLD (Fig. 10.1), we are now at a point well below point A (see also Fig. 10.3). The accretion rate in the QS is very low and the accretion flow inside R_{tr} is radiatively inefficient, hot, geometrically thick, and optically thin. Such flows are described very well by advection dominated accretion flows (ADAF, Narayan and Yi 1994, 1995; Abramowicz et al. 1995). As a rule, when BHTs are in the QS, they exhibit a radio jet. Soft, cyclotron photons from the jet are upscattered by the electrons in the jet (Giannios 2005) or the ADAF (Done et al. 2007) and produce the observed power-law, hard X-ray spectrum. Since the thin disk outside R_{tr} emits optical and infrared radiation, a fraction of it will also be upscattered, but the majority of the soft seed photons is expected to come from the jet, which is directly above and below the ADAF.

Three-dimensional magnetohydrodynamic simulations of astrophysical accreting systems (Machida et al. 2006; Hawley 2009; Romanova et al. 2009; Mignone et al. 2010) have shown that jet formation is due to one of two types of physical mechanisms: either plasma gun/magnetic tower (Contopoulos 1995; Lynden-Bell 1996) or centrifugal driving (Blandford and Payne 1982). In either case, a strong, large-scale magnetic field is absolutely necessary. Such strong magnetic fields can either come from far away, in which case the flow, via advection, brings them in and amplifies them (Igumenshev 2008; Lovelace et al. 2009; Tchekhovskoy et al. 2011) or they are produced *locally*. We favor the second possibility because we consider it simple and elegant. It produces a *strong, poloidal* magnetic field where it is needed and when it is needed (see below). We find it awkward to rely on a random process rather than a direct one to create an ordered magnetic field.

The mechanism for the formation of a strong, poloidal magnetic field is the Poynting-Robertson Cosmic Battery (PRCB, Contopoulos and Kazanas 1998; see also Contopoulos et al. 2006; Christodoulou et al. 2008). The battery works very efficiently, because the inner part of the flow is geometrically thick. Thus, most of the radiation emitted near the ISCO participates in the mechanism of the battery. The formation timescale of the magnetic field, in the various states of a BHT, is discussed in Kylafis et al. (2012). Since BHTs spend most of their time in the QS, there is ample time for the formation of the strong, poloidal magnetic field needed for the jet.

The efficiency of the Cosmic Battery is proportional to the luminosity of the source. Thus, an increase in the luminosity, as the source moves from quiescence to the hard state, results in the PRCB producing a stronger magnetic field, which can support a stronger jet, with an increased radio emission (e.g., Giannios 2005). Thus, it is natural to expect the radio luminosity to be positively correlated with the X-ray luminosity (see Gallo et al. 2006).

10.2.4 Hard State

In the HLD (Fig. 10.1, top panel), we are now at point A (see also Fig. 10.3). As the mass-accretion rate increases, the accretion flow remains ADAF-like, the X-ray luminosity increases, the PRCB works more efficiently, the jet becomes stronger, but the X-ray spectrum (power law in the 2–20 keV band with a photon-number spectral index $\Gamma \sim 1.5$) remains almost the same (almost the same hardness ratio). Thus, in the HLD, the track that the BHTs follow is an almost vertical line (upwards from point A in Fig. 10.1, top panel). There is some softening as the source brightens, but it is much less compared to that in the HIMS, see Motta et al. (2009). Since the jet produces the radio spectrum and also provides seed photons for the hard X-rays, it is natural that the two are correlated (see Gallo 2010 and references therein). Furthermore, since the up-scattering of the seed photons steals energy from the jet and/or the ADAF (Compton cooling; Done et al. 2007), the high-energy cutoff E_c of the X-ray spectrum should decrease. This is exactly what has been observed by Motta et al. (2009).

With increasing mass-transfer rate, the transition radius R_{tr} moves inward (Esin et al. 1997; Liu et al. 1999; Das and Sharma 2013). The thin, outer part of the accretion flow has no direct contribution to the hard X-ray spectrum, though it has an indirect one, because it provides a fraction of the soft seed photons for Comptonization (Sobolewska et al. 2011). The ADAF-like flow at $r < R_{tr}$ is permeated by a poloidal magnetic filed, which is very strong, near equipartition, at the ISCO (Kylafis et al. 2012) and drops off with radius. This poloidal magnetic field not only supports the jet, but also *increases* the strength of the magneto-rotational instability (Bai and Stone 2013). As a result, there is efficient transport of matter inwards, which keeps the density low. Thus, the source goes upwards in the q-diagram (from point A to point B in the upper panel of Fig. 10.1) and *does not turn left*, i.e., towards points E and D. In other words, the poloidal magnetic field, which is created by the PRCB *forces* the source to remain in the hard state and, until something changes, there is no transition to the soft state.

As the source moves from point A to point B (Fig. 10.1, top panel), the luminosity increases, the cutoff of the power-law X-ray spectrum decreases significantly (Motta et al. 2009), the photon-number spectral index increases slightly, and the thin disk begins to contribute to the band measured by RXTE (typically 3 – 20 keV). As a result, the hardness ratio decreases abruptly and the source turns left (point B in the upper panel of Fig. 10.1) in the HLD (see also Fig. 10.3). The source then enters the HIMS.

It has been suggested (Miller et al. 2006a,b) that in the LHS the thin disk is not truncated at several tens R_g, but it extends all the way in, to the ISCO. However, the derived thin-disk inner radius in the LHS is quite sensitive to the assumed continuum shape. Thus, the above suggestion has been challenged by Done et al. (2007). There are recent claims (Reis et al. 2010; Reynolds and Miller 2013) insisting that in the LHS the thin disk extends all the way to the ISCO, but also work with the opposite conclusion (Plant et al. 2013). In our view, the fact that the truncated thin-disk model explains physically not only the X-ray spectra and the observed time variability (Done et al. 2007; Ingram et al. 2009; Ingram and Done 2011, 2012), but also the formation and destruction of jets (Kylafis et al. 2012), makes it more favorable.

During the LHS, the PDS is fitted with a small number (typically 3–4) of very broad Lorentzians plus, in some cases, a low-frequency type-C QPO peak (e.g., Cui et al. 1999). The thin-disk emission can only vary on a viscous timescale, which is of order hundreds seconds. This has been confirmed by Uttley et al. (2011); see also Wilkinson and Uttley (2009). Therefore, the above variability comes from the component that produces the hard X-rays, i.e., from the ADAF and/or the jet. This has been confirmed by Axelsson et al. (2013).

10.2.5 Hard Intermediate State

In the HLD (Fig. 10.1, top panel), we are now at a point to the left of B, say, halfway between points B and C (see also Fig. 10.3). In the HIMS, and as time progresses, $\dot{M}$ increases, R_{tr} decreases, the ADAF part of the accretion flow shrinks, and the thin disk extends to progressively smaller radii. This means that if x is the fraction of the power-law, hard X-ray luminosity in the RXTE band, then $1-x$ is the fraction of the soft X-ray luminosity produced by the thin accretion disk, and x *decreases* monotonically with time. This is exactly what is observed by Munoz-Darias et al. (2011b) for MAXI J1659-152.

As the ADAF part of the accretion flow shrinks, there is an additional reason for the decrease of the power-law luminosity. The smaller the transition radius R_{tr}, the narrower the jet becomes, the smaller its optical depth, the less efficient the Comptonization becomes, and the steeper the high-energy power law (Γ increases to $\sim$2, Kylafis et al. 2008). A similar argument can be made if the Comptonization takes place in the ADAF. In addition, as the ADAF shrinks to smaller and smaller radii, a point is reached where the disk cannot sustain the magnetic field produced by the PRCB (Kylafis et al. 2012), the jet becomes eruptive, emitting discrete blobs with Lorentz factors $\gamma > 2$ (Fender et al. 2004). As the sources reach the so-called jet line (Fender et al. 2004; Fig. 10.1), the thin disk extends all the way to the ISCO.

Since the thin disk may support energetic flares (Poutanen et al. 1997; Gierliński et al. 1999), the high-energy cutoff E_c is no longer determined by the electrons in the jet or the ADAF, but rather by the non-thermal electrons in the flares. This may explain the increase of E_c in the HIMS (Motta et al. 2009). It is a different spectrum, and therefore a different E_c in the HIMS than that in the LHS.

The PDS in the HIMS can be viewed as a high-frequency extension of the PDS in the LHS. It can be decomposed into a number of broad Lorentzian components, which correspond to those found in the LHS (e.g. Belloni et al. 2011), only they have higher frequencies. The characteristic frequencies of all these components vary in unison (see Belloni et al. 2005; in Cyg X-1, not a transient system, an additional Lorentzian component remains fixed, see Pottschmidt et al. 2003), so they are really one frequency! The total fractional rms variability is lower ($10 - 20\,\%$) than in the LHS. The most prominent feature in the PDS is a type-C QPO (e.g., Motta et al. 2011), with centroid frequency varying between $\sim$0.2 and $\sim$20 Hz. As in the LHS, the Lorentzian components and the QPO vary together (Belloni et al. 2005, 2011). However, in the HIMS they are strongly correlated with hardness. Softer spectra correspond to higher frequencies and to lower integrated rms variability. These two correlations are explained naturally in our picture. The spectrum becomes softer, because the thin disk occupies an increasingly larger part of the accretion flow. Since the rms variability of the thin disk is very small (Gilfanov 2010), the total rms variability decreases. Furthermore, since the variability (at least the high-frequency one) appears to come from the jet/ADAF, the higher the energy of observation, the larger the variability, as it is observed (see e.g. Gilfanov 2010; Belloni et al. 2011). We remark here that at high energies we have two contributions to the spectrum: one from the jet/ADAF and one from the energetic flares discussed above.

In addition to the X-ray spectra, an accretion flow consisting of an inner ADAF part and an outer part of the Shakura-Sunyaev type (the so-called truncated disk model) explains very well the broad-band noise and the type-C QPO. Ingram and Done (2012), see also Ingram and Done (2011) and Ingram et al. (2009) have demonstrated that the type-C QPO can be explained as Lense-Thirring precession of the inner, ADAF part of the accretion flow. It is therefore natural to expect a correlation between the high-energy power-law spectrum and the QPO frequency, as they are both produced in the same region, the ADAF. If the high-energy power-law X-ray spectrum is produced in the jet, the characteristic frequency of the QPO has been interpreted in the case of Cyg X-1 (Kylafis et al. 2008) as the inverse of the ejection timescale of the jet, which could be related to the Lense-Thirrring precession frequency.

The broad-band noise spectrum has been interpreted by Ingram and Done (2011, 2012) as propagating mass-accretion rate fluctuations in the ADAF part of the accretion flow. This explains the fact that the features in the power spectrum and the QPO vary in unison.

Our picture makes the prediction that in the upper branch of the q-shaped curve (between points B and C in Fig. 10.1, top panel), the average timelag of the hard X-rays with respect to the soft X-rays should *decrease* as the hardness ratio decreases. This is because the ADAF shrinks in size in the HIMS and therefore the light-travel time of the upscattered photons is reduced.

10.2.6 First Jet-Line Crossing

The jet line (Fig. 10.1, top panel) is a sharp transition between the HIMS and the SIMS. It marks the end of a detectable compact jet. For a detailed study of this transition see Miller-Jones et al. (2012). As discussed above, the thin disk cannot sustain the magnetic field produced by the PRCB or any other mechanism and it is expelled. The transition between HIMS and SIMS is marked only by the timing properties (Belloni 2010). The spectrum shows no noticeable change during the transition, apart from a minor softening. At this time of the outburst, the accretion disk is geometrically thin, the energy spectrum is soft, and the hardness ratio is rather low. The observed variability is detected only at high energies, so it must come from the invoked magnetic flaring activity above and below the thin accretion disk. The characteristic values of the hardness ratio of the jet line depend on the energy bands used for the definition of the hardness ratio.

10.2.7 Soft Intermediate State

Before our discussion of the soft intermediate state, it is important to point out that the HIMS and the SIMS are what was called very high state (VHS) in the *Ginga* era (see Miyamoto et al. 1993; Takizawa et al. 1997). However, now we know that these two states are reached *before* the soft state is entered, which has tremendous implications for our understanding of the physical picture. The *Ginga* coverage was not sufficient to show this and the Esin et al. (1997) model was affected by this interchange of states.

In the HLD (Fig. 10.1, top panel), we are now just to the left of the jet line. The energy spectrum in the SIMS is nearly the same as in the HIMS, only slightly softer. On the other hand, significant changes occur in the PDS. A type-B QPO is now present in the power spectrum, which *is not* an evolution of the type-C QPO observed in the SIMS (see Motta et al. 2011 and references therein). The characteristic frequency of the type-B QPO is in the narrow (compared to the type-C) range of 1–6 Hz (see Motta et al. (2011) and references therein) while at high-flux intervals the range becomes even narrower (4–6 Hz) (Casella et al. 2004; Motta et al. 2011).

Unlike type-C QPOs, which are fitted with Lorentzians, type-B QPOs are fitted with Gaussians. This is due to the fact that type-B QPOs jitter in time on short timescales. Thus, in the average PDS, which is the usual that is fitted, the peak is broadened (Nespoli et al. 2003). As in the HIMS, the fractional rms increases with energy. This can be understood, if we invoke that most of the rms variability comes from the flares above and below the disk.

In view of the qualitative differences between type-B and type-C QPOs, and taking into account that the type-B QPOs are seen only during the SIMS, we are tempted to speculate that the type-B QPO is associated with the last "gasps" of the jet.

10.2.8 Soft State

In the HLD (Fig. 10.1, top panel), we are now at point C (see also Fig. 10.3). In outbursts like those of GX 339-4, our prototype source, the HSS marks the highest reached accretion rate, which then starts to decrease. There are sources where the accretion rate keeps increasing after the HSS is entered. Those are the ones that enter the anomalous state (Belloni 2010). We do not address this state, because it is not seen in GX 339-4 and we consider it as a "first-order effect". We stress, however, that this anomalous state *is not* the VHS of Esin et al. (1997).

In the soft state, the BHTs are characterized by a dominant soft X-ray spectrum and an energetically negligible (about 5 % of the total X-ray luminosity) steep power-law spectrum ($\Gamma > 2.2$) extending to more than 1 MeV (see Grove et al. 1998). As it was mentioned in Sect. 10.2.1 above, it is generally accepted that the soft X-ray spectrum comes from the thin accretion disk and the power-law spectrum comes from Comptonization of the soft disk photons by the non-thermal electrons in the flares above and below the thin disk.

As the mass-accretion rate decreases, the X-ray luminosity decreases and the entire accretion disk remains geometrically thin, emitting progressively less luminosity. Since luminosity decreases, its temperature will also decrease, resulting in a decrease in hardness, which is observed (see Motta et al. 2010; Stiele et al. 2012). This brings the BHTs to point D (Fig. 10.1, top panel, and Fig. 10.3).

Fourier-resolved energy spectra in the HSS (see Churazov et al. 2001) show that the observed variability comes from the power-law spectrum and not from the multi-temperature blackbody of the thin disk. This is consistent with our interpretation that the power-law spectrum comes from magnetic flaring above and below the thin accretion disk. Sometimes, a type-A QPO is observed (see Motta et al. 2011).

10.2.9 Soft Intermediate State Again

In the HLD (Fig. 10.1, top panel), we are now at a point to the right of point D (see also Fig. 10.3). As the mass-accretion rate decreases further, a point is reached (point D in Fig. 10.1, top panel, and Fig. 10.3) where the inner part of the accretion disk becomes radiatively inefficient and puffs up (Das and Sharma 2013). Evaporation also plays a role (Liu et al. 1999; for an interesting alternative see the Chapter of Kazanas in this volume). In any case, the simulations of Das and Sharma (2013) have shown that the transition from a geometrically thick inner flow to a geometrically thin outer disk moves *outward* as the accretion rate decreases. In other words, in the lower branch of the q-shaped curve (i.e., from point D to point E in Fig. 10.1, top panel, and Fig. 10.3) we have the *reverse* of what we had in the upper branch (from point B to point C in Fig. 10.1, top panel, and Fig. 10.3). The transition radius R_{tr} now moves *outwards* with time (Fig. 10.3).

We remark here that, according to the above, the source will *always* turn right at point D (i.e., at a specific accretion rate), independent of how high the luminosity was in the upper branch of the q-shaped curve (points B and C in Fig. 10.1, top panel). This is exactly what has been seen in the multiple outbursts of GX 339-4 (Motta et al. 2011).

With the establishment of an ADAF-like structure at the inner part of the accretion flow, the PRCB operates efficiently and creates the magnetic field needed for the establishment of a jet. Type-B QPOs are seen once again (Stiele et al. 2011; Motta et al. 2011). Extending our previous speculation (Sect. 10.2.7), we propose that the type-B QPOs observed in this part of the HLD are again due to the jet, at its *first appearance*. We note here that a compact jet is formed much later, when the source has approached the hard state and the jet has become strong enough to be detected as such. This gradual development of the jet was recently observed in detail (Corbel et al. 2013).

10.2.10 Hard Intermediate State Again

In the HLD (Fig. 10.1, top panel), we are now at a point, say, halfway between points D and E (see also Fig. 10.3). The transition from the SIMS to the HIMS in the lower branch of the q-shaped curve is uneventful. No eruptive jet, similar to those in the upper branch, has ever been seen and, our prediction is that, one will never be seen. This is because the ADAF-like part of the accretion flow increases continuously, the PRCB continues working, and the jet builds up smoothly. No instability occurs in the lower branch of the q-shaped curve, though variability is naturally expected as the jet forms.

The thick and hot part of the flow inside R_{tr} introduces a hard component in the X-ray spectrum. This comes from up-scattering of soft photons from the thin disk in the hot, thick, inner part of the flow or the forming jet. Thus, the hardness ratio increases with time. As in the HIMS at higher luminosity, if x' is the fraction of the power-law, hard X-ray luminosity in the RXTE band, then $1 - x'$ is the fraction of the soft X-ray luminosity produced by the thin accretion disk, and x' *increases* monotonically with time. This is exactly what is observed by Munoz-Darias et al. (2011b) for MAXI J1659-152.

Type-C QPOs are seen in this HIMS also (Motta et al. 2011). Their explanation is naturally provided by Lense-Thirring precession of the inner ADAF-like part of the accretion flow (Ingram et al. 2009; Ingram and Done 2011, 2012).

Our picture makes the prediction that in the lower branch of the q-shaped curve (between points D and E in Fig. 10.1, top panel), the average timelag of the hard X-rays with respect to the soft X-rays should *increase* as the hardness ratio *increases*. This is because the ADAF, where the upscattering of the soft photons occurs, moves to progressively larger radii in the HIMS (R_{tr} increases; Das and Sharma 2013) and the light travel time of the upscattered photons increases.

10.2.11 Second Jet-Line Crossing

In the HLD (Fig. 10.1, top panel), we are now at a point, say, a bit to the left of point E. As we discussed above, the jet line in the hard to soft transition (transition from HIMS to SIMS) has no counterpart in the soft to hard transition (SIMS to HIMS). Thus, in the "return" track, we define the jet line as the line indicating the establishment of a *compact* radio jet. This line occurs before the sources reach the LHS (Miller-Jones et al. 2012).

By this time, the major part of the accretion flow is geometricaly thick and R_{tr} has increased to tens or hundreds of gravitational radii. Thus, the hardness ratio at this jet line is significantly larger than the one at the jet line at higher luminosity. As a result, the line joing the disappearance of the compact jet (upper branch) and the appearance of a compact jet (lower branch) is slanted (Fig. 10.1, top panel).

10.2.12 Hard State Again

When the geometrically thick accretion flow occupies most of the accretion flow, the sources reach the LHS (point E in Fig. 10.1, top panel). As at the beginning of the outburst, the soft seed photons are mainly cyclotron photons from the jet and the up-scattering occurs in the ADAF and/or the jet. The thin disk has a small contribution of soft photons.

10.2.13 Return to the Quiescent State

As the mass-accretion rate continues to decrease, the flow remains ADAF like, the magnetic field produced by the PRCB weakens, the jet also weakens, but the hard X-ray spectrum (power law) remains approximately the same. Thus, the sources trace a nearly vertical path from point E downwards (Fig. 10.1, top panel). To "zeroth order", this path is identical to the one at the beginning of the outburst, because in both cases the accretion flow inside R_{tr} is ADAF-like.

10.3 Remarks

The q-shaped curve exhibited by BHTs in a hardness – luminosity diagram is a hysteresis curve, like those seen in many branches of science. Such curves are exhibited when the system's future development depends not only on its current state, but also on its past history.

In our case, the poloidal magnetic field that is created by the PRCB in the ADAF during the quiescent state and persists in the hard state, "reminds" the flow that it should remain ADAF-like, despite the fact that the accretion rate has increased significantly. Only at accretion rates near the Eddington limit the thin accretion disk prevails everywhere.

In view of the above, we can describe the stability of the two main spectral states (soft and hard) at the same luminosity as follows. Consider an accretion rate such that the observed luminosity is between the upper and lower branches of the q-shaped curve, i.e., between points D and C or points E and B in Fig. 10.1, top panel. For such an accretion rate, the accretion flow is happy to be either ADAF-like (points between E and B in Fig. 10.1) or Shakura-Sunyaev – type (points between D and C in Fig. 10.1). *Which of the two it will be, depends on its previous history.* In going from point A to point B in Fig. 10.1, the system is "reminded" by the poloidal magnetic field that it is on an ADAF-like solution and that it should remain on it. Similarly, in going from point C to point D in Fig. 10.1, the system is happy with the Shakura-Sunyaev – type solution that it is on and continues on it.

In summary, "memory" of the system plus one parameter, the accretion rate, are enough to explain the q-shaped curve in BHTs.

We remark however that astrophysical models and ideas do not explain all the observational details. Our proposed picture here is no exception. Thus, we leave it to future work by us or others to address "first order" observations, like the anomalous state (Belloni 2010), the fact that the loops in the HLD traversed by GX 339-4 in its four recent outbursts are one inside the other, the repeated crossings of the first jet line, and so on.

Conclusions

By making two rather obvious and generally accepted assumptions, we have been able to explain the major effects observed in the HLD of BHTs. The assumptions are the following:

1. The accretion rate onto BHTs as a function of time has a generic bell-shaped curve. It starts from a very low accretion rate, it increases steadily up to an accretion rate comparable to the Eddington rate, and ends again with a very low accretion rate. This is justified by the fact that BHTs start and end their outbursts at very low luminosity, where the sources are usually undetected within short observing times. About midway into the outburst, the luminosity reaches an approximate plateau and has a value comparable to the Eddington luminosity.
2. At high accretion rates, the accretion disk is optically thick, geometrically thin, and it is described well by the model of Shakura and Sunyaev (1973). At low accretion rates, the accretion flow is radiatively inefficient, geometrically thick, optically thin, and advection dominated (ADAF;

(continued)

Narayan and Yi 1994, 1995; Abramowicz et al. 1995). Both of these pictures are well accepted and have been verified by numerical simulations (Ohsuga et al. 2009).

According to Kylafis et al. (2012), when the accretion flow has an inner ADAF-like part, the PRCB works efficiently. Thus, a strong, poloidal magnetic field is established and a jet forms within hours or days, depending on the luminosity. This is exactly what is observed in a HLD. In the right part of the q-shaped curve in a HLD, the accretion rate is relatively low, part or all of the accretion flow is ADAF-like, and a jet is always present.

At high accretion rates, the accretion disk is of the Shakura and Sunyaev (1973) type, the PRCB works inefficiently, and a strong, poloidal magnetic field that would form a jet is not established. This is consistent with our current knowledge. No jet has ever been detected in the soft state.

Because of the above, it is natural to expect a flaring jet in the first jet-line crossing, but not in the second one. As the sources approach the first jet line, the ADAF part of the accretion flow, that feeds the jet, shrinks. On the other hand, since the thin accretion disk cannot sustain the existing magnetic field, the thin disk becomes unstable to Rayleigh-Taylor-type instability modes, and the accumulated magnetic field escapes to the outer disk in the form of magnetic “strands” (Kylafis et al. 2012). Magnetic reconnection then produces the flare. This is not the case though in the second jet-line crossing. There, due to the low luminosity, the magnetic field builds up slowly in the ADAF part of the accretion flow, and continues increasing as the ADAF part of the flow grows outwards and eventually occupies most of the flow.

In summary, we have shown that the main phenomena observed in the q-shaped curve in a HLD of BHTs can be explained with only one parameter, the accretion rate. Furthermore, we have shown that the q-shaped curved will always be traversed in the counterclockwise direction and we predict that no source will ever be seen to traverse the entire q-shaped curve in the clockwise direction. In addition, we predict that the average timelag of the hard X-rays with respect to the soft X-rays will decrease with time in the upper branch and increase with time in the lower branch of the q-shaped curve.

Regarding jet formation and destruction, our picture is applicable to neutron-star and white-dwarf X-ray binaries as well (Kylafis et al. 2012).

Acknowledgements We thank an anonymous referee for helpful suggestions and comments, which have improved our paper in both content and readability. We have also profited from discussions with Iossif Papadakis. One of us (NDK) acknowledges useful discussions with P. Casella, I. Contopoulos, B. F. Liu, S. Motta, R. Narayan, and A. Zdziarski. This research has been supported in part by EU Marie Curie projects no. 39965 and ITN 215212 (“Black Hole Universe”), EU REGPOT project number 206469, a Small Research Grant from the University of Crete, a COST-STSM-MP0905 grant and a grant from the European Astronomical Society in 2012. TMB acknowledges support from INAF-PRIN 2012-6.

References

Abramowicz, M.A., Chen, X., Kato, S., Regev, O. ApJ, **438**, L37 (1995)
Axelsson, M., Hjalmarsdotter, L., Done, C., MNRAS, **431**, 1987 (2013)
Bai, X.-N., Stone, J.M.: ApJ **767**, 30 (2013)
Belloni, T.: LNP **794**, 53 (2010)
Belloni, T., Psaltis, D., van der Klis, M.: ApJ **572**, 392 (2002)
Belloni, T., Homan, J., Casella, P., van der Klis, M., Nespoli, E., Lewin, W.H.G., Miller, J.M., Méndez, M.: A&A **440**, 207 (2005)
Belloni, T., Motta, S., Muñoz-Darias, T.: BASI **39**, 409 (2011)
Blandford, R.D., Payne D.G.: MNRAS **199**, 883 (1982)
Casella, P., Belloni, T., Homan, J., Stella, L.: A&A **426**, 587 (2004)
Casella, P., Belloni, T., Stella, L.: ApJ **629**, 403 (2005)
Christodoulou, D.M., Contopoulos, I., Kazanas, D.: ApJ **674**, 388 (2008)
Churazov, E., Gilfanov, M., Revnivtsev, M.: MNRAS **321**, 759 (2001)
Contopoulos, I.: ApJ **450**, 616 (1995)
Contopoulos, I., Kazanas, D.: ApJ **508**, 859 (1998)
Contopoulos, I., Kazanas, D., Christodoulou, D.M.: ApJ **652**, 1451 (2006)
Corbel, S., Aussel, H., Broderick, J.W., et al.: MNRAS **431**, L107 (2013)
Cui, W., Zhang, S.N., Chen, W., Morgan, E.H.: ApJ **512**, L43 (1999)
Das, U., Sharma, P.: MNRAS, **435**, 2431 (2013). arXiv:1304.1294
Done, C., Gierliński, M., Kubota, A.: A&ARv **15**, 1 (2007)
Esin, A.A., McClintock, J.E., Narayan, R.: ApJ **489**, 865 (1997)
Esin, A.A., Narayan, R., Cui, W., Grove, J.E., Zhang, S.-N.: ApJ **505**, 854 (1998)
Esin, A.A., McClintock, J.E., Drake, J.J., Garcia, M.R., Haswell, C.A., Hynes, R.I., Muno, M.P.: ApJ **555**, 483 (2001)
Fender, R.P., Belloni, T.M., Gallo, E.: MNRAS **355**, 1105 (2004)
Fender, R.P., Homan, J., Belloni, T.M.: MNRAS **396**, 1370 (2009)
Gallo, E.: LNP **794**, 85 (2010)
Gallo, E., Fender, R.P., Miller-Jones, J.C.A., Merloni, A., Jonker, P.G., Maccarone, T.J., van der Klis, M.: MNRAS **370**, 1351 (2006)
Gilfanov, M.: LNP **794**, 17 (2010)
Giannios, D.: A&A **437**, 1007 (2005)
Giannios, D., Kylafis, N.D., Psaltis, D.: A&A **425**, 163 (2004)
Gierliński, M., Newton, J.: MNRAS **370**, 837 (2006)
Gierliński, M., Zdziarski, A.A., Poutanen, J., Coppi, P.S., Ebisawa, K., Johnson, W.N.: MNRAS **309**, 496 (1999)
Grove, J.E., Johnson, W.N., Kroeger, R.A., McNaron-Brown, K., Skibo, J.G., Phlips, B.F.: ApJ **500**, 899 (1998)
Hawley, J.F.: Ap& SS **320**, 107 (2009)
Homan, J., Belloni, T.: Ap&SS **300**, 107 (2005)
Homan, J., Wijnands, R., van der Klis, M., Belloni, T., van Paradijs, J., Klein-Wolt, M., Fender, R., Méndez, M.: ApJSuppl. **132**, 377 (2001)
Igumenshev, I.V.: ApJ **677**, 317 (2008)
Ingram, A., Done, C.: MNRAS **415**, 2323 (2011)
Ingram, A., Done, C.: MNRAS **419**, 2363 (2012)
Ingram, A., Done, C., Fragile, P.C.: MNRAS **397**, L101 (2009)
Kalemci, E., Tomsick, J.A., Rothschild, R.E., Pottschmidt, K., Kaaret, P.: ApJ **563**, 239 (2001)
Kalemci, E., Tomsick, J.A., Rothschild, R.E., Pottschmidt, K., Corbel, S., Wijnands, R., Miller, J.M., Kaaret, P.: ApJ **586**, 419 (2003)
Kalemci, E., Tomsick, J.A., Rothschild, R.E., Pottschmidt, K., Kaaret, P.: ApJ **603**, 231 (2004)
Kalemci, E., Tomsick, J.A., Buxton, M.M., Rothschild, R.E., Pottschmidt, K., Corbel, S., Brocksopp, C., Kaaret, P.: ApJ **622**, 508 (2005)

Kalemci, E., Tomsick, J.A., Rothschild, R.E., Pottschmidt, K., Corbel, S., Kaaret, P.: ApJ **639**, 340 (2006)
Kylafis, N.D., Papadakis, I.E., Reig, P., Giannios, D., Pooley, G.G.: A&A **489**, 481 (2008)
Kylafis, N.D., Contopoulos, I., Kazanas, D., Christodoulou, D.: A&A **538**, 5 (2012)
Lasota, J.-P.: NewAR **45**, 449 (2001)
Levine, A.M., Bradt, H, Morgan, E.H., Remillard, R.: Adv. Sp. Rev. **38**, 2970 (2006)
Liu, B.F., Yuan, W., Meyer, F., Meyer-Hofmeister, E., Xie, G.Z.: ApJ **527**, L17 (1999)
Lovelace, R.V.E., Rothstein, D.M., Bisnovatyi-Kogan, G.S.: ApJ **701**, 885 (2009)
Lynden-Bell, D.: MNRAS **279**, 389L (1996)
Machida, M., Nakamura, K., Matsumoto, R.: PASJ **58**, 193 (2006)
Markoff, S., Nowak, M.A., Wilms, J.: ApJ **635**, 1203 (2005)
Marsh, T.R., Robinsin, E.L., Wood, J.H.: MNRAS **266**, 137 (1994)
McClintock, J.E., Narayan, R., Garcia, M.R., Orosz, J.A., Remillard, R.A., Murray, S.S.: ApJ **593**, 435 (2003)
McClintock, J.E., Remillard, R.A., Rupen, M.P., Tores, M.A.P., Steeghs, D., Levine, A.M., Orosz, A.: **ApJ** 698, 1398 (2009)
Meyer, F., Meyer-Hofmeister, E.: A&A **104**, L10 (1981)
Meyer, F., Liu, B.F., Meyer-Hofmeister, E.: A&A **361**, 175 (2000)
Meyer-Hofmeister, E., Meyer, F.: A&A **380**, 739 (2001)
Meyer-Hofmeister, E., Liu, B.F., Meyer, F.: A&A **508**, 329 (2009)
Mignone, A., Rossi, P., Bodo, G., Ferrari, A., Massaglia, S.: MNRAS **402**, 7 (2010)
Miller, J.M., Homan, J., Miniuti, G.: ApJ **652**, L113 (2006a)
Miller, J.M., Homan, J., Steeghs, D., et al.: ApJ **653**, 525 (2006b)
Miller-Jones, J.C.A., Sivakoff, G.R., Altamirano, D., et al.: MNRAS **421**, 468 (2012)
Miyamoto, S., Iga, S., Kitamoto, S., Kamado, Y.: ApJ **403**, L39 (1993)
Miyamoto, S., Kitamoto, S., Hayashida, K., Egoshi, W.: ApJ **442**, L13 (1995)
Motta, S., Belloni, T., Homan, J.: MNRAS **400**, 1603 (2009)
Motta, S., Muñoz-Darias, T., Belloni, T.: MNRAS **408**, 1796 (2010)
Motta, S., Muñoz-Darias, T., Casella, P., Belloni, T., Homan, J.: MNRAS **418**, 2292 (2011)
Munoz-Darias, T., Motta, S., Belloni, T.M.: MNRAS **410**, 679 (2011a)
Munoz-Darias, T., Motta, S., Stiele, H., Belloni, T.M.: MNRAS **415**, 292 (2011b)
Narayan, R., Yi, I.: ApJ **428**, L13 (1994)
Narayan, R., Yi, I.: ApJ **452**, 710 (1995)
Narayan, R., McClintock, J.E., Yi, I.: ApJ **457**, 821 (1996)
Narayan, R., Barret, D., McClintock, J.E.: ApJ **482**, 448 (1997)
Nespoli, E., Belloni T., Homan, J., Miller, J.M., Lewin, W.H.G., Méndez, M., van der Klis, M.: A&A **412**, 235 (2003)
Ohsuga, K., Mineshige, S., Mori, M., Kato, Y.: PASJ **61**, L7 (2009)
Orosz, J.A., Bailyn, C.D., Relillard, R.A., McClintock, J.E., Foltz, C.B.: ApJ **436**, 848 (1994)
Plant, D.S., Fender, R.P., Ponti, G., Munoz-Darias, T., Coriat, M.: MNRAS, **435**, 2431 (2013). arXiv1309.4781
Pottschmidt, K., Wilms, J., Nowak, M.A., Pooley, G.G., Gleissner, T., Heindl, W.A., Smith, D.M., Remillard, R., Staubert, R.: A&A **407**, 1039 (2003)
Poutanen, J., Krolik, J.H., Ryde, F.: MNRAS **292**, L21 (1997)
Reig, P., Kylafis, N.D., Giannios, D.: A&A **403**, L15 (2003)
Qian, L., Liu, B.F., Wu, X.-B.: ApJ **669**, 1145 (2007)
Reis, R.C., Fabian, A.C., Miller, J.M.: MNRAS **402**, 836 (2010)
Remillard, R.A., McClintock, J.E.: ARA&A **44**, 49 (2006)
Reynolds, M.T., Miller, J.M.: ApJ **769**, 16 (2013)
Romanova, M.M., Ustyugova, G.V., Koldova, A.V., Lovelace, R.V.E.: In: Tsinganos, K., Ray, O., Stute, M. (eds.) Protostellar Jets in Context. Astrophysics and Space Science Proceedings, vol. 153. Springer, Berlin (2009)
Shakura, N.I., Sunyaev, R.A.: A&A **24**, 337 (1973)
Smak, J.: Acta. Astron. **34**, 161 (1984)

Sobolewska, M.A., Papadakis, I.E., Done, C., Malzac, J.: MNRAS **417**, 280 (2011)
Stiele, H., Motta, S., Munoz-Darias, T., Belloni, T.M.: MNRAS **418**, 1746 (2011)
Stiele, H., Muñoz-Darias, T., Motta, S., Belloni, T.: MNRAS **422**, 679 (2012)
Takizawa, M., Dotani, T., Mitsuda, K., et al.: ApJ **489**, 272 (1997)
Tchekhovskoy, A., Narayan, R., McKinney, J.C.: MNRAS **327**, L79 (2011)
Uttley, P., Wilkinson, T., Cassatella, P. Wilms, J., Pottschmidt, K., Hanke, M., Böck, M.: MNRAS **414**, L60 (2011)
Wilkinson, T, Uttley, P. 397, 666 (2009)

CPSIA information can be obtained at www.ICGtesting.com
Printed in the USA
LVOW05*2229180115

423359LV00001B/58/P